지은이	담디 편집부 엮음
펴낸이	서경원
편집	나진연
디자인	이철주
펴낸곳	도서출판 담디
등록일	2002년 9월 16일
등록번호	제 9-00102호
주소	서울시 강북구 삼각산로 79, 2층
전화	02-900-0652
팩스	02-900-0657
이메일	damdi_book@naver.com
홈페이지	www.damdi.co.kr

Compiler	DAMDI Publishing House
Publisher	Kyongwon Suh
Editor	Jinyoun Na
Art Director	Cheolju Lee
Publishing	DAMDI Publishing House
Office Address	2F, 79, Samgaksan-ro, Gangbuk-gu, Seoul, 01036, Korea
Tel	+82-2-900-0652
Fax	+82-2-900-0657
E-mail	damdi_book@naver.com
Homepage	www.damdi.co.kr

정가 88,000원
Printed in Korea
ISBN 978-89-6801-080-4 (94540)
978-89-6801-079-8 (set)

이 도서의 국립중앙도서관 출판예정도서목록(CIP)은 서지정보유통지원시스템 홈페이지(http://seoji.nl.go.kr)와 국가자료공동목록시스템(http://www.nl.go.kr/kolisnet)에서 이용하실 수 있습니다.(CIP제어번호: CIP2018023911)

Architectural Material 1

BRICK & TILE

담디 DAMDI

Contents

008 **How use the Brick & Tile?**

Interview with Architects

009 Tell us about your favourite project that you used Brick & Tile in or another architect's work
- it can be in the interior, on the facade, doesn't matter where it's used.

022 What are the strengths and weaknesses of Brick & Tile?

036 **Interviewee Profile**

A case study of

046 **Brick & Tile Architecture**

048 **TILE**

Interior

048	534	HWKN
050	'Aktipis' Flowershop	Point Supreme Architects
054	Amasion	Concrete Architectural Associates
056	Kalavrita Café,	Point Supreme Architects
058	Minto Midtown Penthouse	II BY IV Design Associates Inc
062	NINETTE HOUSE	Arenas Basabe Palacios Arquitectos
068	No Visitor House KOSID	jay is working
072	Office and Hotel Room	BOARD
076	TILESCAPE	BudCud
080	Witloof	Maurice Mentjens Design

Housing

082	CD House	Donner Sorcinelli Architecture
090	Pété Mane	architecture paradigm

Multi - Housing

094 Properly breathing house — H&P Architects

102 VM Houses — BIG

Commercial

108 Doosan Art Sqaure — jay is working

116 LEGO House — BIG

124 Warehouse 8B — Arturo Franco

130 wasl Tower — UNStudio

Installation

134 Mountain — TAKK Architecture

Public (Urban & Landscape)

140 Jinzhou — Casanova+Hernandez Architects

146 Library Creativity from Knowledge — ma0

150 Plaza del Salvador Rehabilitation — OOIIO Architecture

158 San Miguel Square Rehabilitation — OOIIO Architecture

Brick

166 Brick

Interior

166 Cafe Grumpy — Z-A studio& Cheng+Snyder
168 Muy Bien — ogisadesign

Housing

172 2 Houses with 6 Homes — nodo17 Architects
178 Another base in 'G' — KimuraMatsumoto Architects
186 Blooming House with wild flowers — studio-GAON
192 Brick Cave — H&P Architects
198 Dr. Sudhakar Jadhav`s Residence — Sunil Patil and Associates
202 iA House — LANDÍNEZ+REY arquitectos
208 LT House — Tropical Space Co.,Ltd
213 Renovation House RS — SUPA Schweitzer Song
218 Seaside Single House — modostudio
226 Termitary House — Tropical Space Co.,Ltd
232 Villa - Villa — nARCHITECTS
238 Y House — Hironori Matsubara

Multi - Housing

244 Asma Bahçeler Residences — M artı D Mimarlık
252 Black & White Twins — Casanova+Hernandez architects
258 BO5 Tin Tin Tower — NL Architects
266 Fairyland Guorui — UNStudio
278 Gapyong Space Invader — NL Architects
280 Kameleon — NL Architects
288 VILLA TORPED — PERIPHERIQUES

Education

296 COLÉGIO DOS PLÁTANOS — murmuro
306 Library and Cafeteria for the Technical University of Applied Sciences in Wildau — Rebecca Chestnutt, Robert Niess
314 MARIA GRAZIA CUTULI SCHOOL — 2A+P/A, IaN+, ma0, Mario Cutuli

Commercial

320 Binet Business Center — AZC
330 Gapyong Space Invader — NL Architects
332 Labels Clothes Shop, Sittard — Maurice Mentjens Design
336 Medi Flower Oriental Hospital — jay is working
342 Terra Cotta Studio — Tropical Space Co.,Ltd
348 WAP Art Space — DAVIDE MACULLO ARCHITECTS
358 Warehouse 17C — Arturo Franco, Fabrice van Teslaar
362 Entrepreneur Offices. Madridejos — OOIIO Architecture
378 FACTORY IN BINH DUONG — NISHIZAWA ARCHITECTS
388 Namsan Patio — Architects Group Raum

Installation

394 Dissolving Arch — stpmj Architecture

Public (Urban & Landscape)

398 Antwerp Zoo — STUDIO FARRIS ARCHITECTS
404 Artena Archaeological Park — 2TR ARCHITETTURA
406 Community Centre Aussersihl Part2 — EM2N
408 Former Haystack Refurbishment — OOIIO Architecture
420 Home Outside — BOARD
422 New National Museum of Afghanistan — Donner Sorcinelli Architecture
426 Gymnasium Hausburgviertel — Rebecca Chestnutt, Robert Niess

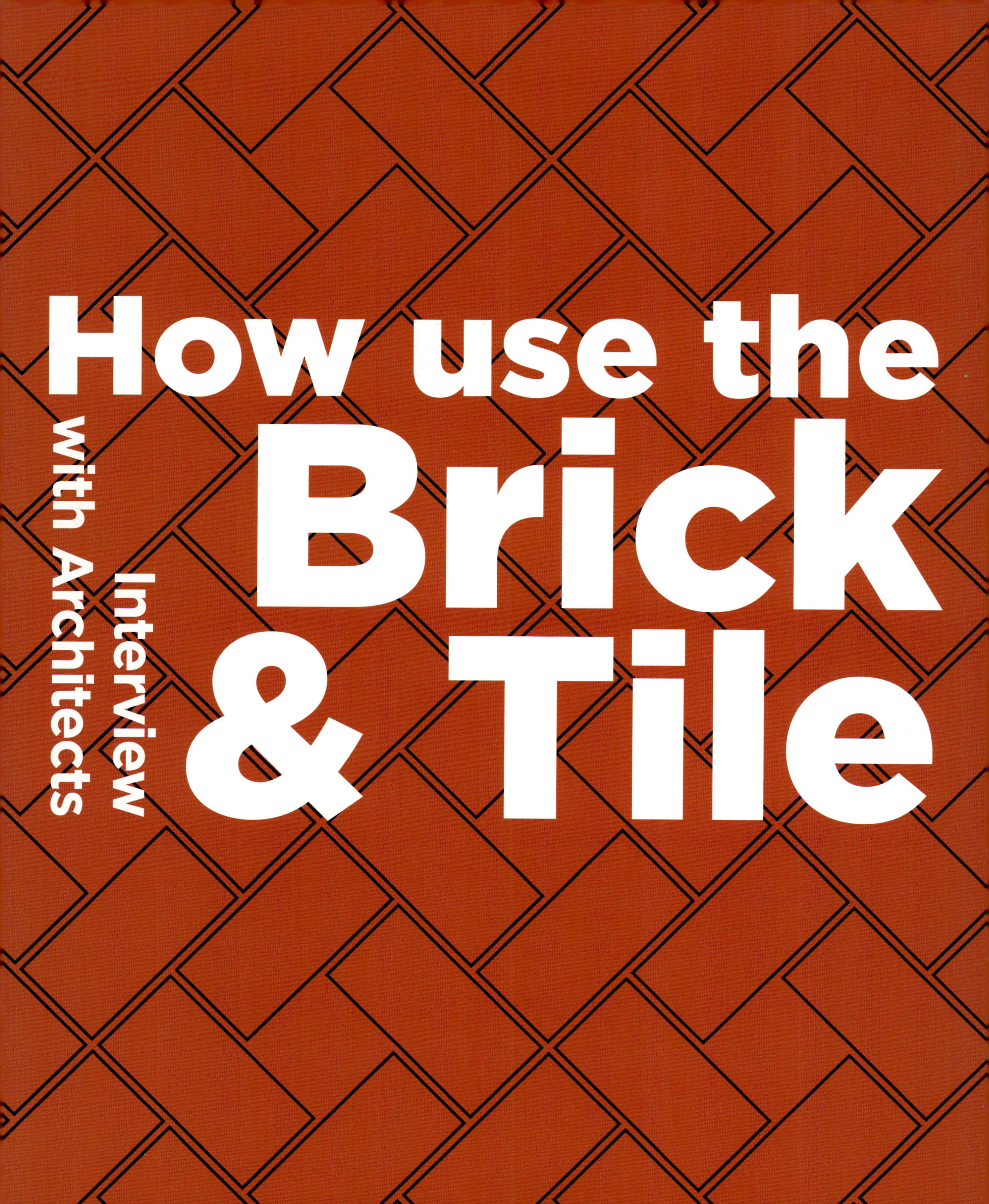
How use the
Brick
& Tile
Interview
with Architects

Q1
Tell us about your
favourite project
that you used Brick & Tile in
or another architect's work
- it can be in the interior, on the facade,
doesn't matter where it's used.

Brick

ARPHENOTYPE

Bricks are an ancient material, which even today can be re-invented. One project which immediately pops up in my mind is the Kolumba Museum(***A1***) in Cologne by Peter Zumthor in Cologne. However in this context, I rather would like to mention another architects, who used bricks in a new way in his time. His name is Eladio Dieste. One of the few architects who managed to blur the boundaries between tradition and innovation. The building I am referring to is the church "Iglesia de Atlántida Cristo Obrero y Nuestra Señora de Lourdes"(***A2***).

He designed the church in 1958, long before the applied use of computers in architecture. Eladio Dieste created double curved walls and ceilings by bricks, which are still today stunning and unique. This creates amazing light conditions and a friendly space, driven by the look and color of the bricks. Lastly I would like to refer to a book by the architect Koen Mulder "Het Zinderend Oppervlak", a book which highlights the new possibilities in pattern and curvature for walls made out of bricks.

A1 Kolumba Museum © HOWI

AZC

We used brick on the last project of an office building: "Hotel industriel Binet(**P.320**)" in Paris.

The white glazed brick façade fits with the north area of Paris where the building belongs and with its function of an office building.

BOARD

Until today, we have done only one project in which we used brick, which is the design of a square with urban furniture, a project that we called "Home Outside(**420p**)". There, we proposed to use brick for the surface of the square to connect it to its context. In that way the typical red brick surface of the neighbourhood could be continued and strong ties to the surroundings created. However, the main element of the square was not the surface, but 3 elements that we added, that are both artistic and functional: a fireplace, a table, and a boat. With these 3 elements we aimed to create a unique square as a work of art, where the inhabitants of the site, called the "Kraanbolwerk", but also its visitors, can feel at home outside and meet, talk, cook, eat, warm up, and play.

A2 Iglesia de Atlántida Cristo Obrero y Nuestra Señora de Lourdes © Andrés Franchi Ugart

Carlos Lampreia

One of my favourites houses made in brick, is the experimental Muuratsalo House by Aalto in Finland, where in the patio we can watch a range of different textures made by bricks working together against the external white painted bricks that helps defining the edge of the building.

Casanova+Hernandez Architects

In our project for a collective housing block in Blaricum(**252p**), the Netherlands, we used black brick as a means to create a cubic pure volume used as a canvas on which to perforate a series of varied openings framed with a white aluminum strip in order to give importance to the void instead of to the architectural mass.

The 'extraction of mass' in this project from the ideal black brick pris-

A3 Elbek & Vejrup

matic volume, especially when it is extracted from the four corners of the building, reduces its massive appearance. The white color in the voids contrasts with the dark color of the brick facade to create what, according to the reification property of the gestalt theory, is called a subjective or illusory contour. As happens in many carved wooden sculptures created by Barbara Hepworth during the 1940s, color helps to render visible the invisible and the void becomes an important constructive element of the work. The black brick reinforces the appearance of the outer skin of the building and the white frames underline the presence of the voids perforated into it.

CEBRA(Mikkel Frost)

We´ve used bricks in many of our projects. In Denmark, most houses are built from this material so it´s deeply rooted in our culture. I think we managed to reinterpret the material and building system in our office building for the tech company Elbek & Vejrup(***A3***). Here the bricks are applied like digital pixels and have multiple bright colors. It´s almost like a Minecraft aesthetic.

A4 Melnikov House © Sergei Arssenev

Davide Macullo Architects

The WAP ART SPACE(***A4***) gallery that we recently built in Gangnam Seoul is an interesting use of the bricks as we intend it. The bricks are like a silk cloth you wear and give the feeling of keeping the body warm or cool. The impact of bricks on the elevations of a building is connected to the reduction of scale: when I see the building from far it has a homogeneous appearance, the closer I come to the building the more I can distinguish the characteristics of the cladding material. This holds my attention during the different phases of the approach and reduces the scale from urban to domestic, from public to private. The more I get closer, the more I can nurture my feelings. Bricks have the extraordinary potential in expressing the human work. Is like seeing a fabric woven by hand instead of by machine. It is alive and expresses a care for Man.

Donner Sorcinelli Architecture

Two types of Brick Masonry have been used for the New National Museum of Afghanistan(**422p**). Stone bricks have been used for the exterior while clay bricks for the interior layer of the building envelope.

The reason was to exploit only local materials in order to reduce the CO2 footprint of the building as well as to use local workforce.

Keiichi Hayashi Architect

Melnikov House(***A4***) / Konstantin Melnicov,

I was surprised to know a house, such as the Cylinder wall modern house with its many hexagonal shape windows was made of bricks, rather than reinforced concrete. At that time, I felt good because I could instantly comprehend the mystery of impressive design. Masonry could easily create curved wall plastically hexagonal shape windows with laid bricks by forty five degrees angles. The placement of the windows fits to a brick construction process and structural mechanics. I assume that Melnikov used masonry bricks because of financial constraints, but I would like to give high evaluation to his free-wheeling design and effort which he put into small residence that he elevated the once sublimated, old masonry technique to a structure which embodies modernism in its composed structure and design.

Kimura Matsumoto architects office

Brick masonry architectural projects that I admire are too numerous to mention. Needless to say, the interesting feature of brickwork is that it is a fundamental unit which when combined can create a wall, floor, ceiling and evening furniture such as chairs and tables. I think the best example of this is "Church of St Peter" by Sigurd Lewerentz.

LANDÍNEZ+REY arquitectos

In the office we appreciate any project that investigates the possibilities of each material to get its architectonical premises. In fact, i do not remember a project in which we have not used bricks and we never make a difference in their disposition for being inside or outside. That´s why we, as an architectural office, have grown up accompanied by our progressive knowledge in its use, technology and expressiveness.

M artı D Mimarlık

Brick is used as cladding material in Asmabahçeler Residences(***A5***). For many users, perception of brick is considered as sympatic. Brick is consitent with notion of "home". For this reason we have used brick in this project. Furthermore, with developing application techniques and technology, it provides heat insulation.

murmuro

We have worked with brick on the project for the Plátanos School. We have used brick on the facade as a constructive and plastic solution for a technical necessity(***A6***).

The client had very specific needs: we could not reduce the outside sports area, they did not want HVAC solutions only natural ventilation taking advantage of the existent wind corridor, we could not open windows towards the sports area for security reasons, the building had to connect two different pre-existent buildings plus two outside spaces all at different heights, had to be built during the functioning of the school and in two phases. Also, the program demanded we used the facade facing the sports area. That is why we came up with the solution of a complete facade in brick, on a hit and miss design. This solution allowed us to have a facade that would stand the games and balls thrown at it, would allow us to open windows behind to introduce natural light into the secondary program of the school, and allow us to disguise several ventilation grids, the ones allowing to inflate fresh and cold air into the building.

NISHIZAWA ARCHITECTS

Katzden Factory(**378p**) is the project we used brick like a main material for the whole building.

OOIIO Architecture

Probably the project where I have used brick with more success is in the refurbishment of an ancient Haystack that was abandoned and

A5 Asmabahçeler Residences © ZM Yasa Photography

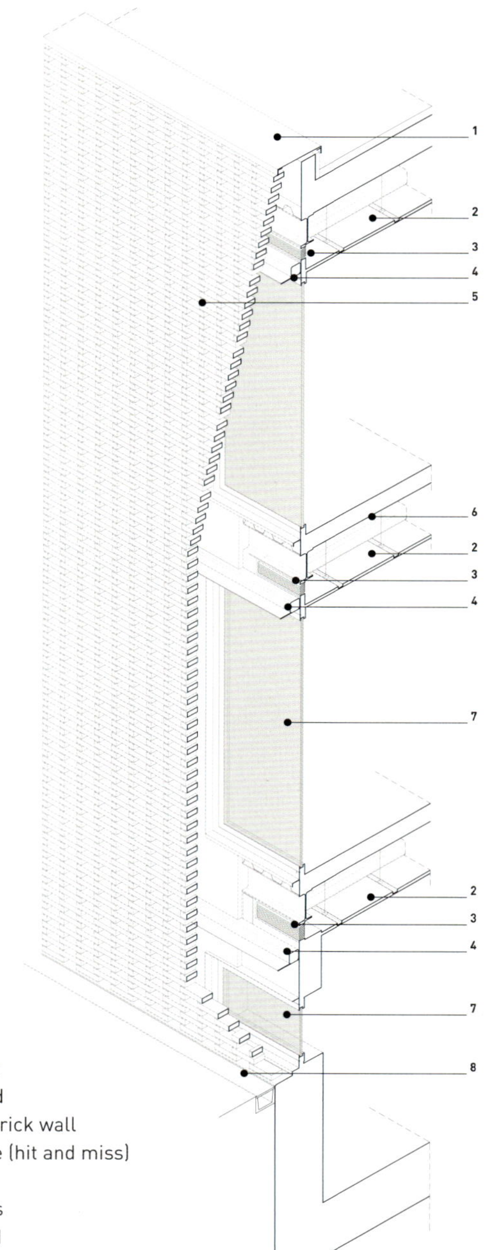

1 zinc sheet flashing
2 ceilling ventilation duct
3 metallic ventilation grid
4 metallic structure for brick wall
5 permeable brick facade (hit and miss)
6 metallic main frames
7 exterior window frames
8 courtyard drainage grid

A6 the Plátanos School Section Detail

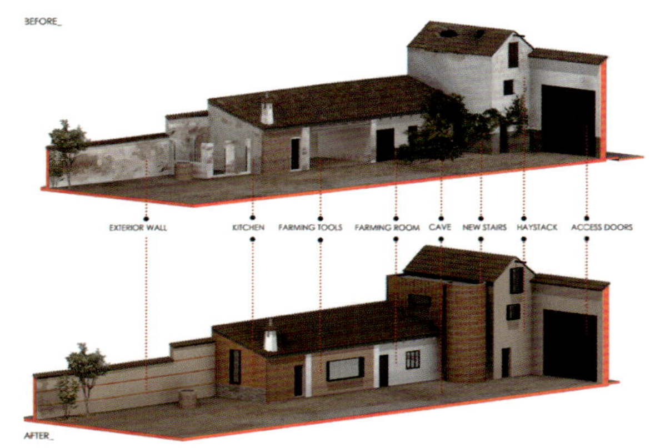

A7 Former Haystack Refurbishment

A8 Brick Townhouses

A9 St. Mark Church © Håkan Svensson Xauxa

A10 The Masonry

falling apart in a village in central Spain(***A7*, 408p**).

There were some poor ancient construction used in old times to keep agricultural tools and animals to work in the countryside.

One of the old walls was made from beautiful ancient big bricks and the client wanted to demolish it because it was falling down. We convinced him to keep it and thanks to this special interest in recovering this brick wall we decided to work always with brick in all the new elements introduced in the buildings, with different patterns.

The result of all this architectonic surgery work has been a patchwork of nuances and details rethinking the old and mixing it with the new, turning the result into a mosaic of nuances and details.

Brick here was definitely the right material to merge old with new!

Piuarch

We've not worked with exposed bricks very often, but the redevelopment project of the Caproni factory in Milan has allowed us to deal with this material and to enhance its characteristics and aesthetic properties.

The brick represents, in this case, the value of industrial archaeology, tells the story of the place, its link with Milan, which has always been a city built with bricks since ancient times. The industrial sheds, covered with metal shed roofs, have brick facades that have been maintained and preserved: they are the memory of the past, in sharp contrast with the new, represented instead by the tower placed between the buildings of the twentieth century and covered with a glass facade and a system of sunshades in anthracite-coloured metal.

We worked again with the brick in the project for the competition for the enlargement of the hospital of Tambacounda, in Senegal: there we hypothesized to realize the facades in blocks of raw clay, that the same population could have produced exploiting the local resources and the climate.

SLOT STUDIO

It is definitely Brick Townhouses(***A8***). In this case the material became a conceptual accomplice. We reached a special treatment in which both conceptuality and shape merged; Line pursues texture and texture integrates space that is the formula.

Stefano Corbo STUDIO

One of the best examples of how brick can be applied to architecture in an evocative and poetic fashion is Sigurd Lewerentz's St. Mark Church(***A9***) in Stockholm, Sweden. This church was built in 1960, and everything is made out of brick: floor, walls, ceiling. Brick and light shape the atmosphere of this sacred space.

stpmj Architecture

The Masonry(***A10***) / stpmj

By using two different kinds of brick in a single brick façade, it exposed the size and scale differences along with the division of the units.

Studio Farris Architects

We used bricks for the project of the Antwerp Zoo(**398p**) and for the Farmhouse Lennik project.

SUPA architects schweitzer song

Brick is most fascinating for us when it still was used structurally and visibly as such. Unfortunately brick today became a finishing material, an aesthetic surface without deeper meaning as an architectural element. Basically today the brick is reduced to a tile.

TAKK Architecture

Sant Pau Hospital in Barcelona(***A11,12, Next Page***) by Lluís Domenech I Montaner is an icon of the modernist catalan movement which inspires us everyday living in bacelona.

TheeAe Architects

There quite many projects we like. If we want to refer one of them, we would like to refer ABC building in South Korea designed by Wise Architecture. The use of brick was well utilized to bring the contrast of density through the gap of the space through the installation of bricks. Also the massing of the building is well expressing of this contrast. That is what we believe how architecture simply can be done well by the use of materials.

TOUCH Architect

"The Rest Avenue"(***A13***) is a neighborhood center in Surin, which is located at North-Eastern part of Thailand, well-known as an ancient city with Prasart; a load-bearing wall structure castle which using all stone-brick. This place consists of food and souvenir retail shops originated from characteristic of Surin.

The 3 elements from 'Prasart' which are PRASART's LAYOUT, BRICK PATTERN, and GRID SYSTEM are being used, and combine them together in order to created an overall plan. The 4 main entrances also have their features related to Surin which are;

NORTH | An Elephant gate
EAST | A Rice gate
SOUTH | A Silk gate
WEST | A Khmer gate.

This West gate is the one that use all brick pattern transmitted from the ancient pattern of stone-brick masonry. It integrates modern pattern with an ancient character.

UNStudio

For the Villa Wilbrink single family house that we designed in the early 1990s, we used a technique for the facade where we glued the bricks together, instead of using cement. Because the layer of glue was so thin, the bricks were much more closely stacked together and it gave a really wonderful, solid effect to the house.

A11 Sant Pau Hospital © Nicky Boogaard

A12 Panorama view Sant Pau Hospital © Jesús Arpón

A13 West Gate of the Rest Avenue

Tile

Arenas Basabe Palacios Arquitectos

The project that best employs tiles in order to build an idea is 'Ninette House(**062p**)', a single family house located in the outskirts of Madrid. On the ground floor level, the flooring, a pixelated design made of micro-terrazzo tiles(***B1***), extends beyond the glazing in a subtle color gradient, blurring the boundaries between the building and the surrounding landscape.

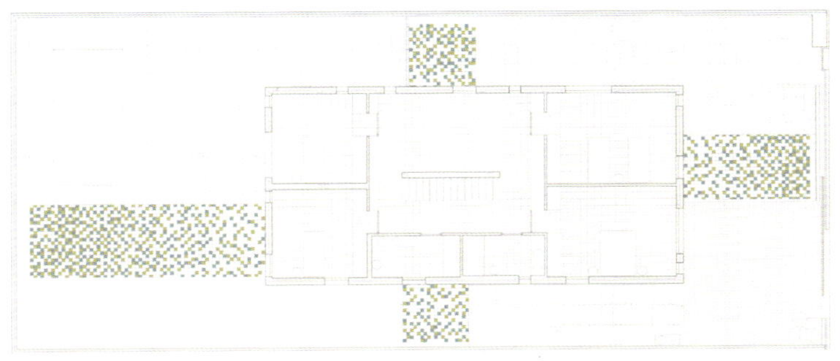

B1 First Floor of Ninette House

ARPHENOTYPE

Tiles are amazing in the past and today. They can have special surfaces that can create a unique look. Most people probably think about bathrooms or floors by mentioning tiles. But what if they are used on facades? They are durable and easy to clean. Architect Frank Lloyd Wright innovatively used three-dimensional tiles in such a way for the Enis house (1923, ***B2***), that the director Ridley Scott shot a scene of the famous futuristic film Bladerunner (1982) in that house. The entire system of the house is based on the tile pattern, which is located both inside and in the facade. However, I would like to point out another architect: Manuel Herz. He used a special tile for the façade of the Jewish Community Center in Mainz, Germany. The building is clad with three-dimensionally riffled, green glazed ceramic tiles, which underlines the sculptural appearance of the center. A special assembly plan has been created which creates a pattern of different standardized lengths of tiles. Lastly, a pavilion, which very impressed me is the Austrian pavilion(***B3***) for the Expo by SPAN & Zeytinoglu Architects in 2010. The organic shape and the gradient by using white and red tiles was an amazing idea. It also shows, that organic shapes can be covered by tiles. In summary, can be applied to facades, ceilings etc. and can have different forms, which in the age of Industry 4.0 and mass-customization creates new possibilities for new forms.

B2 Enis House © Mike Dillon

AZC

We used glazed tiles on the exterior cladding of "Terrasse 9"(***B4, Next Page***) housing project.

BOARD

One of my favourite projects in which we used tiles is our design for an "Office and Hotel Room(**072p**)" that was recently completed in the centre of Rotterdam. There, since the entrance of the project was already slightly set apart from the main staircase towards a little niche, we saw an opportunity to create a "secret" and hidden door to the office and hotel room, which would disconnect the space magically and as a hideaway from the business of the city and 'real life' in general. To emphasize the hiddenness of the place, we continued the existing white ceramic tiles from the corridor into the niche of the entrance and covered the entire door with the same tiles too, which camouflages the entrance to such an extent that until today the postman is having trouble finding it.

B3 Austrian Pavilion © Kimon Berlin

Carlos Lampreia

Tiles are a very popular material in Portugal, and i like to use them

B4 Facade View of Terrasse 9

B5 Chiado Building in Lisbon by Alvaro Siza © carlos lampreia

B6 Elbek & Vejrup, Mikkel Frost

B7 Church of San Marco de Canavezes © Manuel Anastácio

as a panel to reflect light and protect the walls from water like in the Chiado Building in Lisbon by Alvaro Siza(***B5***), where soft coloured tiles in the facades bring light and sun reflections from the river to the streets around.

Casanova+Hernandez Architects

For our project in Jinzhou(**140p**), China, we have used broken tiles to create a hybrid landscape consisting of a park and a museum where the limits between landscape and architecture are erased. By using the same broken tiles of 4 different colors to create the pavement and the benches of the park on the one hand, and the facades and the roof of the museum on the other, a three-dimensional landscape is created, understood as a complete work of art where colors and reflections unify the complex.

In this case, the material, its properties and its possibilities of use, have been at the very core of the entire design concept. The commission required from all invited designers to reflect on their own national identity. Instead of entering the narrow field of national identities, we preferred to make a reflection on the concept of cultural hybridization, a phenomenon representative of all commercial and cultural relations between West and East during centuries already. The development and use of ceramics as mosaics in the West on the one hand and the development of the crackled glaze of the Chinese porcelain on the other offered us an interesting conceptual starting point to make research on and to base our design. The material defined the design since the beginning of the design process.

CEBRA(Mikkel Frost)

We´ve used tiles graphically in a few of our projects – like digital pixels(***B6***). At the Elbek & Vejrup project, we also made black space invaders on the bathroom floors and in the Experimentarium project we created optical illusions with tiles. It´s a cheap but visually very powerful way to twist spaces and beef up dull spaces.

Davide Macullo Architects

The only tiles we consider as quality product are the hand produced ones. I think that besides the beauty of the historical buildings around the world that have seen wonderful tile works, in the recent history Alvaro Siza has been able to use the tiles in a simple but extraordinary way. For example in the church of San Marco de Canavezes in Portugal(***B7***).

Donner Sorcinelli Architecture

CD House(**082p**) roof has been built with large size flat dark grey ceramic tiles not so usual for the local environment which is usually characterized by light brown Spanish colonial ones.

Keiichi Hayashi Architect

La maison de Jean Pierre Raynaud(***B8, Next Page***) / Jean Pierre Raynaud, Raynaud completely covered the interior of his house with white tiles and created whole space as his artwork. New space stands as bulk material of accumulation when all materials were restored into a single tile scale. Thus, He represented how masonry joints play an important role in scale to fill newly-opened space. I was attracted by his artwork because it creates freedom in space that is confined by the restriction of surficial, divisible nature of tiles, transforming it into super-monomaniac space that compulsively creates a feeling of minimalism. Furthermore, the style of tiling itself attracts me more as his artwork.

LANDÍNEZ+REY arquitectos

We still fascinated by the use of the tile in the fifth facade that results from the project of Mercat de Santa Caterina (Barcelona, Spain, ***B9***) by Enric Miralles and Benedetta Tagliabue, 2005.

M artı D Mimarlık

We use tiles generally at wet spaces like bathrooms, wc due to its hygenic and water-proofing characteristics. In Automative Exporters Vocational High school, we have used a large dimensioned tile (100x300cm) for interior wall cladding.

object-e architecture

Our project for the redesign of the Koum Kapi waterfront competition(***B10,11***) (that won the second prize) is to a large extend a study of a tiling system. Aim of the proposed intervention is the design of a public space that will be able to address both the local (at the scale of the neighborhood) and the city level. The proposal is organized through the subdivision of the ground plane, that follows an approach that departs from what we usually expect in similar situations, that is, linear tracings along the seaside that are defining zones of different movements and activities. Instead the ground plane is subdivided in rectangular shapes of different sizes and materiality. Therefore the design is structured around different ways according which tiles can be arranged and manipulated in order to form paths, define movement and direction, differentiate functions, create continuities and signify important areas. The subdivision and tiling is controlled through a packing algorithm. The parameters of the algorithm helped us control density, direction and position of the tiles. The maximum and minimum size of the packing process is altered locally according to the different events that each area will host. The result is a patchwork of sizes and materials, that apart from its main organizational properties it also functions as a reference to both the nature of the Koum Kapi area as a patchwork of diverse elements and to patterns that can be found throughout the city of Chania (for example the masonry of the city's walls). The size of the rectangulars range from a few centimeters to 10 meters, allowing this way for a large variation.

Along this patchwork various 'spacial events', of larger or smaller scale, are taking place that are usually signified with the presence of the larger rectangular shapes. The tiling functions more like a field, whose elements are of varying and locally differentiated size and density. Therefore the result of the proposal is an urban field that at the same time promotes local differentiation and provides a recognizable vocabulary throughout the seaside.

Out project for the Souda Ferry Terminal competition(***B12***) (honorable mention), aims to create a unique, narrative experience for the user by challenging what we expect to encounter in buildings with such 'heavy' uses, which need to accommodate large numbers of visitors in transit. The detailing and character of the proposal, that wouldn't be out of place in very different architectural typologies, references a time when travel for leisure was a form of luxury and

B8 Sculptures in gardens of Palauet Albéniz © Jordiferrer

B9 Mercat de Santa Caterina © Tony Hisgett

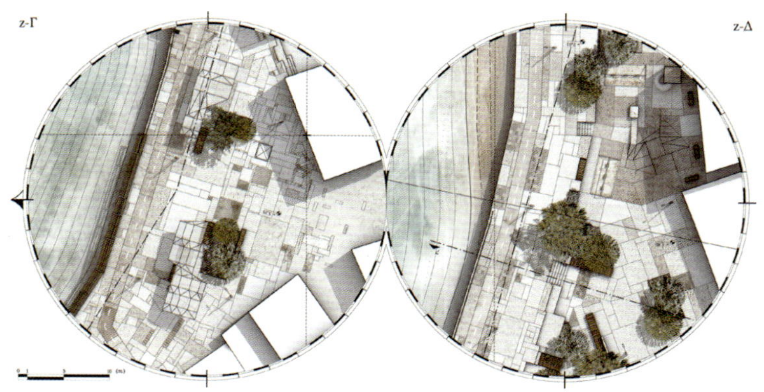

B10 the Koum Kapi Waterfront Competition

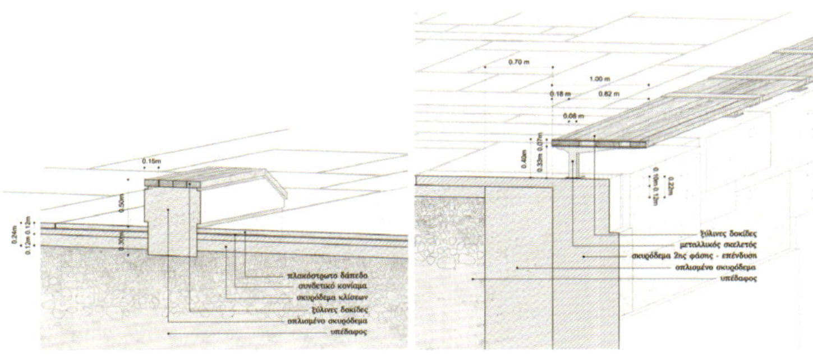

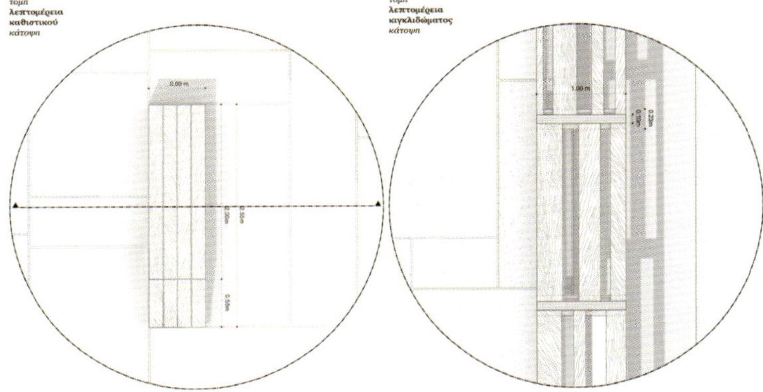

B11 the Koum Kapi Waterfront Detail

B12 the Souda Ferry Terminal Competition

B14 Castel Vecchio Museum, © Paolo Monti

therefore the buildings that were meant to support it were echoing that character. At the same time, the simple functional layout of the building and the materials used assure that the proposed design will be able to sustain the very high load of use that results from the number of visitors of a ferry terminal. However, the building tries to transform a place that you are supposed to move through in order to go somewhere else into one that you could probably want to be – by affirming the contradictions that may arise from such a reversal.

The ways in which we employ tiles therefore echoes the above: Colored concrete tiles are employed both on the interior and exterior of the station in order to provide a filling of 'luxury' to a place that is connected to mass tourism. We are trying to imagine our proposal as a building that will recreate the filling of the station as a place where something special is happening; where slowness can maybe take the place, if only as a feeling and not an actual condition, of the speed that we usually encounter in such places. Using concrete tiles, along with the other material selections, enhance that feeling.

OOIIO Architecture

Definitely the refurbishment of two squares in Talavera de la Reina, a 75.000 inhabitant town in Spain(**150,158p**).

I decided to use local Talavera´s traditional pottery as the main material, as it is a product of excellent plastic qualities and deeply rooted in local culture. Usually it used with drawings and motifs from the Renaissance in small-scale domestic objects. I propose greatly increase the size of those drawings that adorn vases and dishes up to the urban scale, showing new possibilities for this fabulous craft material.

Piuarch

In our architecture we've never worked with ceramics as the architectural envelope of the project. Ceramics have been used exclusively as a horizontal and vertical coating. On this subject we have often experimented with customised solutions in collaboration with companies.

In several occasions, we've designed surfaces to best integrate with the philosophy of the project, restoring a global aesthetic: this is the case, for example, with the concept store for the women's lingerie chain Yamamay, where the ivory square flooring, with geometric inserts in marble ton sur ton, allows you to communicate the ideals of lightness, softness, brightness, geometry and symmetry of the point of sale.

SLOT STUDIO

We designed and built an assembly plant for Quin(***B13, Next Page***), the leading automotive wooden parts producer and we needed to achieve a very special aesthetic standard: a trendy-elegant-consistent and refreshing effect. Tiles helped us with this task by covering up the main building's exterior.

Stefano Corbo STUDIO

Castelvecchio Museum(***B14***), a project of adaptive reuse designed by Carlo Scarpa in Verona, Italy, between 1958 and 1974, is a brilliant example of how tiles, along with other materials, can contribute to create a mosaic of colors and spatial effects within the same build-

B13 Quin

ing.

Tiles, in fact, are always combined with other materials –stone, concrete, wood– and respond to the architect's idea to take the visitor along an intense journey along the medieval castle; tiles help emphasize the relationship with the existing structure by analogy or by contrast.

stpmj Architecture

Concert hall and Music school (MUCA) / COR ASOCIADOS,

A simple treatment with ceramic tile as exterior finish, provides various outlooks depending on its reflection and where you stand.

SUPA architects schweitzer song

A tile is the little pretty brother of the brick. It is still the best solution to certain functional problems within a building (e.g. in the bathroom). We try to apply it very reduced, white, minimalistic.

TAKK Architecture

Casa Vicens(***B15***), which was the first house built by Antoni Gaudí. again an icon of catalan modernism where we find that the architect in this case really managed to take advantage of the properties of tile.

TheeAe Architects

Tile is most widely used interior materials in building industry. We use tile for floors and walls. Even tiles can be easily customized with any pattern or images. The floor pattern(***B16***) we designed for one of our interior project was baked in the factory and installed on floor. The process of ordering the product from the manufacturer was not simple, since we needed to make numerous times of reproduction due to unsatisfactory on color differences. However, overall quality after installation was remarkable.

TOUCH Architect

Sukpisan SHRINE is a 'Brahma shrine' combining with a 'house of Guardian spirit'. Since the overall context is modern contemporary architecture, the shrine was designed by following the context which finally becomes a modern-like shrine but still keep traditional function as it should be.

The architecture has 3 gables, big one, medium one, and small one which is for MAIN ENTRANCE, BRAHMA, and GUARDIAN SPIRIT respectively. Floor level has been elevated for 1 meter because the shrine should be placed on higher level than the house. It is located in a large green space in between each building which can easily access and worship. Moreover, Brahma has 4 faces, in our belief, 4 faces can see things and protect us in 360 degrees.

The big gable for main entrance was designed to make the people who walking up to the shrine, should 'BOW THEIR HEADS' since the gable roof has been elevated down at the last step of staircase, in order to force everyone to salute the Brahma.

Material using of this shrine is mainly 'white marble tile' cladding on reinforced concrete structure. It has been used in all floor, wall, and roof, since the continuity of massing was created. Moreover, tile material and lighting design shows different effects between DAYTIME and NIGHTIME. During daytime, golden Brahma can easily be seen from 'white marble tile' background, while 'recessed lighting' makes the golden Brahma be seen in darkness. The shrine shines and becomes a lightness during nighttime.

UNStudio

The wasl Tower in Dubai(**130p**) I think is the most exciting use of tile that we have developed to date.

In this project the specially designed tiles principally have sustainable functions, whereby they lessen the heat load on the building. But they also give the facade an unusual visual effect, which relates to the scale of product design.

B15 Casa Bicens ©jorapa

B16 The Artist House

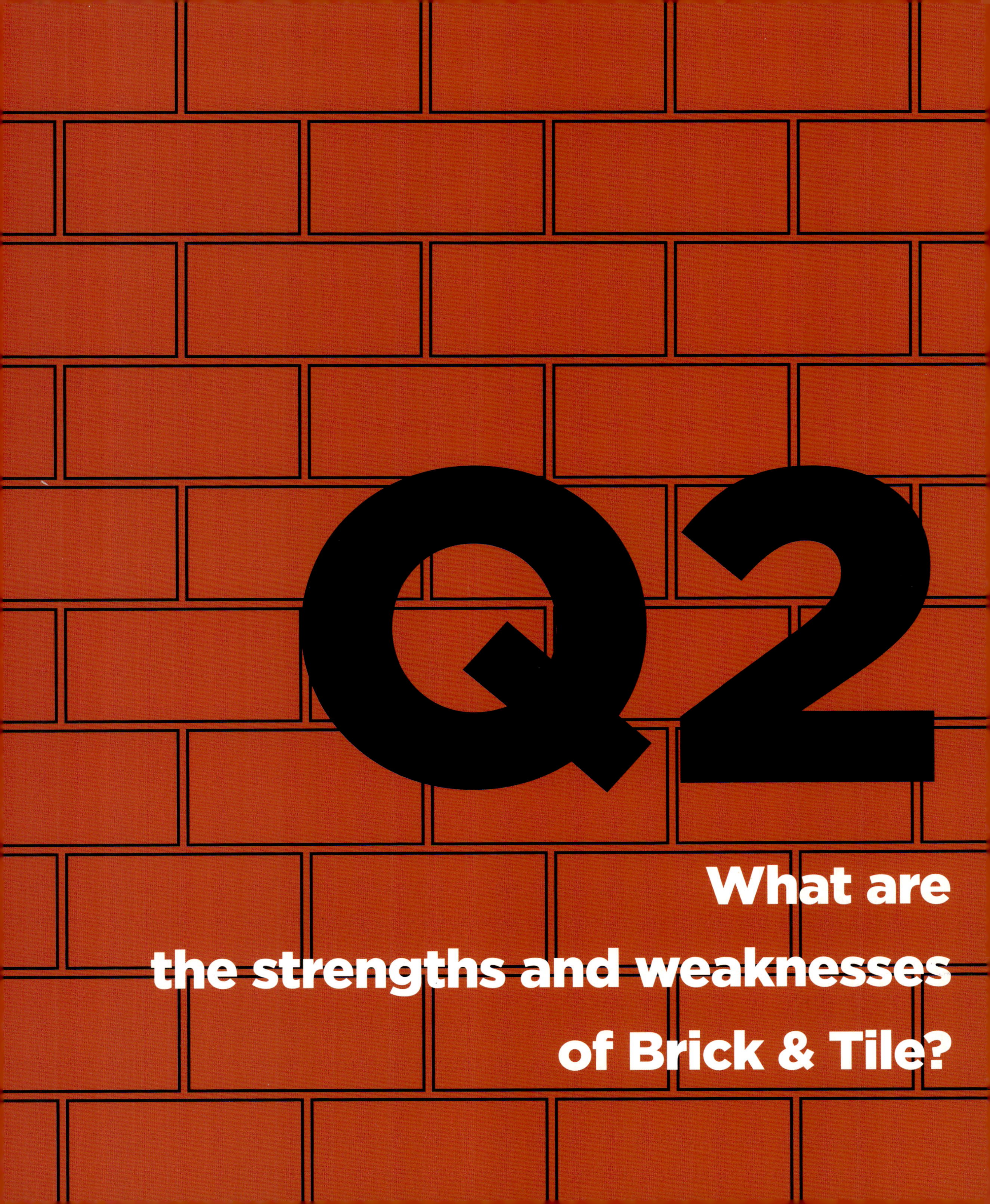
Q2
What are
the strengths and weaknesses
of Brick & Tile?

Brick

ARPHENOTYPE

Bricks can be seen an nearly any culture. They can be mass-produced, or customized. They give architecture a native look. They can with assistance by algorithms and robotics create wonderful patterns, if we refer to the works by Gramazio & Kohler(***A1***), which have been possible since ages by bricklayers, but most of the techniques are forgotten. Nowadays, robots are applying these techniques faster and even with a high accurately.

AZC

Brick is a very graphic material as the light showcases every brick and draws every joint. It also is a long-lasting material with no need for refurbish or clean.

It would be nice if we could use it as a structural material for columns or beams also. It would be interesting to use it as an insulating material also.

BOARD

Some of the strengths of brick are certainly its robustness, longevity, affordability, and ease of use and application. For the project "Home Outside" we wanted to use all of these qualities. But since the square was located on a former industrial site - a former city port, trading, and production location – where a renowned factory called "de Volharding"(***A2***), that was mainly constructed with brick, once stood, we wanted to connect the project through the use of brick for the square's surface with the industrial past of the place. In order to deal with one of the weaknesses of brick - namely its lack of reinforcement, making brick easy to assemble but also easy to disassemble – we placed the 3 elements of the square on concrete slabs with engraved in them information about the meaning behind the elements.

Carlos Lampreia

The strength of brick to me is about surface and order in construction, and also because when someone put them together gets a finished surface and a texture.

Has Louis Kahn once said a brick 'likes an arch', and in that sense it's a material that works well producing massive walls, but needing arches or other types of substructures in order to achieve voids between those walls.

Casanova+Hernandez Architects

In our opinion, brick is a good means to create a neutral base on top of which other materials can get the predominant expressive role.

Brick can be also an elegant, silent material to express serenity and neutrality in architecture if its quality is good and its texture, proportions, and color are carefully chosen.

The weakness of brick is that it can also be understood as a quite vulgar material if it is not chosen and used carefully.

CEBRA(Mikkel Frost)

It doesn't quite make sense to me to talk about weaknesses when it comes to materials. Different materials simply have different properties – like musical instruments. Drums are amazing for rhythm, but if you need to play a melody you might need something like a guitar. So, materials are all great, but they need to be applied properly and used in the right situation. For bricks specifically, I´m often faced with economical challenges. If I manage to find a stone that is just right in terms of color and texture, it is often over budget and I have to settle for something else. Making beautiful details with brick also cost labor hours and is, unfortunately, a rare possibility.

A2 de Volharding © Joris

Davide Macullo Architects

In our practice we build up an awareness of emphasizing the strength of the materials or using the materials for their adaptability to our goals. So we don't care about weakness.

Donner Sorcinelli Architecture

Natural properties are pros as well as the intense work required is cons.

Keiichi Hayashi Architect

Strength: It can become ruins.

Weakness: It can not be roofed.

Kimura Matsumoto architects office

During the modernization in Japan, brick masonry buildings were constructed to prevent fires in urban areas crowded with timber houses. In the project "Another base in G" we tried to reduce the sound transmission by focusing on the inherent sound insulating qual-

A1 Acoustic Bricks ©Gramazio Kohler Research, ETH Zurich

ities resulting from the weight of brick.

Both of these cases were positively affected by the characteristics of brick masonry.

As I mentioned above, although a brick's hardness can make it difficult to process, its form and size (designed to be carried in the hand) make it possible to form various shapes. This ambiguity is something I find attractive and the balance between a material's strengths and weaknesses is something I pay close attention to.

For example, by using brick masonry as a structural wall we lose the ability to create openings at will. However rather than see this as a negative, we can except this and celebrate its unexpected results.

Not just in the case of brickwork, but with any material, I think it is important to understand and accept both it's strengths and weaknesses and attempt to work with them.

LANDÍNEZ+REY arquitectos

It is evident that the use of any construction system is intimately linked to what is its price. In the case of brick this depends directly on the price of labor and the tradition of using the material in each country. In Spain it is a technology with a huge industrial experience and technological development, which has led to a constant experimentation by architects.

The strength of the brick is its expressive capacity and experimentation capabilities, as well as its local tradition. It weakness is only the result of its misuse.

M artı D Mimarlık

Widespread use of brick enables builders to find it, also it is economical and builders are familiar with it.

murmuro

As every material it has its strengths and weaknesses, but also depends on how we look at it and what we make of it. Brick is normally associated with masonry walls, sturdy and robust walls, however we have designed an entirely perforated wall with bricks. It is also very durable and has very low maintenance. As weakness, we would point out that a brick façade as complex as the one we've designed requires highly skilled handwork, which can easily transform a very competitive solution into an expensive one.

NISHIZAWA ARCHITECTS

Strength: Help the temperature inside the building become stable, it will be warm in rainy season and cool in hot seasons.

Second-hand brick is easy for us to show off the traditional Vietnam colonial architecture's atmosphere under the French domination, combine with new materials make the building look more contemporary than ever.

Weakness: Cause the wall need the thickness so we have to use the concrete-beam like a main-factor, brick at the time become like a finishing-layer for exterior wall of the building.

OOIIO Architecture

Strengths: low maintenance, cheap material (in Spain), easy to work with, can solve structure and façade at the same time.

Weakness: not the best climatic behavior with sun heat, it works as a heat accumulator.

Piuarch

It's difficult, not to say impossible, to make general statements about a material's ability to perform. Each material makes sense when applied in a given context and each context pushes to the search and identification of particular solutions.

SLOT STUDIO

It is a rough hard-sensed material, and this is good for certain purposes but once you are looking to invoke fresh, light, transparency signs it won't be very easy to handle. Presence is definitely on its side, so is traditionality and warmth.

Stefano Corbo STUDIO

Defining pros and cons in the use of brick is difficult, as every material should be considered in relation to the overall logic of the project, and based on the relationship between its different components.

Bricks are normally associated to a traditional vision of architecture, which is somehow disconnected by technological advances.

stpmj Architecture

Strengths: Brick assembly can give diverse result of the surface with its various sizes, colors, patterns of assembly, and colors of grouts. Also, it requires relatively low maintenance over time.

Weakness: It is not impossible but there is limit to actualize certain size or shape of space with its stack assembly, it is not strong against seismic wave.

Studio Farris Architects

The most interesting advantages are: It is economical, it is hard and durable, it is low maintenance, reusable and recyclable, it produces less environmental pollution during the manufacturing process.

The disadvantages could be: construction with bricks is time consuming. Since bricks absorb water easily, they cause flaking when not exposed to air.

SUPA architects schweitzer song

A brick is very raw and fundamental. It is a minimal building block out of which almost everything can be generated.

TAKK Architecture

Brick is a very cheap material, that can be used both as a structure or as a façade, that is something we find very interesting. On the other hand it is very linked nowadays in our context to a certain architectural style with certain political implications that are not exactly followed by us.

TheeAe Architects

For answering of advantage of brick, it is almost maintenance free material. No need paintings, caulking, or staining for years and years. In addition, it protects sound from the outside wall, and it is fire resistant, looking very natural and comfortable, and most importantly environmental friendly.

For disadvantages, it can be expensive due to much involvement on labor. Thus, too heavy to require strong foundation. Also, it takes time to build well.

TOUCH Architect

Brick material, in general, is durable and highly fire resistant. Since it is a very strong material, that can be used as a load-bearing wall and helps to support the building structure. Moreover, it is easy to construct which does not require any finishing or painting if not necessary, while still have their own aesthetic by creating different orientations and patterns. However, it does have some weakness. Heavy weight will increase stress to the piling and foundation which is needed to support the bricks. Moreover, it consumes more time to create such a wall.

UNStudio

Brick laying of course doesn't require huge machinery, it can be done by hand, which makes it great for smaller scale projects. However, brickwork is not strong enough to be used for very large spans, unless the walls are very thick - which is why we don't ever see brick skyscrapers.

***B1** Micro-terrazzo tiles © Shlomo katzaw*

Tile

Arenas Basabe Palacios Arquitectos

The versatility of the micro-terrazzo tiles(***B1***) allows us to use them both inside and outside, varying only in finish (either polished or rough). This allows us to reinforce the idea of a seamless continuity between interior and exterior spaces. The main 'weakness' of this material could be associated to the irregularity of its edges, which could stand in the way of the sharpest detailing. Nonetheless, we do not perceive it as a weakness rather than one of its inherent qualities. No tile is identical to another one, turning imperfection into one of the sources of its appeal.

ARPHENOTYPE

As a building material, ceramic tiles are very ecologically friendly to the environment. They are really just clay, some minerals, water and fire. Ceramic tiles usually have a dense surface such as window glass (eg. porcelain stoneware) and are often very smooth. For this reason, dirt and germs on ceramic tiles have little chance. In general they are standardized rectangular patterns, which can be positive or negative.

AZC

Tile is a material that does not need maintenance. It lasts for a long time in excellent condition.

BOARD

Apart from being a hard-wearing material with an ability to cover surfaces in a simple, solid, and affordable way, we consider the ability of tiles to create complex mosaics - and with it narratives and multifaceted meanings - as one of its greatest strengths. This quality we wanted to use when we designed one of the lavatories for the "Office and Hotel Room" project. Because, while the main space of the project is white, open, neutral, and flexible, this lavatory contains an impressive hand-made piece of art depicting a palm tree, a 1:8 reproduction of a design originally made by STAR strategies + architecture and intended for a public square in the city of Elche, Spain. Thus, the clients, admirers of Adolf Loos, preferred to keep the most excessive ornamental space for private use. In order to realize the time-consuming mosaic, which, in our particular case, could be viewed as a weakness of the use of tiles for this project, we flew in an expert and artisan from Spain.

Carlos Lampreia

Tiles are a coating material, usually used to cover walls or floors inside and outside while defending them from water. Since its a ceramic product it can improve the light and receive a full range of colour and textures. The weakness about them is firstly their thickness, and secondly that they can became fragile with the use and constructive tensions.

Casanova+Hernandez Architects

By using broken tiles, in the same way that Gaudi(***B2, Next Page***) and the Catalan modernist architects did, to cover pavements, facades, and urban elements, we have been able to create unique hybrid landscapes with complex geometries in which all horizontal, diagonal and vertical planes work synchronically as a stage where urban life occurs. By breaking the tiles into small pieces, they can be adapted easily to almost any surface, which gives the designer a lot of freedom of expression.

The versatility in colors and patterns of the broken tiles provides as well a high degree of freedom of design and of chromatic richness. This has allowed us to play with the perception of the whole by defining the relation between the different parts of a design in terms of colors and patterns, thus providing the opportunity of creating perspective effects and rich views throughout buildings and landscapes.

The weakness of tiles is that they can be broken easily under certain circumstances, but if they are applied already broken on purpose, this point becomes irrelevant.

B2 Gaudi's Tile work

CEBRA(Mikkel Frost)

In cold climates like here in Denmark applying tiles to an exterior façade can prove technically difficult. In the very few places I know where architects have stuck their neck out the tiles simply come clattering down.

Davide Macullo Architects

The handmade tiles are simply a pleasure for the senses. The industry produced tiles are aggressive and don't belong to the human space because it is a product that express an industrial process without soul and this is psychologically negative for the environment of people to live within.

Donner Sorcinelli Architecture

They are easy to assembly on the roof but easily affected by the local weather conditions.

Keiichi Hayashi Architect

Strength: You feel like being in the water when you are in a tiled room.

Weakness: It slips easily.

LANDÍNEZ+REY arquitectos

The tile industry in Spain is one of the most powerful in the world, with cutting-edge technological development. Undoubtedly, the good use in its tradition, together with this representation of the future, constitutes its best potential.

M artı D Mimarlık

Characteristic of being modular material is useful for design process but, seams and joints may causes some hygene problems. However, larger sizes of tiles have been produced for a few years and it minimises hygene problems while increasing need of precise work at construction process.

object-e architecture

We tend to think of materials in architecture not in terms of strengths or weaknesses, but rather of possibilities. Therefore we are interested in the possibilities that tiles can offer to us. Tiles of course are based on the idea of repetition. An element has to be repeated, one way or another, in order for the tiling system to be created. Therefore we find a lot of possibilities in the rules that we can apply to the repetition of the tiles. By incorporating difference in the repetition process, many unique condition can arise. Difference in the case of the tiles can be related to position, size and material properties/color; or – maybe more interestingly – to a combination of the above. Through such techniques tiles can stop being a single element on a grid that can create a surface and become many more.

OOIIO Architecture

Strengths: amazing plastic possibilities, good to keep rain water out of your walls, in some places like the city were we work with it, there is a deep tradition with it and great craft artisans.

Weaknesses: You need to glue it really well on the walls to avoid tiles jumping out.

SLOT STUDIO

Tiles are versatile and economic. They can go from simulating to conceptualizing, but one has to be very careful when it comes to choosing and justifying the use of it cause you may fall into pretentiousness or cheapness.

Stefano Corbo STUDIO

Defining pros and cons in the use of tiles is difficult, as every material should be considered in relation to the overall logic of the project, and based on the relationship between its different components.

stpmj Architecture

Strengths: Tile can be applied various cases, it is light and has wide variety of selection.

Weakness: For the use of exterior, some loss or crack could be happened because of the contraction and expansion of the material itself as exposed on sunlight and water directly.

SUPA architects schweitzer song

We are dreaming of a seamless tile, the beautiful surface of a ceramic tile without the joints. If it would be possible to bake and glaze a whole bathroom or kitchen we would be the first to apply it.

TAKK Architecture

Tile cannot work as a structure, but on the other hand the aggregation of tiles can be very beautiful worked in terms of patterns and geometries, thank to the multiple possibilities of colors and shapes that it can develop.

TheeAe Architects

Most of our interior projects were done with the use of (ceramic) tiles for the finish materials. There are truly tons of choices designers can choose from and it is also easy installation by semi-skilled cement workers. Also, its application is quite simple compare to other building material installation.

In contrast, the weakness of using tiles might its strength. It is easily broken when it is dropped on floor and it requires time and labour to install one by one as same as brick installation.

TOUCH Architect

The tile itself, has a huge variety of sizes, shapes, colours, patterns, and textures. It can be used for flooring, wall cladding, and for some furniture such as, a counter top for kitchen. It creates different types of an architectural character through its colour and pattern, since it is a simulation form of natural elements, such as wood, concrete, granite, and marble. Besides, it helps protect the concrete floor and wall from cracking and dirt since it is easy to clean, as well as, being respect from most clients because of trusting in its quality with reasonable price. However, there is some cons using the tile. Even if the tile is not affected by humidity, it may cause some mold on tile grout which is not easy to be replaced.

UNStudio

Their lack of strength is literally their weakness. Although the strength is improving all the time, we still need to be careful with sizes, as these materials can break very easily.

Q1
본인이 벽돌이나 타일을 사용한 프로젝트 중 가장 마음에 드는 작품 또는 다른 건축가의 작품을 소개해 달라.
(인테리어, 파사드 등 어디에 사용했든 관계없다.)

Brick

ARPHENOTYPE

벽돌은 고대의 재료로 오늘날에도 재발명할 수 있다. 바로 떠오르는 프로젝트 중 하나는 피터 줌터(Peter Zumthor)가 독일 쾰른(Köln)에 지은 콜럼바 박물관(Kolumba Museum)이다. 그러나 이 맥락에서 또 다른 건축가를 언급하고 싶다. 그는 당시에 벽돌을 새로운 방식으로 사용했으며 그의 이름은 일라디오 디에스테(Eladio Dieste)이다. 전통과 혁신의 경계를 흐리게 만든 몇 안 되는 건축가 중 한 명이다. 내가 생각하고 있는 건물은 이글레시아 데 아틀란티다 크리스토 오브레로 이 누에스트라 세뇨라 데 로르데스(Iglesia de Atlántida Cristo Obrero y Nuestra Señora de Lourdes) 교회이다. 그는 건축에서 컴퓨터를 사용하기 훨씬 전인 1958년에 이 교회를 디자인했다. 일라디오 디에스테가 벽돌로 만든 이중 곡선 모양의 벽과 천장은 오늘날에도 여전히 놀랍고 독특하다. 이는 벽돌의 모양과 색상에 의해 구동되는 놀라운 빛과 우호적인 공간을 만든다. 마지막으로 나는 건축가 코엔 멀더(Koen Mulder)의 "Het Zinderend Op'pervlak"이라는 책을 언급하고 싶다. 이 책은 벽돌로 만든 벽에 패턴과 곡선의 가능성을 강조한다.

AZC

파리의 '비네 오피스 지구(Hôtel Indsutriel Binet)'에 있는 마지막 사무용 건물에 벽돌을 썼다. 하얀 유약처리가 된 벽돌 파사드는 건물이 있는 파리의 북부 지역과 사무용 건물이라는 용도에 잘 맞았다.

BOARD

오늘까지 우리는 우리 프로젝트에 단 한 번 벽돌을 썼었다. 이는 도시 가구가 있는 광장의 디자인으로, 우리가 "바깥의 집"이라고 부르는 프로젝트이다. 여기서 우리는 광장의 표면에 벽돌을 써서 그 콘텍스트에 연결할 것을 제안했다. 그렇게 하면 그 동네의 전형적인 붉은 벽돌 표면이 계속되면서 광장과 주변 환경 사이에 강한 유대 관계를 형성할 수 있었다. 그러나 광장의 주요 요소는 표면이 아니라 우리가 추가한 예술적이면서 기능적인 3가지 요소였다. 이는 벽난로, 테이블 및 보트였으며 이 세 가지 요소로 우리는 하나의 예술 작품인 독특한 광장을 만드는 것을 목표로 삼았다. "크란볼베르크(Kraanbolwerk)"라고 불리는 사이트의 주민들뿐만 아니라 방문객까지 바깥에서 집처럼 편안히 느끼면서 만나고, 얘기하고, 요리하고, 먹고, 몸을 따뜻이 하고, 놀 수 있게 하고 싶었다.

Carlos Lampreia

내가 가장 좋아하는 벽돌로 된 집 중 하나는 핀란드에 있는 알토(Aalto)의 무라찰로 실험집(Muuratsalo Experimental House)이다. 테라스에서 벽돌이 만드는 다양한 질감을 볼 수 있고 이는 외부의 흰색 벽돌에 맞서 건물의 윤곽을 더 뚜렷이 보이게 한다.

Casanova+Hernandez Architects

네덜란드 블라리쿰(Blaricum)에 있는 공동 주거 단지 프로젝트에서 우리는 건축적 매스보다 보이드(void)에 중요성을 부여했다. 가느다랗고 하얀 알루미늄 테가 둘린 다양한 구멍을 뚫을 캔버스로써 검은 벽돌로 순수한 용적을 만들었다.

이 프로젝트의 '매스 추출'은 검은 벽돌로 된 이상적인 용적, 특히 건물의 네 모서리에서 추출할 때 그 거대한 외관을 감소시킨다. 보이드의 흰색은 벽돌 외관의 어두운 색과 대조를 이루며 게슈탈트 이론의 구체화 특성이 말하는 주관적 또는 환상적인 윤곽을 만든다. 1940년대 바바라 헵워스(Barbara Hepworth)가 만든 많은 나무 조각에서와 마찬가지로, 색상은 보이지 않는 것을 볼 수 있게 해주고 보이드는 작품의 중요한 건설적 요소가 된다. 검은 벽돌은 건물의 바깥 외피 모양을 강화하고 흰색 프레임은 구멍이 뚫린 보이드의 존재를 강조한다.

CEBRA(Mikkel Frost)

우리는 많은 프로젝트에서 벽돌을 사용했다. 덴마크에서는 주택 대부분을 벽돌로 지으며 우리 문화에 깊이 뿌리를 둔다. 우리는 테크 회사인 엘벡 앤드 베이럽(Elbek&Veirup)의 사무실 건물에서 벽돌과 그 건축 시스템을 재해석할 수 있었다. 여기서 벽돌을 디지털 픽셀처럼 적용했으며 여러 가지 밝은 색상을 썼다. 거의 마인크래프트의 미학 같다.

Davide Macullo Architects

최근 서울 강남에 지은 WAP ART SPACE 갤러리는 우리의 의도대로 벽돌을 흥미롭게 사용할 수 있었다. 벽돌은 몸에 걸치는 실크 천과 같고 몸을 따뜻하게 혹은 시원하게 유지하는 느낌을 준다. 건물 입면에 벽돌이 미치는 영향은 축척의 축소와 연관된다. 멀리서 건물을 보면 균질한 외관이지만, 건물에 가까이 올수록 클래딩 재료의 특성을 더 구별할 수 있다. 이는 다른 단계의 접근을 거치는 동안 계속해서 나의 관심을 끌고, 도시에서 가정, 공공에서 개인으로 축척을 축소한다. 가까이 다가갈수록 내 감정을 더 잘 느낄 수 있다. 벽돌은 인간의 작품을 표현할 수 있는 특별한 잠재력을 가지고 있다. 기계가 아닌 손으로 짠 천을 보는 것과 같다. 벽돌은 살아 있고 인간에 대한 보호를 잘 표현한다.

Donner Sorcinelli Architecture

아프가니스탄의 새로운 국립 박물관에 벽돌 쌓기의 두 가지 유형을 사용했다. 건물 외부에 조적 벽돌을, 건물 외피의 내부 층에는 점토 벽돌을 사용했다.

지역 재료만을 활용한 이유는 건물의 CO2 footprint를 줄이고 지역 인력을 활용하기 위해서였다.

Keiichi Hayashi Architect

멜닉코브 하우스 / 콘슨탄틴 멜니코브

나는 육각형 모양의 창문이 많은 원통형 벽이 있는 현대식 주택이 철근 콘크리트가 아닌 벽돌로 만들어졌다는 사실에 놀랐다. 그 당시 나는 이 인상적인 디자인의 수수께끼를 바로 이해할 수 있었기 때문에 기분이 좋았다. 벽돌을 45도 각도로 배치함으로써 육각형 모양의 창문이 있는 유연한 곡선의 벽을 쉽게 만들 수 있었다. 창문의 배치는 벽돌 건설 과정과 구조 역학에 적합하다. 멜니코프가 재정적 제약 때문에 벽돌을 사용했을 거로 추측하지만, 나는 이 작은 주택에 넣은 자유분방한 디자인과 노력에 대해 높은 평가를 하고 싶다. 멜니코브는 한때 승화되고 오래된 벽돌 기술을 차분한 구조와 디자인을 통해 모더니즘을 구현하는 구조로 향상시켰다.

Kimura Matsumoto architects office

내가 동경하는 벽돌 건축 프로젝트는 너무 많아서 하나만 언급하기 너무 어렵다. 벽돌의 가장 흥미로운 특징은 벽, 바닥, 천장 그리고 의자나 탁자 같은 모든 가구를 만들 수 있는 기본 단위라는 것이다. 이것의 가장 좋은 예는 Sigurd Lewerentz의 "성 베드로 교회"라고 생각한다.

LANDÍNEZ+REY arquitectos

우리 사무소에서는 각 재료가 건축학적 전제를 얻을 가능성을 조사하는 모든 프로젝트를 환영한다. 사실, 내 기억에 우리가 벽돌을 사용한 프로젝트 중 내부든 외부든 그 사용방식에 변화를 주지 않았던 프로젝트는 없었다. 우리가 건축 사무소로서 벽돌의 사용, 기술 및 표현력에 대한 진보적인 지식과 함께 성장해온 이유이다.

M artı D Mimarlık

벽돌은 아스마바흐첼라 주택(Asmabahçeler Residences)에서 클래딩 재료로 사용했다. 많은 이가 벽돌을 호의적으로 인식하고, 벽돌은 '집'이라는 개념과 일치한다. 이런 이유로 우리는 이 프로젝트에서 벽돌을 사용했다. 또한 벽돌은 건설 기법과 기술의 발달로 이제 단열재 역할도 한다.

murmuro

우리는 플루타노스 학교 프로젝트에 벽돌을 사용했다. 기술적인 필요성을 위한 건설적이고 가소성이 좋은 해결책으로 파사드에 벽돌을 썼다.

이 프로젝트의 의뢰주는 아주 구체적인 요구 사항이 있었다. 우리는 외부 스포츠 구역을 축소할 수 없었고, HVAC 없이 기존의 바람이 통하는 복도를 이용하여 자연 환기를 해야 했으며 보안을 위해 스포츠 구역 쪽으로 창을 열 수 없었다. 새 건물은 현존하는 각기 다른 높이의 다른 두 건물과 바깥 공간 두 곳을 연결해야 했다. 학교가 운영되는 동안 두 단계에 걸쳐서 지어져야 하는 점도 있었다. 그리고 프로그램 때문에 스포츠 구역을 향하는 파사드를 써야 했다. 이것이 무계획적으로 보이는 디자인에 파사드 전체를 벽돌로 만드는 해결책을 제시하게 된 이유이다. 이 해결책으로 각종 스포츠 시합과 던져지는 공을 견딜 수 있는 파사드를 지었고, 학교의 부차적인 프로그램에 자연채광을 들일 수 있도록 뒤쪽에 창문을 열 수 있었으며, 건물에 신선하고 차가운 공기를 불어 넣을 환기구를 숨길 수 있었다.

NISHIZAWA ARCHITECTS

카츠덴 공장(Katzden Factory)은 우리가 건물 전체에 주요 재료로 벽돌을 사용한 프로젝트이다.

OOIIO Architecture

아마도 내가 벽돌을 좀 더 성공적으로 사용했던 프로젝트는 스페인 중부의 한 마을에서 버려지고 붕괴하고 있는 오래된 헛간의 보수 공사일 것이다.

옛날에 시골에서 일할 때 필요한 농업 도구와 동물을 위해 사용된 오래된 건축물이었다.

오래된 벽 중 하나는 아름다운 옛날식 큰 벽돌로 되어있었다. 이 벽이 무너지고 있어서 의뢰인은 철거하고 싶어했지만, 우리는 벽을 보존하는 쪽으로 설득했다. 이 벽돌 벽을 복구하는데 특별히 흥미가 있었기 때문에 건물에 도입하는 모든 새로운 요소 역시 전부 벽돌로 하기로 했다.

이 모든 건축적 수술 작업의 결과는 오래된 것을 다시 생각하고 새로운 것과 혼합하여 낳은 뉘앙스와 디테일의 패치워크였다.

여기서 벽돌은 오래된 것과 새로운 것을 결합하는 데 확실히 적합한 재료였다!

Piuarch

우리는 노출된 벽돌을 자주 쓰지는 않지만, 밀라노에 있는 카프로니(Caproni) 공장의 재개발 프로젝트로 이 재료를 다뤄보고 그 특징과 미적 특성을 향상할 수 있었다.

이 경우, 벽돌은 산업 고고학의 가치가 있으며, 고대부터 벽돌로 지어진 도시인 밀라노와의 연결 고리와 그 장소에 대한 이야기를 전한다. 금속 지붕으로 덮인 산업 창고에는 잘 유지되고 보존된 벽돌 외관이 있다. 20세기의 건물 사이에 서서 유리 외관과 무연탄색 금속으로 된 차양 시스템으로 덮인 타워가 상징하는 새로움과는 대조적으로 이 건물은 과거의 기억이다.

우리는 세네갈의 탐바쿤다(Tambacounda) 병원 확장을 위한 공모전에서 벽돌을 다시 연구했다. 여기서 우리는 원시 점토 블록의 파사드를 만들고 같은 지역의 인구가 지역 자원과 기후를 이용해서 이를 생산할 수 있었을 거라고 가정해보았다.

SLOT STUDIO

단연코 '벽돌 타운하우스'를 꼽겠다. 이 경우 재료가 디자인 컨셉의 공범이 되었다. 디자인 컨셉과 모양이 합쳐지는 특별한 경우였다. 선은 텍스쳐를 추구하고 텍스쳐가 공간을 통합한다. 그것이 공식이다.

Stefano Corbo STUDIO

벽돌을 연상적이고 시적인 방식으로 건축에 적용할 수 있는 가장 좋은 사례 중 하나는 스웨덴 스톡홀름에 있는 시구르드 레베렌츠(Sigurd Lewerentz)의 세인트 마크 교회(St. Mark Church)이다. 이 교회는 1960년에 지어졌으며 바닥, 벽, 천장을 비롯한 모든 것이 벽돌로 만들어졌다. 벽돌과 빛이 이 신성한 공간의 분위기를 형성한다.

stpmj Architecture

조적집/stpmj- 다른 크기의 벽돌을 사용하여 하나의 조적 facade에서 그 스케일 변화가 드러나게 계획되었으며 내부 세대의 구분을 의미하기도 한다.

Studio Farris Architects

우리는 벨기에의 앤트워프 동물원과 레니크 농장 프로젝트에 벽돌을 사용했다

SUPA architects schweitzer song

벽돌은 구조적으로, 혹은 구조로서 눈에 띄게 사용되었을 때 가장 매혹적이다. 불행히도 벽돌은 오늘날 건축 요소로서의 깊은 의미가 없는 미적 표면 뿐인 마감재가 되었다. 한마디로 오늘날 벽돌은 타일이 되어버렸다.

TAKK Architecture

루이스 도메네 이 몬타네(Lluís Domenech i Montaner)가 지은 바르셀로나의 산트 파우 병원(Sant Pau Hospital)은 바르셀로나에서 매일 생활하는 우리에게 영감을 주는 카탈로니아 모더니스트 운동의 아이콘이다.

TheeAe Architects

우리가 좋아하는 프로젝트가 꽤 많다. 그중 하나를 꼽자면, 우리는 와이즈 건축(Wise Architecture)이 디자인한 한국의 ABC 건물을 언급하고 싶다. 여기서 벽돌은 공간의 간격을 통해 밀도의 대비를 일으키는 데 잘 사용되었다. 건물의 매싱 또한 이러한 대비를 잘 표현하고 있다. 이 프로젝트는 재료를 사용하여 건축이 어떻게 단순하게 잘 지어질 수 있는지 우리가 믿는 바를 보여준다.

TOUCH Architect

"레스트 애브뉴 (The Rest Avenue)"는 쁘라삿이 있는 고대 도시로 잘 알려졌고 태국의 북동부에 위치한 수린의 이웃 센터이다. 쁘라삿은 돌과 벽돌을 사용하는 하중 지지 벽 구조의 성이다. 레스트 애브뉴는 수린의 특색을 살린 음식점과 기념품 가게로 이루어져 있다.

쁘라삿의 세가지 특징인 배치, 벽돌 패턴과 그리드 시스템을 사용했고 전반적인 평면을 위해 이들을 결합했다. 4개의 주요 입구에도 수린과 관련된 특징이 있다.

북쪽 입구 = 코끼리 문
동쪽 입구 = 쌀 문
남쪽 입구 = 비단 문
서쪽 입구 = 크메르 문

이 서쪽 문은 석공을 쓰는 고대 패턴에서 전해지는 벽돌 패턴을 사용한 문이며 현대적 패턴과 고대의 특징을 통합했다.

UNStudio

1990년대 초에 우리가 설계한 빌라 윌브링크(Villa Wilbrink) 단독 주택을 위해 우리는 파사드에 시멘트 대신 풀로 벽돌을 붙이는 기법을 사용했다. 풀이 매우 얇았기 때문에 벽돌이 훨씬 더 밀착되어, 집에 정말 훌륭하고 견고한 효과를 낳았다.

Tile

Arenas Basabe Palacios Arquitectos

아이디어 하나를 건설하는데 타일을 가장 잘 사용한 우리 프로젝트는 마드리드 외곽에 있는 단독주택인 니네트 하우스(Ninette House)이다. 1층에 마이크로 테라초 타일로 된 픽셀 디자인의 바닥재가 미묘한 색상 그라디언트를 따라 유리를 건너서 건물과 주변 경관 사이의 경계를 흐린다.

ARPHENOTYPE

타일은 과거에도 오늘날에도 놀라운 재료이다. 독특한 외관을 만들 수 있는 특별한 표면도 있다. 대부분의 사람은 타일이라고 하면 욕실이나 바닥을 생각할 것이다. 하지만 타일을 파사드에 사용한다면 어떨까? 타일은 내구성이 있고 청소하기 쉽다. 건축가 프랭크 로이드 라이트(Frank Lloyd Wright)는 에니스 하우스(1923년 - Enis House)에 3차원적인 타일을 굉장히 혁신적으로 사용했으며, 영화감독 리들리 스콧(Ridley Scott)이 여기서 유명한 초현대적인 영화 블레이드 러너(1982년)의 한 장면을 촬영할 정도였다. 집 전체의 시스템은 내부와 외관 모두에 있는 타일 패턴을 기반으로 한다. 하지만, 여기서 나는 또 다른 건축가인 마누엘 헤르츠(Manuel Hertz)를 언급하고 싶다. 그는 독일 마인츠(Mainz)에 있는 유대인 커뮤니티 센터의 파사드에 특별한 타일을 사용했다. 잔물결을 3차원적으로 만든 녹색 유약 세라믹 타일로 건물을 덮었으며, 이 타일은 센터의 조각 같은 형태를 강조한다. 이 프로젝트를 위해 특별 생산 계획을 세워 표준화된 여러 길이의 패턴을 만들 수 있었다. 마지막으로, 2010년 SPAN & Zeytinoglu Architects가 엑스포를 위해 디자인한 오스트리아 파빌리온에서도 깊은 인상을 받았다. 흰색과 빨간색 타일을 사용하여 만든 유기적인 모양과 그라디언트는 놀라운 아이디어였다. 또한 타일로 유기적인 모양을 덮을 수 있다는 점을 보여준다. 요약하자면, 타일은 파사드, 천장 등에 적용할 수 있으며 다양한 형태를 띨 수 있다. 4차 산업 및 대량 맞춤화의 시대에는 새로운 형태에 대한 새로운 가능성을 창출한다.

AZC

우리는 '테라스 9 (Terrasse 9)'라는 주택 프로젝트의 외장 마감재로 채유 타일(표면에 유약을 입힌 타일)을 썼다.

BOARD

타일을 사용한 내가 가장 좋아하는 프로젝트 중 하나는 로테르담 중심부에서 최근에 완성된 "사무실과 호텔 방(Office and Hotel Room)" 디자인이다. 프로젝트의 입구가 이미 작은 벽감을 향해 메인 계단과 약간 떨어져 있었고 우리는 이를 사무실과 호텔 방으로 향하는 "비밀"스럽고 숨겨진 문을 만들 기회로 보았다. 덕분에 마술처럼 공간을 분리하고 사무실과 호텔 방을 도시의 일상과 일반적인 '진짜 인생'으로부터 은신처로 만들 수 있었다. 이 공간을 더 잘 숨기기 위해 복도에 있는 기존의 흰색 세라믹 타일을 입구의 벽감까지 연결하고 문 전체도 같은 타일로 덮어서 입구를 위장했다. 오늘날까지도 우체부가 찾기 어려워할 정도다.

Carlos Lampreia

타일은 포르투갈에서 매우 인기 있는 재료이다. 알바로 시자(Alvaro Siza)가 지은 리스본의 치아도(Chiado) 건물과 같이 빛을 반사하고 물에서 벽을 보호하기 위해 타일을 패널로 사용하는 것을 좋아한다. 파사드에 있는 부드러운 색의 타일은 빛과 강에서 반사되는 햇빛을 건물 주변의 길로 불러온다.

Casanova+Hernandez Architects

중국 진저우의 프로젝트를 위해 우리는 깨진 타일을 사용하여 풍경과 건축 사이의 경계가 지워지는 공원과 박물관으로 구성된 하이브리드 풍경을 만들었다. 4가지 색상의 깨진 타일을 사용하여 한편으로는 길과 공원 벤치를, 다른 한편으로는 박물관의 정면과 지붕을 만들어 3차원의 풍경을 만들었다. 타일의 색상과 반사가 이 모두를 통합하여 프로젝트는 완전한 예술 작품이 된다.

이 경우 재료, 그 특성 및 사용 가능성은 전체 디자인 컨셉의 매우 중요한 핵심이었다. 이 프로젝트는 모든 초청 디자이너가 각자의 국가 정체성을 반영해야 했다. 우리는 국가 정체성이라는 좁은 영역에 들어가는 대신, 이미 수 세기 동안 이루어져 온 서구와 동양 간의 모든 상업 및 문화 관계를 대표하는 현상인 문화적 하이브리드화의 개념을 반영하는 것을 선호했다. 서양의 모자이크로서 도자기의 개발 및 사용과, 중국 도자기에서 금이 간 유약의 개발은 우리의 조사를 시작하고 디자인의 기초를 둘 흥미로운 개념적 출발점을 제공했다. 여기서 재료는 디자인 과정의 시작부터 전체적인 디자인을 정의했다.

CEBRA(Mikkel Frost)

우리는 몇 가지 프로젝트에서 타일을 디지털 픽셀처럼 그래픽으로 사용했다. 엘벡 앤드 베이럽 프로젝트에서 우리는 욕실 바닥에 흑색의 침입자를 만들었고, 엑스메리멘타리움 프로젝트에서는 타일로 착시 현상을 일으켰다. 타일은 저렴하지만, 공간을 비틀고 따분한 공간을 강화하는 시각적으로 매우 강력한 재료이다.

Davide Macullo Architects

우리가 고품질 제품으로 간주하는 유일한 타일은 수제 타일이다. 전 세계 역사적인 건물의 아름다움에서 볼 수 있었던 멋진 타일 작품 외에도 최근 역사에서 알바로 시자(Alvaro Siza)가 단순하지만 특별한 방식으로 타일을 사용했다고 생각한다. 예를 들면 포르투갈의 산 마르

코 데 카나베즈(San Marco de Canavezes) 교회가 있다.

Donner Sorcinelli Architecture

CD 하우스 지붕은 일반적으로 밝은 갈색의 스페인 식민지풍 지붕이 특징인 그 지역에서는 평범하지 않은, 크기가 크고 평평한 진한 회색 세라믹 타일로 지었다.

Keiichi Hayashi Architect

레스페이스 제로 / 쟝 피에르 레노, 쟝 피에르 레노의 집 / 쟝 피에르 레노

레노는 흰색 타일로 집 내부를 완전히 덮고 공간 전체를 자기 작품으로 만들었다. 이 공간에 있던 모든 재료가 타일 한 장의 축척으로 복원될 때 새 공간은 축적의 대량자재가 된다. 따라서 그는 석조 조인트가 새롭게 열린 이 공간을 채우기 위해 축척 면에서 어떻게 중요한 역할을 하는지 보여준다. 나는 타일의 표면적이고 분열 가능한 특성으로 제한되는 공간에서 자유를 만들어 내면서 이를 강박적으로 미니멀리즘의 느낌을 주는 초 편집광적인 공간으로 변형시키는 그의 작품에 매료되었다. 게다가, 타일링 그 자체의 스타일은 그의 작품으로서 더욱 매력적이다.

LANDÍNEZ+REY arquitectos

우리는 2005년 엔릭 미라예스(Enric Miralllas)와 베네데타 타글리아부에(Benedetta Tagliabue)가 스페인 바르셀로나에 있는 산타 카테리나 시장(Mercat de Santa Caterina)의 다섯 번째 파사드에 타일을 사용한 방식에 여전히 매료되어 있다.

M artı D Mimarlık

우리는 일반적으로 위생 및 방수 특성 때문에 화장실과 같은 젖은 공간에서 타일을 사용한다. 우리는 자동차 수출자 직업 고등학교의 내부 벽 클래딩에 치수가 큰 타일(100x300cm)을 사용했다.

object-e architecture

쿰 카피(Koum Kapi) 해안가 재설계 프로젝트 공모전(2등 상 수상)은 거의 타일링 시스템에 대한 연구이다. 우리가 제시한 디자인의 목표는 지역(동네 규모)과 도시 모두를 맡을 수 있는 공공 공간의 설계이다. 이 프로젝트는 지상 평면의 세분하여 구성된다. 여기서 우리는 유사한 상황에서 일반적으로 기대하는 것, 즉 해변을 따라 각기 다른 움직임과 활동의 영역을 정하는 접근법을 피했다. 대신 지상 평면은 크기와 물질성이 다른 직사각형 모양으로 세분된다. 따라서 경로를 형성하고, 이동과 방향을 정하고, 기능을 차별화하고, 연속성을 만들고, 중요한 영역을 나타내기 위해 타일을 배열하고 조작할 수 있는 다양한 방식으로 디자인이 구성된다.

세분화와 타일 공사는 포장 알고리즘으로 조절했다. 알고리즘의 변수는 타일의 밀도, 방향 및 위치를 제어하는 데 도움이 되었다. 포장 프로세스의 최대 및 최소 크기는 각 영역이 개최할 여러 이벤트에 따라 국소로 변경했다. 그 결과는 다양한 크기와 재료의 패치워크이며, 주요 구조적 속성과는 별도로 코움 카피 지역의 다양한 요소의 패치워크 같은 자연과 차니아(Chania)시 전역에서 발견할 수 있는 패턴을 나타낸다 (예: 도시 벽의 석조). 타일 직사각형의 크기는 몇 cm에서 10m까지이며, 덕분에 타일 패턴에 폭 넓은 변형이 가능하다. 이 패치워크를 따라 더 크거나 작은 규모의 다양한 '공간 이벤트'가 발생하며, 일반적으로 더 큰 직사각형이 있음을 뜻한다. 타일 패턴은 요소가 다양하고 국부적으로 차별화 된 크기와 밀도를 가진 들판과 비슷하다. 따라서 디자인의 결과는 동시에 디자인의 국부적인 차별화를 촉진하고 해변 전체에 눈에 띄는 어휘를 만드는 도시의 들판이다.

수다 페리 부두(Souda Ferry Terminal) 공모전(장려상)을 위한 우리 프로젝트는 많은 방문객을 수용해야 하는 유동인구가 많은 건물에 보통 예상하는 것과 다르게 방문객에게 독특하고 서술적인 경험을 주는 것을 목표로 한다. 매우 다른 건축 유형에서는 그다지 특별하지 않을 이 디자인의 디테일과 특성은 여가를 위한 여행이 사치였던 시대를 나타내기 때문에 그를 위한 건물도 그 특성을 반영한다. 동시에 건물의 간단하고 기능적인 레이아웃과 재료는 이 디자인이 페리 터미널 방문자 수에서 오는 매우 높은 사용량을 지탱할 수 있음을 보장한다. 하지만 이 건물은 사람들이 다른 곳으로 이동하기 위해 통과해야 할 장소를 머물고 싶은 곳으로 바꾸려고 하고 이러한 반전으로 생길 수 있는 모순을 긍정적으로 해결하려 한다.

따라서 우리가 타일을 사용하는 방식은 위의 내용을 반영한다. 색이 있는 콘크리트 타일은 대중적이고 관광과 연결된 장소에 '고급스러운 느낌'을 주기 위해 역의 내부와 외부에 사용된다. 우리는 이 프로젝트를 특별한 일이 일어나는 장소였던 역의 느낌을 재현할 건물로 상상한다. 실제가 아닌 느낌만으로라도 이런 장소에서 볼 수 있는 일반적인 속도를 대신하여 느림이 있을 수 있는 곳으로 본다. 다른 재료와 함께 콘크리트 타일을 사용하여 그 느낌을 향상한다.

OOIIO Architecture

단연코 7만5천명의 주민이 사는 스페인의 탈라베라 델 라 레이나(Talavera de la Reina) 시에 있는 광장 두 개의 보수 공사이다.

여기서 나는 현지 탈라베라 전통 도자기를 주요 재료로 사용하기로 했다. 타일은 훌륭한 가소성을 지닌 제품이며 지역 문화에 깊이 뿌리를 두고 있기 때문이다. 일반적으로 타일은 르네상스 시대의 그림과 모티프로 소규모 가정적 물체에 쓰이지만 나는 화병과 접시를 장식하는 그림을 도시 규모로 크기를 크게 늘려 이 멋진 공예품에 대한 새로운 가능성을 제안했다.

Piuarch

건축에서 우리는 세라믹을 건축의 외피로 사용한 적이 없다. 세라믹은 오로지 수평 및 수직 코팅으로 사용했다. 이 주제를 위해 우리는 종종 여러 회사와 협력하여 맞춤형 방안을 실험해왔다.

여러 차례에 걸쳐 우리는 프로젝트의 철학과 가장 잘 통합할 수 있는 표면을 설계하여 전체적인 미학을 복원했다. 예를 들어, 여성 란제리 체인 야마메이를 위한 컨셉 스토어에서 톤 위에 톤을 써서 상아색 사각형 바닥재와 대리석의 기하학적인 인서트로 이상적인 가벼움, 부드러움, 밝기, 기하학과 매장의 대칭을 전할 수 있었다.

SLOT STUDIO

우리는 자동차 목재 부품 생산 업체인 퀸(Quin)을 위한 조립 공장을 설계하고 건설했다. 매우 특별한 미적 표준을 달성해야 했는데 트렌디하고 우아하며 상쾌한 느낌을 이루어야했다. 타일로 본관의 외관을 가려서 이룰 수 있었다.

Stefano Corbo STUDIO

1958년부터 1974년까지 카를로 스카파(Carlo Scarpa)가 디자인한 건물 재사용 프로젝트인 이탈리아 베로나의 카스텔 베키오(Castel Vecchio) 박물관은 타일이 다른 재료와 함께 같은 건물 내에서 색상과 공간 효과의 모자이크를 만드는 데 어떻게 기여할 수 있는지에 대한 훌륭한 예이다.

여기서 타일은 사실 석재, 콘크리트, 목재 같은 다른 재료와 항상 결합돼있고 방문객을 이 중세의 성을 따른 강렬한 여행에 초대하는 건축가의 아이디어에 답한다. 또한 타일은 유추적 또는 대조적으로 기존 구조와 관계를 강조하는 데 도움이 된다.

stpmj Architecture

Concert hall and Music school (MUCA) / COR ASOCIADOS

세라믹타일로 마감하여 보는 위치에 따라 건물의 외관이 달리 보이는 효과를 내고 있다.

SUPA architects schweitzer song

타일은 벽돌의 예쁜 동생이다. 이것은 여전히 건물 내의 화장실과 같은 특정 기능적 문제에 대한 최선의 해결책이다. 우리는 타일을 흰색에 매우 간결하게 사용하려고 한다.

TAKK Architecture

카사 비센스(Casa Vicens)는 안토니 가우디가 지은 최초의 집이었다. 이 역시 카탈로니아 모더니즘의 아이콘으로, 건축가가 실제로 타일의 특성을 아주 잘 이용한 경우이다.

TheeAe Architects

타일은 건축 산업에서 가장 널리 사용되는 실내 자재이다. 우리는 바닥과 벽에 타일을 사용한다. 타일은 패턴이나 이미지에 맞춰 쉽게 맞춤제작까지 할 수 있다. 우리가 디자인한 인테리어 프로젝트의 바닥 패턴은 공장에서 구워서 바닥에 설치했다. 색상 차이가 만족스럽지 않아 여러 번 재생산해야 했기 때문에 제조업체에서 제품을 주문하는 과정은 간단하지 않았지만 설치 후 전반적인 품질은 놀라웠다.

TOUCH Architect

숙피산 신사는 수호신의 집과 결합된 브라마*의 신사이다. 전반적인 콘텍스트는 현대 건축이기 때문에 신사는 그에 따라 설계됐고 마침내 현대적이지만 여전히 전통적인 기능을 유지

하는 신사가 되었다.

이 신사에는 정문, 브라마, 수호신용으로 각각 대,중,소의 박공지붕이 있다. 신사는 집보다 높은 곳에 배치해야 하므로 신사의 바닥이 집보다 1미터 높으며 쉽게 접근하고 예배할 수 있는 각 건물 사이의 큰 녹지에 위치한다. 게다가 브라마는 4개의 얼굴을 가지고 있고 그 얼굴은 360도에서 사물을 보고 우리를 보호할 수 있다고 믿는다.

신사에 오는 모든 사람이 브라마에게 경의를 표하도록 하기 위해서 정문의 큰 박공지붕은 일정한 높이로, 마지막 계단에 오르면 머리를 숙이게 디자인되었다.

이 신사는 주로 철근 콘크리트 구조에 흰색 대리석 타일로 지었다. 용량감의 연속성을 위해 모든 바닥, 벽 및 지붕에 타일을 사용했다. 또한 타일 및 조명 디자인은 낮과 밤 간의 다른 효과를 만든다. 낮에는 황금 브라마가 흰색 대리석 타일 배경으로 눈에 띄며 밤에는 매입형 조명으로 황금 브라마를 어둠 속에서도 잘 볼 수 있다. 숙피산 신사는 밤에 반짝이고 빛이 된다.

*역주 - 브라마: 힌두교의 창조신

UNStudio

두바이의 탑(Tower in Dubai)은 지금까지 우리가 개발한 가장 흥미로운 타일 사용법이라고 생각한다. 이 프로젝트에서 특별히 디자인된 타일은 주로 건물의 열 부하를 줄이는 지속 가능한 기능을 한다. 거기에 파사드에 특이한 시각 효과를 더하며 이는 제품 디자인의 규모와 관련된다.

Q2
본인이 생각하는 벽돌과 타일의 장단점에 대해 얘기해달라.

Brick

ARPHENOTYPE

거의 모든 문화에서 벽돌을 찾아볼 수 있다. 대량 생산하거나 맞춤형으로 만들 수 있고, 건축에 태생의 모습을 준다. 그래마지오와 콜러(Gramazio & Koehler)의 작품에서 볼 수 있듯이 알고리즘과 로봇공학의 도움을 받아 훌륭한 패턴을 만들 수도 있다. 이는 벽돌공이 오랜 세월 동안 해온 일이지만, 기법 대부분이 잊혔다. 요즘은 로봇이 이러한 기술을 더 빠르고 정확하게 적용하고 있다.

AZC

벽돌은 빛이 비추면 모든 벽돌이 잘 보이고 모든 이음매가 눈에 띄기 때문에 매우 그래픽한 재료이다.

또한 보수나 청소없이도 오래가는 재료이다. 기둥이나 들보 같은 구조에도 사용할 수 있다면 좋을 것 같고 절연재로 사용하는 것도 흥미로울 것 같다.

BOARD

벽돌의 장점 중 일부는 확실히 견고함, 수명, 경제성, 사용 및 적용 용이성이다. 프로젝트 '바깥의 집'을 위해 우리는 이 모든 특성을 사용하고 싶었다. 하지만 그 외에도 그 광장은 이전에 산업 현장으로 - 전 도시 항구, 무역 및 생산 장소였다 - 주로 벽돌로 지어진 "드 볼하딩(de Volharding)"이라는 유명한 공장이 있었기 때문에 벽돌을 사용하여 프로젝트를 사이트의 산업 과거와 연결하고 싶었다. 벽돌의 단점 중 하나는 보강물이 부족한 점이며, 이 덕분에 벽돌은 조립이 쉽지만 분해하기도 쉽다. 우리는 광장의 세 가지 요소를 콘크리트 슬래브에 배치하고 각 요소가 의미하는 바에 대한 안내정보를 새겼다.

Carlos Lampreia

벽돌의 장점은 건설할 때의 표면과 질서이고 누군가 다 쌓으면 얻을 수 있는 마감된 표면과 질감이다.

루이스 칸(Louis Kahn)은 벽돌이 '아치를 좋아한다'고 말했다. 그런 의미에서 거대한 벽을 잘 만드는 재료이지만 벽 사이의 공간을 달성하기 위해 아치 또는 다른 유형의 하부 구조를 필요로 한다.

Casanova+Hernandez Architects

우리 생각에 벽돌은 다른 재료가 주된 표현적 역할을 할 수 있는 중립적인 기반을 만들기에 좋은 수단이다.

벽돌은 품질이 좋고 질감, 비율 및 색상을 신중하게 선택한다면 건축에서 평온함과 중립성을 표현하는 우아하고 고요한 재료가 될 수 있다.

벽돌의 단점은 신중하게 선택하고 사용하지 않으면 매우 저속한 재료로 보일 수 있다는 점이다.

CEBRA(Mikkel Frost)

재료를 논할 때 단점을 이야기하는 것은 말이 되지 않는다. 여러 재료는 악기와 같이 다른 속성을 가지고 있을 뿐이다. 드럼은 리듬에 아주 적합하지만, 멜로디를 연주해야 한다면 기타와 같은 것이 필요할 수 있다. 따라서 재료는 모두 훌륭하지만 올바르게 적용되고 적절한 상황에서 사용되어야 한다. 특히 벽돌을 쓸 때 나는 종종 경제적 어려움에 직면한다. 색깔과 질감 면에서 딱 맞는 벽돌을 찾으면 종종 예산을 넘어서 다른 제품으로 만족해야 한다. 벽돌로 아름다운 디테일을 만드는 것은 또한 노동 시간이 많이 소요되며 불행히도 달성할 가능성이 매우 작다.

Davide Macullo Architects

우리 사무소는 재료의 장점을 강조하거나 우리의 목적에 맞는 적응력을 가진 재료를 쓰는 것에 대한 인식을 쌓기 때문에 재료의 단점에 대해서는 신경 쓰지 않는다.

Donner Sorcinelli Architecture

자연스러운 특성이 장점이며, 손이 많이 가는 점이 단점이다.

Keiichi Hayashi Architect

장점: 유적이 될 수 있다.

단점: 지붕으로 쓸 수 없다.

Kimura Matsumoto architects office

일본의 근대화 과정에서 목조 가옥의 화재를 예방하기 위해 벽돌 건물을 지었다. "Another base in G" 프로젝트에서 우리는 벽돌의 무게로 인한 고유한 방음 특성에 초점을 맞추어 음향 전달을 줄이려고 했다.

이 두 사건 모두 벽돌 석조의 특성에 의해 긍정적으로 영향을 받았다.

내가 위에서 언급했듯이, 벽돌의 경도는 처리를 어렵게 할 수 있지만, 벽돌의 형태와 크기는 다양한 모양을 만드는 것을 가능하게 한다. 이 모호함은 내가 매력적이라고 생각하는 것이고 재료의 강점과 약점 사이의 균형은 내가 세심하게 주의를 기울이는 것이다.

예를 들어, 벽돌 석조를 구조 벽으로 사용함으로써, 우리는 마음대로 개구부를 만들 수 있는 능력을 잃게 된다. 하지만 이것을 부정적으로 보기보다는, 그것을 제외시키고 다른 예상치 못한 결과를 축하할 수 있다.

벽돌의 경우뿐만 아니라, 어떤 재료에 대해서도, 그것의 장점과 단점을 이해하고 그것들을 가지고 작업하려고 시도하는 것이 중요하다고 생각한다.

LANDÍNEZ+REY arquitectos

어떤 건설 방식을 사용하든 가격이 얼마인지와 밀접하게 관련되어 있다는 점은 명확하다. 벽돌의 경우 이것은 각 국의 인건비와 전통 재료에 직접 연관이 있다. 스페인에서 벽돌은 거대한 산업 경험과 기술의 발달을 갖춘 기술이므로, 건축가가 지속해서 여러가지로 실험해 올 수 있었다.

벽돌의 장점은 표현 능력과 실험성뿐만 아니라 이 지역의 전통인 점이다. 단점이 있다면 단지 오용의 결과일 뿐이다.

M artı D Mimarlık

벽돌의 광범위한 사용으로 건설업자가 재료를 찾기 쉬우며 경제적이고 현장 업자에게 익숙하다.

murmuro

모든 재료는 각기 장단점이 있지만 보는 관점과 사용하기에 따라 달라진다. 벽돌은 보통 축벽이나 튼튼하고 강력한 벽이 연상되지만 우리는 벽돌로 완전히 구멍이 뚫린 벽을 디자인했었다. 벽돌은 또한 내구성이 아주 좋고 관리에 손이 많이 가지 않는다. 단점으로, 우리가 디자인했던 것처럼 복잡한 디자인의 벽돌 파사드는 고도로 숙련된 수세공이 필요하다. 이 점은 아주 경쟁력 있는 디자인을 비싼 디자인으로 쉽게 바꾸기도 한다.

NISHIZAWA ARCHITECTS

장점: 건물 내부의 온도가 안정되도록 돕는다. 우기에는 따뜻하고 더운 계절에는 시원하다.

중고 벽돌로 프랑스 지배하에 있었던 베트남 식민지 시대의 전통적인 분위기를 보여주기 쉽다. 이를 새로운 재료와 결합하면 건물은 그 어느 때보다 현대적으로 보인다.

단점: 벽돌 벽은 두께가 필요하기 때문에 주된 요소마냥 콘크리트 들보를 사용해야 한다. 그럴 때 벽돌은 건물의 외벽을 위한 마무리 층처럼 되어버린다.

OOIIO Architecture

장점: 낮은 유지 보수, (스페인에서는) 저렴함, 다루기 쉬운 점, 구조와 외관을 동시에 해결할 수 있는 점.

단점: 태양열에 알맞은 열적 거동이 없고 열을 축적하는 점.

Piuarch

불가능한 것은 아니지만, 재료의 수행 능력에 대한 일반적인 진술을 하는 것은 어렵다. 각 재료는 주어진 상황에 맞춰 적용될 때 의미가 있으며 각 컨텍스트는 특정 방안의 검색과 식별이 필요하다.

SLOT STUDIO

벽돌은 거칠고 단단한 재료이며 특정 목적에는 적합하지만 신선하고 가볍고 투명한 느낌을 원한다면 다루기 쉽지않다. 존재감은 분명히 벽돌의 장점 중 하나이며 전통성과 따뜻함도 있다.

Stefano Corbo STUDIO

벽돌에 대한 장단점을 정의하는 것은 어렵다. 모든 재료는 프로젝트의 전반적인 논리는 물론 다른 요소 간의 관계에 따라 고려되어야 하기 때문이다.

벽돌은 일반적으로 건축의 전통적인 비전이 연상된다. 이 비전은 기술적 진보로 인해 어쩐지 단절되어 버렸다.

stpmj Architecture

장점: 사용되는 벽돌의 크기와 색깔, 줄눈의 패턴과 깊이 및 쌓는 방법 등의 세밀한 변화에 따라 전체의 면을 이룰 때 다채로운 입면을 구성할 수 있으며 시간에 따른 재료의 오염도가 안정적이다.

단점: 쌓아 올라가는 조적의 방식으로는 (불가능하지는 않지만) 공간의 형태나 크기에 따라 제약이 있고 인장력의 부재로 내진에 취약하다.

Studio Farris Architects

가장 흥미로운 장점은 경제적이며, 단단하고, 내구성이 있으며, 유지 보수가 낮고, 재사용 및 재활용을 할 수 있으며, 제조 과정에서 환경 오염이 적다는 점이다.

단점은 벽돌로 건축하는 것은 시간이 많이 소요되고, 물을 쉽게 흡수하기 때문에 공기에 노출되지 않으면 박리가 생긴다는 점이다.

SUPA architects schweitzer song

벽돌은 매우 노골적이고 근본적이다. 거의 모든 것을 생성 할 수 있는 최소한의 재료이다.

TAKK Architecture

벽돌은 매우 값싼 재료로, 구조나 외관으로 사용할 수 있으며 이는 우리가 매우 흥미롭게 생각하는 점이다. 반면에 오늘날 벽돌은 우리 주변에서 우리가 특별히 따르지 않는 특정 정치적 의미를 지닌 특정 건축 양식과 매우 연관되어있다.

TheeAe Architects

벽돌의 장점은 관리가 거의 필요 없는 재료라는 점이다. 오랜 시간동안 페인트나 코킹, 스테인 같은 관리가 필요 없다. 또한, 벽돌은 외부로부터 소리를 차단하고, 내화성이 있고, 매우 자연스러우며 편안하지만, 가장 중요한 장점은 친환경적인 점이다.

벽돌의 단점은 건설하는데 필요한 노동량이 많아 비용이 많이 들 수 있다는 점이다. 그리고 매우 무겁기 때문에 튼튼한 토대가 필요하다. 또한 잘 지으려면 시간이 걸린다.

TOUCH Architect

벽돌 재료는 일반적으로 내구성이 뛰어나고 내화성이 높다. 매우 튼튼한 재료이기 때문에 하중 지지 벽으로 사용할 수 있으며 건물 구조를 지지하는데 도움이 된다. 더욱이, 디자인에 필요하지 않으면 특별히 따로 마감이나 페인팅이 필요하지 않다. 다양한 방향과 패턴을 만들 수 있기 때문에 그만의 미학도 있다. 하지만 벽돌에도 약점은 있다. 무게가 무겁기 때문에 그를 지탱하는데 필요한 말뚝과 건물의 기초에 부담이 늘어난다. 더욱이 그런 벽을 만드는데 더 많은 시간이 소모된다.

UNStudio

벽돌을 놓는 일에는 물론 거대한 기계가 필요하지 않고, 손으로 할 수 있기 때문에 소규모 프로젝트에 적합하다. 그러나 벽이 매우 두껍지 않은 한, 벽돌은 매우 큰 경간에 사용할만큼 강하지 않다. 그래서 벽돌로 된 고층 건물이 없는 것이다.

Tile

Arenas Basabe Palacios Arquitectos

마이크로 테라초 타일은 그 유동성 덕분에 내부와 외부 모두에 사용할 수 있으며 마무리 (폴리싱 혹은 거침)만 다르다. 이는 내부와 외부 공간 사이의 원활한 연속성을 강화할 수 있게 해준다. 이 재료의 주 '단점'은 가장자리가 고르지 못한 점이며, 아주 깨끗한 디테일에는 방해가 될 수 있다. 하지만 우리는 이 점을 고유한 특성 중 하나로 생각하지 단점으로 보지 않는다. 완전히 같은 타일은 하나도 없으며 바로 이 불완전함이 타일의 매력 중 하나이다.

ARPHENOTYPE

건축 자재로서 세라믹 타일은 매우 친환경적이다. 타일은 따지고 보면 그저 점토, 미네랄, 물, 그리고 불이다. 세라믹 타일은 일반적으로 유리창과 같은 고밀도 표면을 가지고 있고 (예: 도자기 석기) 종종 아주 매끄럽다. 이런 이유로 세라믹 타일에는 흙과 세균이 거의 없다. 일반적으로 타일은 포지티브이든 네거티브이든 직사각형 패턴으로 표준화되어있다.

AZC

타일은 특별한 관리가 필요 없는 재료이다. 훌륭한 상태로 오랫동안 지속된다.

BOARD

단순하고 견고하며 저렴한 방법으로 표면을 덮을 수 있는 단단한 소재인 것 외에도 복잡한 모자이크를 만드는 능력과 내러티브 및 다각적인 의미를 타일의 가장 큰 장점 중 하나로 간주한다. 우리가 '사무실과 호텔 방' 프로젝트의 화장실 중 하나를 설계할 때 사용하고 싶었던 것이 이 특성이다. 프로젝트의 주요 공간은 흰색에 개방형이고, 중립적이고 유연한 반면, 이 화장실에는 야자수를 묘사한 인상적인 수제 예술 작품이 포함되어 있다. 이는 STAR Strategies + Architecture가 원래 스페인 엘체(Elche)에 있는 공공 광장을 위해 만든 디자인으로, 1 대 8의 축척으로 재현했다. 따라서 아돌프 루스(Adolf Loos)의 찬양자인 의뢰인은 이 가장 과도한 관상용 공간을 사적인 용도에 쓰는 것을 선호했다. 시간이 오래 걸리는 모자이크는 타일의 단점으로 간주 될 수 있지만 우리 프로젝트의 경우 스페인에서 전문가이자 장인을 데려왔다.

Carlos Lampreia

타일은 코팅된 재료로, 일반적으로 벽이나 바닥을 안팎으로 덮고 물에서 보호하는 데 사용된다. 세라믹 제품이기 때문에 빛을 개선할 수 있고 모든 색상과 질감이 가능하다. 단점은 첫째로 그 두께이며, 둘째로 사용과 건설적인 장력으로 깨지기 쉽다는 점이다.

Casanova+Hernandez Architects

가우디(Gaudi)와 스페인 카탈루냐(Catalonia)의 모더니스트 건축가들이 그랬던 것처럼, 우리는 도로, 파사드, 도시 요소를 덮기 위해 부서진 타일을 사용했다. 덕분에 모든 수평, 대각선 및 수직면이 도시 생활이 일어나는 무대로 동기화하는 복잡한 기하학의 독특한 하이브리드 풍경을 만들 수 있었다. 타일을 작은 조각으로 부수면 거의 모든 표면에 쉽게 적용할 수 있어 디자이너에게 표현의 자유가 많이 부여된다.

깨진 타일의 다양한 색상과 패턴은 디자인할 때 많은 자유와 풍부한 색채를 제공한다. 덕분에 우리는 색상과 패턴면에서 각기 다른 디자인 요소 사이의 관계를 정의하여 건물 전체에 대한 인식으로 여러가지 실험할 수 있었다. 이는 건물과 풍경 전체에 원근감 효과와 풍부한 전망을 창출할 기회를 만들었다.

타일의 단점은 특정 상황에서 쉽게 깨진다는 점이나, 이미 의도적으로 깨진 채로 적용되면 이 점은 무관하다.

CEBRA(Mikkel Frost)

덴마크의 추운 기후에서는 외부 파사드에 타일을 적용하는 것은 기술적으로 어렵다. 타일을 쓰려고 건축가가 목을 내놓았던 내가 아는 매우 적은 경우에서 타일은 간단히 달그락거리며 떨어져 버렸다.

Davide Macullo Architects

수제 타일은 단순히 감각을 위한 즐거움이다. 대량 생산된 타일은 공격적이며 영혼이 없는 산업 과정을 표현하는 제품이기 때문에 인간의 공간에 속하지 않는다. 이는 사람들이 살 환경 내부에 쓰기에는 심리적으로 부정적이다.

Donner Sorcinelli Architecture

지붕에서 조립하기 쉽지만, 기상 조건에 쉽게 영향을 받는다.

Keiichi Hayashi Architect

장점: 타일로 된 방에 있으면 물 속에 있는 것 같다.

단점: 미끄러지기 쉽다.

LANDÍNEZ+REY arquitectos

스페인의 타일 산업은 최첨단 기술 개발로 세계에서 가장 뛰어나다고 말할 수 있다. 말할 것도 없이, 최첨단 기술이라는 미래의 표현과 전통적인 활용법이 만나 최고의 잠재력을 만든다.

M artı D Mimarlık

모듈형 재료의 특성은 설계 과정에는 유용하지만 이음매와 조인트가 위생상 문제를 일으킬 수 있다. 하지만 지난 몇 년간 더 큰 크기의 타일은 생산되면서 위생 문제는 최소화했지만, 건설 과정에서 정밀한 작업의 필요성이 높아졌다.

object-e architecture

우리는 건축에서 재료를 강점이나 약점의 관점에서가 아니라 가능성의 관점에서 생각하는 경향이 있다. 그러므로 우리는 타일이 우리에게 제공할 가능성에 관심이 있다. 물론 타일은 반복이라는 개념에 기초를 두고 있다. 타일 시스템을 만들려면 한 요소를 어떠한 방식으로든 반복해야 한다. 그래서 우리는 타일의 반복에 적용할 수 있는 규칙에서 많은 가능성을 찾는다. 반복 과정에 특색을 포함하여 독특한 상태가 많이 발생할 수 있다. 여러 타일 사이의 차이점은 위치, 크기 및 재료 특성/색상과 관련될 수 있다. 또는 더 흥미롭게, 위의 조합에서 나올 수도 있다. 이러한 기술을 통해 타일은 그리드의 단일 요소가 아니라 표면을 생성하고 더 많은 것이 될 수 있다.

OOIIO Architecture

장점: 놀라운 가소성, 빗물을 잘 막는 점, 우리가 일하는 도시에서와 같이 일부 장소에서는 깊은 전통과 훌륭한 장인이 있다는 점.

단점: 타일이 튀어 나오지 않도록 벽에 잘 붙일 필요가 있다는 점.

SLOT STUDIO

타일은 다재다능하고 경제적이다. 시뮬레이션에서 개념화까지 폭넓게 쓸 수 있지만, 허세나 저렴함에 빠질 수도 있기 때문에 타일을 선택하고 사용을 정당화하는데 매우 조심해야 한다.

Stefano Corbo STUDIO

타일에 대한 장단점을 정의하는 것은 어렵다. 모든 재료는 프로젝트의 전반적인 논리는 물론 다른 요소 간의 관계에 따라 고려되어야 하기 때문이다.

stpmj Architecture

장점: 타일은 디자인 적용의 폭이 넓다. 제품의 다양성이 확보되어 있고 가볍다.

단점: 외기에 면하는 경우 기후에 따라 제품 자체가 갖는 재료의 수축 팽창에 영향을 주기 때문에 탈락 및 제품의 하자로 이어질 확률이 있다.

SUPA architects schweitzer song

우리는 매끄러운 타일, 조인트가 없는 아름다운 세라믹 타일 표면을 꿈꾼다. 만약 화장실이나 부엌 전체를 구워서 유약을 칠 수 있다면, 우리가 제일 먼저 쓸 것이다.

TAKK Architecture

타일은 구조로 쓸 수 없지만 타일의 집합은 패턴과 기하학 측면에서 매우 아름답다. 이는 개발할 수 있는 색상과 모양의 다양한 가능성 덕분이다.

TheeAe Architects

우리 인테리어 프로젝트의 대부분은 마감재로 (세라믹) 타일을 사용했다. 디자이너가 고를 수 있는 선택의 폭이 정말 넓으며 적당히 숙련된 시멘트공이라면 쉽게 설치할 수 있다. 또한, 다른 건축 자재와 비교하여 설치가 매우 간단하다.

대조적으로, 타일을 사용하는 단점은 내구성이다. 바닥에 떨어졌을 때 쉽게 부서지며 벽돌과 마찬가지로 하나씩 설치하는 데 시간과 노동이 필요하다.

TOUCH Architect

타일은 크기, 모양, 색상, 패턴 및 질감이 아주 다양하며 바닥재, 벽 혹은 주방 카운터와 같은 일부 가구에 사용할 수 있다. 목재, 콘크리트, 화강암이나 대리석과 같이 자연을 가장하기 때문에 색상과 패턴을 통해 다양한 유형의 건축적 특징을 만든다. 게다가 콘크리트 바닥과 벽을 균열과 먼지로부터 보호하는데 도움이 되며 청소하기 쉽다. 의뢰인 대부분은 타일이 합리적인 가격으로 품질을 제공하는 것을 믿고 쓴다. 하지만 타일을 사용하는데 몇 가지 단점이 있다. 타일이 습도에 영향을 받지 않더라도 타일 그라우트에 곰팡이를 약간 일으킬 수 있고 그라우트는 교체하기 쉽지 않다.

UNStudio

힘의 부족은 말 그대로 타일의 약점이다. 비록 내구력이 항상 향상되고 있지만, 이런 재료는 매우 쉽게 깨지기 때문에 여전히 크기에 주의해야 한다.

Interviewee
PROFILE

Arenas Basabe Palacios Arquitectos

Overview

Arenas Basabe Palacios arquitectos is a young architecture and urbanism studio based in Madrid. Its partners Enrique Arenas Laorga, Luis Basabe Montalvo and Luis Palacios Labrador have been working together since 2006 and have won more than thirty prizes in architecture and urbanism competitions. They have given lectures in diverse institutions and presented their work and investigation in several international exhibitions. Their work has been published in Spain, France, Italy, Switzerland, UK, Austria, Germany, Cyprus, India and Korea.

Directors

Enrique Arenas Laorga (1974), architect (ETSAM, Madrid) and Doctor Cum Laude (UPM Madrid, 2016). He has developed projects in very different areas: rehabilitations, housing, institutional and events. He has held lectures at several academic institutions, and taught as professor at the European Institute of Design in Madrid (IED).

Luis Basabe Montalvo (1975), architect graduated at the TU Graz. He has been visiting professor at Dipartimento di Architettura e Studi Urbani (DAStU) at Politecnico di Milano. Since 2003 he teaches design studio at ETSAM as an associate professor. He has been guest researcher and guest lecturer at various Universities: RWTH Aachen (Germany), Cambridge (UK) and CEPT Ahmedabad (India).

Luis Palacios Labrador (1983), architect graduated at ETSAM (Madrid, 2009), Master in Advanced Innovation and Technology (ETSAM, 2011) and Doctor Cum Laude (UPM, 2017). He currently teaches design studio as an associate professor. He has worked in the Netherlands, investigated in Berlin and held lectures and workshops in India and UK.

ARPHENOTYPE

Dietmar Köring, Dipl.-Ing.(FH) M.Arch. Architect BDA, is an architect, researcher, and educator living in Cologne. He is head of the architectural research office Arphenotype, where he focuses on blurring the boundaries of different artistic disciplines. From 2012 to 2017 he was a research fellow at TU Berlin / CHORA City & Energy and Dietmar has taught Digital Design at TU Braunschweig from 2010 to 2012, he was Guest Professor for Virtual Realities & Experimental Architecture at the University Innsbruck ./Studio3 in 2011, Technology and Design Lecturer at the Cologne Institute for Architectural Design / C-I-A-D and visiting lecturer for digital design at the DeMontfort University Leicester. From 2011 to 2012 he was assistant professor for Smart Grid research (Smart City Concepts 2022) at the Institute for Corporate Architecture at the Cologne Technical University.

He studied architecture at the University of Applied Sciences Cologne, the University of Western Sydney and at the Muthesius Academy of Fine Arts, where he graduated as in 2005 as Dipl.-Ing. (FH). Dietmar received his MArch in 2007 at the Bartlett School of Architecture University College London, under Prof. Neil Spiller and Phil Watson. Since 2008 he is a registered Architect at the AKNW and ARB.

Through his career he has worked internationally for offices such as Coop Himmelblau, Graft, 3deluxe and Andrew Wright Associates. His research has been awarded by the Jaap Bakema Fellowship / NAI and his works have been internationally published and exhibited, including MoMa New York, Heide Museum of Contemporary Arts Australia and Deutsches Technikmuseum Berlin. Dietmar has given international lectures, guest critiques and workshops. Since 2013 he is collaborating with Simon Takaski as Takasaki Koering Architects.

Dietmar is member of the narrative research network .horizon.com.

AZC

AZC was founded in 2001 with the idea that exploring architecture and its techniques could help to improve our built environments. Our interest does not lie in inventing concepts, we have always sought to realize buildings for real life's needs.

Through competitions and direct commissions, our office has worked on over a hundred projects of varied scales and uses. Most of our built projects are intended for a wide audience; sports facilities, lecture halls, office buildings and residential, some of which very specific for vulnerable populations. We also have, eight metro stations under construction, including four in Paris and four in Rennes and studies for a new station in Lyon, are ongoing.

Through some recently completed buildings, which have different purposes, we want to share our current concerns of coherence with global and local contexts which today represent the major issues of architecture.

We are not alone in the projects process, our clients and our partners share this common experience, which is engaging and meaningful; they allow us to reflect on our own actions that relate to the projects. We aspire to a high quality in any form of collaboration.

Most of our work has been published, displayed, sometimes awarded and we have often been given the opportunity to speak on topics of sustainability, diversity and innovative techniques, which all illustrate our commitments.

BOARD

BOARD (Bureau of Architecture, Research, and Design) was founded in Rotterdam in 2005 and is active in many fi elds: as an architecture, urban design, and design practice, as a research board and as a platform for comparative analysis on urban issues through its bi-annual journal MONU – Magazine on Urbanism. BOARD won several prizes recently in prestigious international architecture and urban design competitions.

Bernd Upmeyer is the founder of BOARD and editor in chief of MONU – Magazine on Urbanism. He studied architecture and urban design at the University of Kassel(Germany) and the Technical University of Delft (Netherlands). From 2004 until 2007 he taught and did research as Assistant Professor at the department of Architecture, Urban Planning and Landscape Planning at the University of Kassel. In 2010 he taught as Adjunct Professor at the department of Urban Design at the Hafen-

City University Hamburg. In 2012 he was a guest critic at the Berlage Institute's fi rstyear postgraduate research studio "Anarcity".

In 2013 he lectured and participated in a discussion about architecture, urbanism and media at Strelka's Urban Studies Session in Moscow. Upmeyer frequently writes for international publications and magazines. He holds a PhD (Dr.-Ing.) in Urban Studies from the University of Kassel(Germany). Upmeyer is the author of the book Binational Urbanism – On the Road to Paradise. The book examines the way of life of people who start a second life in a second city in a second nation-state, without saying goodbye to their fi rst city.

Upmeyer coined the term "binational urbanism".

BOARD employs an international team of architects and planners and collaborates with national and international external consultants and specialists.

Carlos Lampreia

Carlos Lampreia, architect (1990), is architecture design teacher at FAA-Universidade Lusíada de Lisboa since 1994, studied at OPorto Architecture School and at Lisbon Technical University FA-UTL. Master in architecture theory, 'towards an objective architecture', 2002. Phd about, strategy, site and material, concerning architecture and arts, 'concept site and material, a strategy in architecture and arts, 1960-2000', 2017. His Lisbon based office, carloslampreia[x]arquitectos, works on an experimental way with young architects and students towards architectural materialisation, participating both in international competitions and individual private requests.

Casanova +Hernandez Architects

Casanova+Hernandez, founded in 2001 by Helena Casanova and Jesus Hernandez, is a design and research studio based in Rotterdam. It focuses on rethinking and designing our urban habitat in order to create vibrant cities while promoting environmental and social sustainability.

Working with an interdisciplinary team and with experience developing projects in very different cultural contexts in Europe, South America and Asia, the office has expanded its capabilities and its international network through close and fruitful collaboration with experts in different continents.

Casanova+Hernandez is structured in two complementary platforms: C+H Projects and C+H Think Tank. C+H Projects is the design platform of Casanova+Hernandez. It operates in the fields of architecture, landscape architecture and urban design, often combining them to create hybrid architectural landscapes.

C+H Think Tank works as an independent platform that analyses urban and social problems and proposes innovative design solutions, new urban strategies and advice on the implementation of new policies.

www.casanova-hernandez.com

CEBRA

CEBRA is a Danish architectural office founded in 2001 by the architects Mikkel Frost, Carsten Primdahl and Kolja Nielsen. In April 2017, architect MAA Mikkel Hallundbæk Schlesinger entered the group of partners.

Based in Aarhus in Denmark and in Abu Dhabi in the UAE, CEBRA employs a multidisciplinary international staff of 50 architects, constructing architects, urban planners and landscape architects, who all share a strong passion for architecture.

CEBRA has gained recognition through award-winning projects such as The Iceberg at the habour front in Aarhus and the Experimentarium science centre in Copenhagen and has a growing international portfolio in Europa and the MENA region.

At CEBRA we want to change the way to think, design and build architecture. We are always pushing artistic and architectural boundaries - pushing these boundaries with a CEBRA attitude and a Nordic mindset that combines our artistic approach to architecture with an understanding of its cultural context.

We design architecture by listening to and understanding our users and clients and studying their context, culture and climate. Our services cover all project phases - from client advisory and user involvement and concept and project development to project and construction management as well as technical supervision.

Most CEBRA projects are within the fields of education, culture and housing - thought, designed, and built in line with our mantra - Architecture with attitude.

Davide Macullo Architects

Davide Macullo (b. Giornico, CH, 1965) lives and works in Lugano, Switzerland. Studied art, architecture and interior design. For 20 years (1990-2010) he was project architect in the atelier of Mario Botta with responsibility for over 200 international projects worldwide. He opened his own atelier in 2000.

The ethos of the studio is one of 'drawing from context' and the various contributions promote a dialogue between the specificity of the project and the universality of the contexts. His work has been published and awarded both at home and abroad. Selected realized projects include the WAP ART foundation mixed use gallery and apartment in Gangnam Seoul, South Korea, the Assuta Hospital in Ashdod, Israel, 5* Hotel and SPA facilities in Greece,the headquarter Jansen AG in Oberriet, Switzerland, Private Museum in Jeju South Korea, Sino-Swiss centre in Tianjing China, several houses and housing in Switzerland and abroad.

Current projects include a new Health and Wellness Hotel in Weggis, Switzerland and Marbella, Spain, houses and residential buildings in Switzerland, a beachfront villa in Heraklion, Greece, a Medical SPAin Baku, Azerbaijan. The work of the studio includes masterplanning, graphic design, branding consulting and custom designed furniture, now in production and spans to the creation of contemporary art collections for clients.

In Rossa Calanca Valley in the Grison Canton, Davide Macullo has started an urbanistic program to promote the intervention in situ of international artists to influence daily life through contemporary art. The first building realized in collaboration with Daniel Buren will be followed by other ten artists.

Donner Sorcinelli Architecture

Donner Sorcinelli Architecture is an international architectural design office based in Italy.

Founded by architects Luca Donner and Francesca Sorcinelli, the firm pays particular attention to the theme of sustainable and affordable architecture in all its variants, based on experimentation and research in various fields like Architecture, Urban Design, Interior and Product Design.

Their projects have been awarded in International competitions:

"Social Housing Dev."- Piazzola sul Brenta 1st prize; " Social Housing Dev." -Presina, 1st prize; "Design Beyond East and West"- Seoul, 1st prize; "International Design Competition for Modern Saudi Houses, Affordability and Sustainability"-Riyadh, 1st prize; Urban Retrofitting of S.Elena's - Silea, 1st prize; Sansovino Masterplan -Montebelluna, 3rd prize; School Campus in Carbonera, 3rd prize;"Your Absolute" for a residential Tower, Mississauga, Honorary Mention; "Daejeon Urban Renaissance"- Daejeon, Honorable Mention.

They have been published in many international magazines, books and presented in several exhibitions in Italy and abroad.

DoSo are winners of the "Cityscape Architectural Review Award 2006", "SAIE selection Awards 2009" and the "20+ 10+ X World Architecture Award 2012". They have received an Honorary Mention at Modern Atlanta Prize 2011, an Acknowledgement Prize at Holcim Awards 2005 for sustainable constructions (MENA region) and they have been nominated by Korean Institute of Architects among "100 Architects of year 2017".

Luca Donner and Francesca Sorcinelli have been teaching at International Universities in Dubai after previous academic experiences in Italian Universities.

www.doso.it

Keiichi Hayashi Architect

Office information
1-6-10 Kumochi-cho Chuou-ku
Kobe-city Hyogo
651-0056 JAPAN
Phone +81.78.221.1868
hayashi @8107.net
Owner: Keiichi Hayashi
Established: 1997

Biography
1967 Born in Osaka
1991 Graduated from Metal Engineering, Kansai University
1993 Graduated from Architecture, Kansai University
1997 Established Keiichi Hayashi Architect

Design Philosophy
It is important for me to make architecture using basic materials and uncomplicated construction methods. I try to create a system that is based on pure architecture but becomes complex when people use it.

www.haya-at.com

Kimura Matsumoto architects office

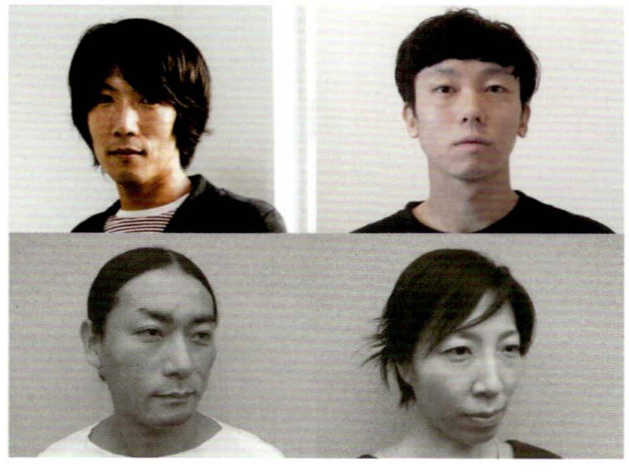

Yoshinari Kimura
Born in Wakayama, Japan in 1973
1996 Graduated from Osaka University of Arts, Department of Architecture (studied under Negishi Kazuyuki)
2003 Established Kimura Matsumoto Architects
Naoko Matsumoto
Born in Kyoto, Japan in 1975
1997 Graduated from Osaka University of Arts, Department of Architecture (studied under Negishi Kazuyuki)
2003 Established Kimura Matsumoto Architects
Katsunobu Tasho
Born in Osaka, Japan in 1973
1997 Graduated from Kyoto institute of technology, Bachelor of Architecture
1999 Graduated from Graduate School of Kyoto Institute of Technology, Master program of Architecture
2003 Established katsunobu tasho / architecture
Masaki Kato
Born in Hiroshima, Japan in 1986
2011 Graduated from Graduate School of Science and Engineering Research, Kindai University, Master program of Architecture
2013 Established masakikato

LANDINEZ+REY arquitectos

LANDINEZ + REY is an architectural practice co-founded in 2000 and based in Madrid(Spain) by the architects David Landinez González-Valcácel (Madrid, 1973) and Mónica González Rey (Paris, 1973). Both are formed as M.Arch (1999) in ETSAM-UPM(Faculty of Architecture of the Polytechnical University of Madrid, UPM) and both are also graduated as Building Engineers by UEM-Madrid (2013). David also is M.Arch in Efficient Buildings and Rehabilitation by UEM (Universidad Europa de Madrid) and Mónica has also postgraduated studies in Analysis and Real Estate Management by Colmillas University (ICAI-ICADE)

LANDINEZ + REY arquitectos [el2gaa] develops its activity as a working plaform where architecture is sought from its capacity as a system generator. Systems capable of providing answer to both the place and the rest of the dimensions, scales and techniques demanded by each draft. Its work is based on the research that arises from the proposals presented to architectural competitions, many of them has been rewarded with awards and honorable mentions, others won and built.

Among these works we can find the Badajoz, Coria and Plasencia secondary schools and the Malpartida de Plasencia and Jaraíz de la Vera Gym Plvillions for the Regional Government of Extremdura, Spain and the RivasFutura Underground Station for Metro de Madrid.

His works have been published in different specialized magazines and exhibitied by different national and foreign insititutions.

www.landinez-rey.com

M artı D Mimarlık

M artı D Mimarlık was founded in 1987 in İzmir by Metin Kılıç and Dürrin Süer. They design various types of projects in various scales such as residential, commercial, healtcare, educational and urban design. With their intention that unites academic and practical skills, they contribute to today's architecture culture.

Metin Kılıç - Partner - Founder(Architect)
He was born in 1962. In 1985 he graduated from 9 September University, Faculty of Architecture. He has been working and conducting M artı D Mimarlık since 1987 as founder and partner. He has many projects sucha as hospitals, educational instutions, hospitals, residences, commercial buildings. He has been principal of İSMD between 2013-2015. He has many prizes in architectural competitions.

Dürrin Süer - Partner - Founder (Phd. Architect)
She was born in 1965 in Ankara. In 1987 she graduated from 9 September University, Faculty of Architecture. She has completed Masters and PhD Degrees in same university. Between 1987 - 2007 she has worked as academician in 9 September University, Faculty of Architecture. She has many articles published in architectural magazines about architecture education, architecture and utopia, architecture and technology, consumption spaces and residental spaces. She has been working in M artı D Mimarlık as founder and partner. She also takes place as jury member in architectural competitions and writes academical researchs and articles.

Ali Can Helvacıoğlu (Architect)
He was born in İzmir in 1989. He was graduated from Izmir University of Economics, Faculty of Architecture in 2011. He has been working in M artı D Mimarlık since 2011.

murmuro

João Caldas (Braga, 1981) is the co-founder, with Rita Breda (Estarreja, 1981) of the office murmuro, working from Braga and Porto, in Portugal. They have started their joint path while still at the university, having graduated from DARQ-FCT, University of Coimbra. As Erasmus students, at the NTNU and the Fine Arts Academy in Trondheim, Norway, they had the opportunity to taken part of an interdisciplinary and collaborative program for architecture and art students, pivotal in their education.

The success of some of their projects led to the foundation, in 2015, of the office murmuro, where they develop projects regardless of their scale, program or budget. From their body of work they highlight the multifamily housing project Doze Casas, in Braga, shortlisted for the PREMIS FAD award in 2016, the VII ENOR Award in 2017, and the 2nd Prize on the Serralves Foundation's Pavilion Competition. Recently murmuro has received EUROPE 40 UNDER 40 award that spotlights the most promissing and the best emerging young architects in europe.

NISHIZAWA ARCHITECTS

SHUNRI NISHIZAWA
was born in 1980. He obtained his B.Arch (2003) and M.Arch (2005) degrees from Tokyo University. He worked from 2005 to 2009 for Tadao Ando Architect & Associates (Osaka). Then, he worked with Vo Trong Nghia (Ho Chi Minh City, 2009-11). He was a Partner in Sanuki+Nishizawa Architects (2011-15), before founding Nishizawa Architects in 2015. His work includes the Binh Thanh House (with Vo Trong Nghia Architects, Ho Chi Minh City, 2013); Thong House (Ho Chi Minh City, 2014); Katzden Factory (Binh Duong City, 2016); House in Chau Doc (2017; published here); Pizza 4P's Ben Thanh (Ho Chi Minh City,2017), all the completed buildings are in Vietnam.

object-e architecture

Object-e architecture is an architectural practice currently based in Thessaloniki, Greece and directed by Katerina Tryfonidou and Dimitris Gourdoukis. It started in 2006, in St. Louis, USA, as a platform with the intention to explore new territories in architecture with the aid of computational tools and techniques. Through time, object-e moved beyond the borders of computation and engaged design at large, trying to graft the new media with the social, political and ecological issues that architecture is facing today. It has won a number of prizes in international competitions, and its works has been published, exhibited and presented internationally.

Object-e is based on several collaborations with people coming from different backgrounds, with different design intentions and agendas. The outcome of object-e, being in most cases collaborative, is therefore defying any concept of style; Identity is formed through difference and constant transformation.

Object-e engages architecture and design in three different - but always connected - levels: Through specific design projects that are answers to specific design questions; private projects or competitions. Through experimental research projects that are aiming to extend our understanding of space, digital media and fabrication. Through teaching; either in established institutions or independently, sharing of knowledge is always the core that allows the rest to happen.

http://object-e.net
https://www.facebook.com/object.e

OOIIO Architecture

OOIIO is an international team of architects, designers and engineers engaged in finding this special "I don´t know what it is" that makes a work unique, exciting and able for transmit sensations on a way that a vulgar work will never get.

In OOIIO we know that not every construction is architecture.
We do architecture.

OOIIO creative process is completely open and random. When we start a project we never know how it will end, what will come out, that's why we introduced throughout the design process countless stimulus and references from any kind that may improve the final design. These are usually unexpected and we find them everywhere, like in the geometry of a mineral that seems attractive, a traditional dish of the place where the project is built, the shape and colors of a vase, a tree or a graffiti that we found painted on a wall.

This constant search for poetical links to architecture through everything around us makes us look at the world with wide open eyes and helps us to get every project as the result of a unique creative process, making each OOIIO project special and standed out.

www.ooiio.com

Piuarch

Francesco Fresa, Germán Fuenmayor, Gino Garbellini and Monica Tricario formed the Piuarch studio in 1996 out of a desire to merge different experiences into a shared architectural project.

The studio is located in an open space in a former industrial building that once hosted a typography business in Brera, in the centre of Milan. Here, Piuarch designs public buildings, office and residential complexes, commercial spaces, boutiques, shopping malls and even urban plans, with the contribution of consultants from various disciplines.

Piuarch has pursued these themes participating in competitions, developing projects from the planning to the final construction phase, elaborating interior design projects.

In recent years Piuarch has developed a number of projects abroad. It is active in China, Algeria, Russia, where it has recently opened an operational office, and in Ukraine, with ongoing and realized projects.

SLOT STUDIO

SLOT is an active and interdisciplinary architectural design studio. People from diverse professional disciplines contribute to this project.

Our work has reached a profound understanding of the human needs, enabling us to intertwine the constructive and philosophical sides of building.

As architects, we push design to its ultimate material consequences and aim for cultural connectivity: sense of usability, mathematics of space and a wide aesthetic research.

Juan Carlos Vidals, Founder and Director of SLOT, graduated from the UNAM - National Autonomous University of Mexico in 2002 and has worked in the LCM/Fernando Romero and collaborated with Rojkind Arquitectos in several projects.

slot.mx

Stefano Corbo Studio

Stefano Corbo (1981) is an Italian architect, researcher, and Assistant Professor at RISD (Rhode Island School of Design).

He holds a Ph.D and an M.Arch. II in Advanced Architectural Design from UPM-ETSAM Madrid (Escuela Técnica Superior de Arquitectura).

Stefano has taught at several academic Institutions: Nanjing University; LAU Beirut (Lebanese American University); The Faculty of Architecture in Alghero, Italy; ETSAM Madrid; he has also been a guest lecturer at SAC Städelschule Frankfurt, Deakin University in Melbourne, College of Design Minnesota, ESALA Edinburgh, The University of Miami, and The University of Wisconsin.

Stefano has contributed to several international journals and has published two books: "From Formalism to Weak Form. The Architecture and Philosophy of Peter Eisenman." (Ashgate-Routledge, 2014), and "Interior Landscapes. A visual atlas." (Images, 2016).

In 2012, after working at Mecanoo Architecten, Stefano founded his own office SCSTUDIO (www.scstudio.eu), a multidisciplinary network practicing architecture and design, preoccupied with intellectual, economic and cultural contexts.

www.scstudio.eu

stpmj

stpmj is an award winning design practice based in New York and Seoul. The office is founded by Seung Teak Lee and Mi Jung Lim with the agenda, "Provocative Realism". It is a series of synergetic explorations that occur on the boundary between the ideal and the real. It is based on simplicity of form and detail, clarity of structure, excellence in environmental function, use of new materials, and rational management of budget. To these we add ideas generated from curiosity in everyday life as we pursue a methodology for dramatically exploiting the limitations of reality.

We design identity, brand and value.
We design creative and innovative culture.
We design unique vision of architecture.

We have been recognized with architectural awards including,

Architectural Record Design Vanguard

American Institute of Architects New York Design Award

Korean Ministry of Cultre, Sports and Tourism Young Architects Award

Kim Swoo Geun Preview Award

American Institute of Architects New Practices New York

Architectural League Young Architects + Designers

Studio Farris Architects

Studio Farris Architects is an architectural practice based in Antwerp, Belgium, founded by Italian architect Giuseppe Farris in 2008.

The studio's goal is to discover the intrinsic potential in every project, questioning the obvious, exploring the surroundings and cultural heritage. Architecture, interior architecture, furniture design, lighting and graphic design are all part of the same overall concept and are designed as a coherent whole, with particular attention to detail.

The work of Studio Farris Architects has been widely featured in international publications such as Architectural Record (USA), AZURE (Canada), TASARIM (Turkey) and Wallpaper* (United Kingdom). Studio Farris Architects is the recipient of Architectural Record's 2015 Design Vanguard Award.

www.studiofarris.com

SUPA architects schweitzer song

Ryul Song was born in Taejon, Korea, and graduated from Hong-Ik University Seoul and Technical University of Dortmund, Germany. Christian Schweitzer was born in Linz, Austria, and graduated from Kaiserslautern University of Technology, Germany. Both met during their master's degree studies at the Frankfurt Staedelschule Academy of Fine Arts and established in 2000 their studio SUPA architects schweitzer song in Frankfurt, Germany. In 2005 they branched out to Seoul, Korea, teaching at Korea National University of Arts, Seoul National University, Korea University, and Hanyang University.

SUPA is working within the narrow intersection of conceptual design, art, theory and education. They are focused on the conceptual approach towards architectural design, finding new ways to expand the vocabulary of architecture. Their special interest lies in the exploration of the specific sociocultural context inherent in a task as the interface between everyday life, art and architecture.

TAKK Architecture

Takk is a space for architectural production focused in the development of experimental and speculative material practices in the intersection between nature and culture in the contemporary framework, with a special attention on the overcoming of anthropocentrism on its different ways (political, ecological, cultural, on gender), and on the definition of new notions of beauty through the articulation of the difference by assembling a multiplicity of materials from different origins and conditions.

Additionally to this profesional practice, Takk is developing a framework in the field of research and teaching. Mireia Luzárraga and Alejandro Muiño are teachers in the Projects Department of the Universidad de Alicante (UA), in BAU Barcelona Design University Centre, and Master Tutors in the Institute of Advanced Architecture of Catalonia (IAAC). They have also participated as teachers in different workshops and summer schools and have explained their work in several lectures internationally.

At the present time, Mireia and Alejandro combine their profesional and teaching labour with the development of their respective PhD Thesis on the politics of ornament and self sufficient micro-communities. They have been granted for them with the scholarship "Junior Faculty - La Caixa".

TheeAe Architects

TheeAe is abbreviation of the evolved architectural eclectic. Its name is about the effort and dedication to the value of architectural aesthetic which shall be laid on place, history and culture of surrounding environment. TheeAe pursues re-searching and re-finding the elements that have been embedded into the context of environment, so as to define the beauty of the architecture within the given context.

In this passion, TheeAe has begun its practice, since 2011 in Hong Kong, to continue to explore and learn of culture and beauty of environment through the design. TheeAe's service has been extensively covered in various areas of architecture & interior design. The projects include, one of landmark projects, Mumbai Airport (Chhatrapati Shivaji International airport), and Mumbai Hotel (Bellagio & Mandarin) in India, Mardina Residence in Saudi Arabia. As well as, projects are developed and proposed in many other places around world, those of which are Afghanistan Museum in Kabul, Afghanistan, Guggenheim Museum in Helsinki, Finland, Science Center in Kaunas, Lithuania, Arch-lay Contemporary Museum in Shenzhen, China, Elevated Elf Land, horse theme park, in South Korea, Dubai Heart, Dubai, UAE., etc.

Our service will continue to serve the clients who are seeking for superior design quality not only to increase the value of the properties but also to bring the meanings of places for people, who will live, visit and pass by.

TOUCH Architect

TOUCH Architect Co.,Ltd. was first established since 2014. It was changed from TOUCH STUDIO Architect partnership, with four years experiences into a company. With a great chance of an improvement, we have two main co-founders consist of Mr. Setthakarn Yangderm as an architect and leader of our firm and Ms. Parpis Leelaniramol as an architect, which will corporate together in design, construction, and management.

SETTHAKARN YANGDERM
Architect // Managing Director // Founder
Bachelor of Architecture
(Department of Architecture, Major in Thai Architecture)
Faculty of Architecture, Chulalongkorn University. (1st top rank of Thailand's university)
Honor in best architectural design in 2007.

PARPIS LEELANIRAMOL
Architect // Co-founder
Bachelor of Science
(International Program in Design and Architecture),
Faculty of Architecture, Chulalongkorn University. (1st top rank of Thailand's university)
Master of Science in Real Estate Business (MRE), Faculty of Commerce and Accountancy, Thammasat University. (1st top rank of business program)
Teaching Assistant - MRE Thammasat Personal Consultant - Mini-MRE at Ananda Development (Real Estate Listed Company in Thailand)

UNStudio

© Inga Powilleit

Ben van Berkel studied architecture at the Rietveld Academy in Amsterdam and at the Architectural Association in London, receiving the AA Diploma with Honours in 1987.

In 1988 he and Caroline Bos set up an architectural practice in Amsterdam, extending their theoretical and writing projects to the practice of architecture. UNStudio presents itself as a network of specialists in architecture, urban development and infrastructure. Current projects include the design for Doha's Integrated Metro Network in Qatar, 'Four' a large-scale mixed-use project in Frankfurt and the Wasl Tower in Dubai.

With UNStudio he realised amongst others the Mercedes-Benz Museum in Stuttgart, Arnhem central Station in the Netherlands, the Raffles City mixed-use development in Hangzhou, the Canaletto Tower in London, a private villa up-state New York and the Singapore University of Technology and Design.

In 2018 Ben van Berkel founded UNSense, an Arch Tech company that designs and integrates human-centric tech solutions for the built environment.

Ben van Berkel has lectured and taught at many architectural schools around the world. Currently he holds the Kenzo Tange Visiting Professor's Chair at Harvard University Graduate School of Design, where he has led a studio on health and architecture. In 2017, Ben van Berkel also gave a TEDx presentation about health and architecture. In addition, he is a member of the Taskforce Team / Advisory Board for the Dutch Construction Industry. He is also currently a member of the Taskforce Team / Advisory Board Construction Industry for the Dutch Ministry of Economic Affairs.

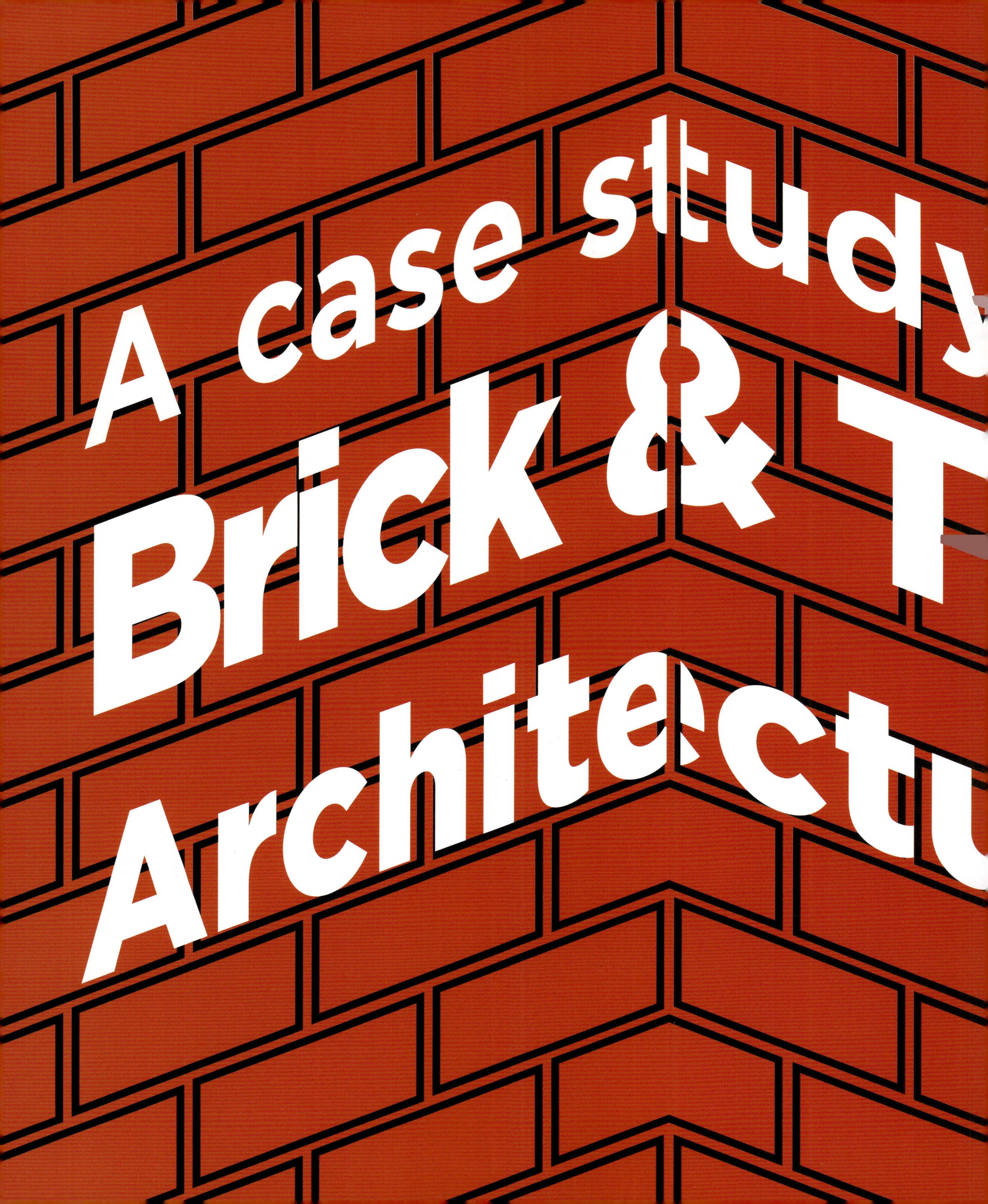
A case study
Brick & T
Architectu

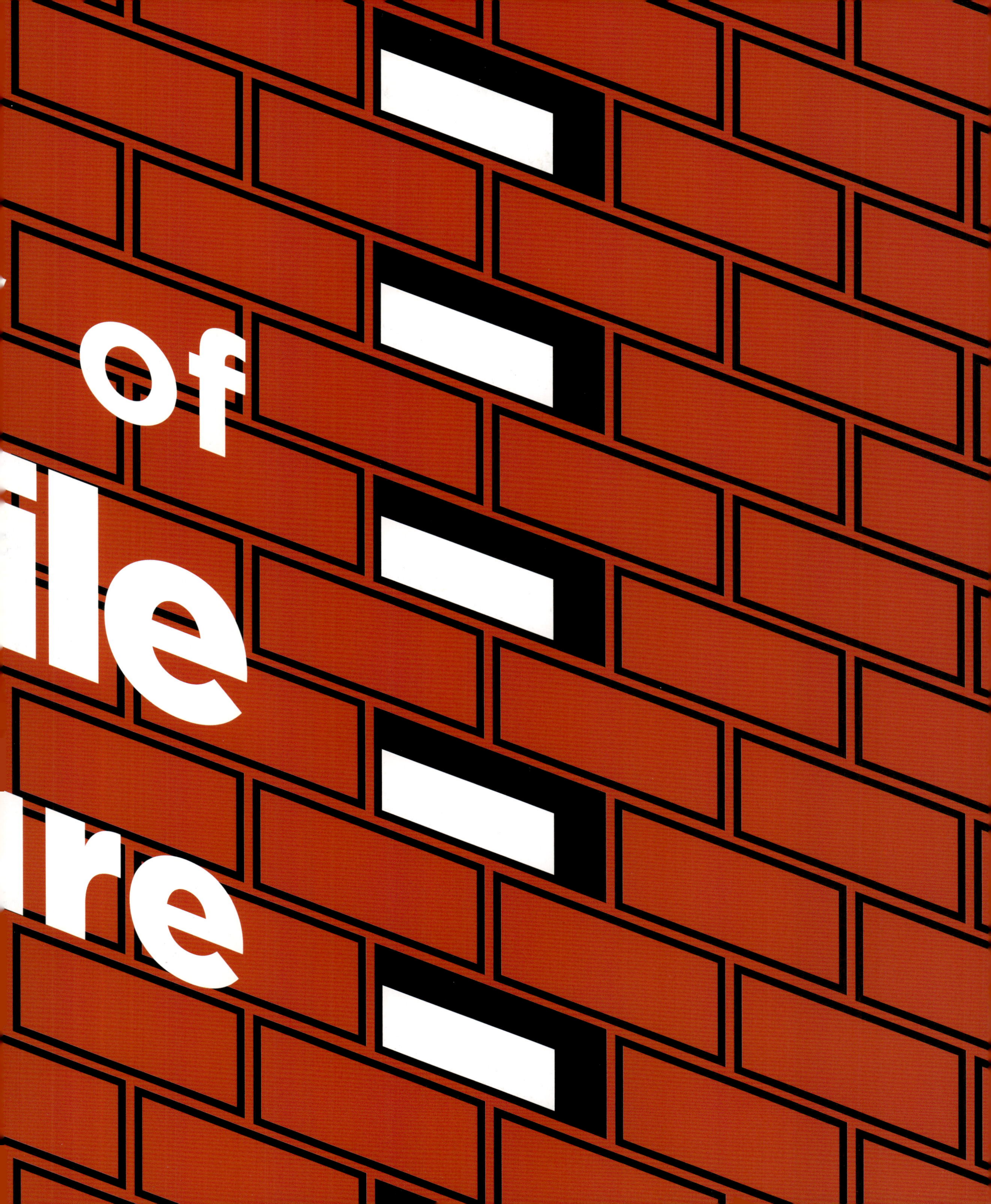

of
ile
re

534, New York, USA
HWKN

© Michael Moran/OTTO

© Michael Moran/OTTO

© Michael Moran/OTTO

© Michael Moran/OTTO

© Michael Moran/OTTO

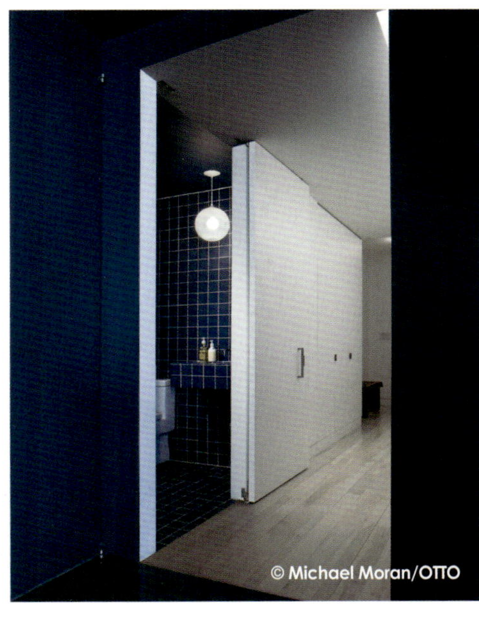

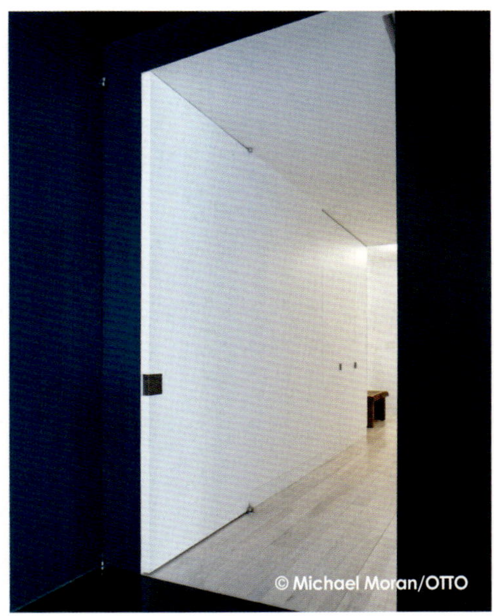

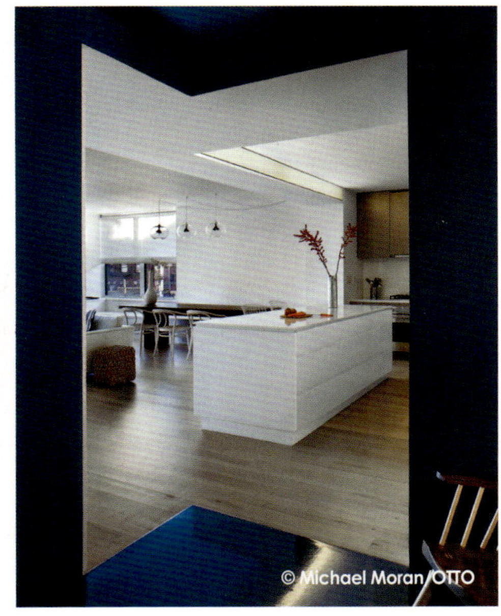

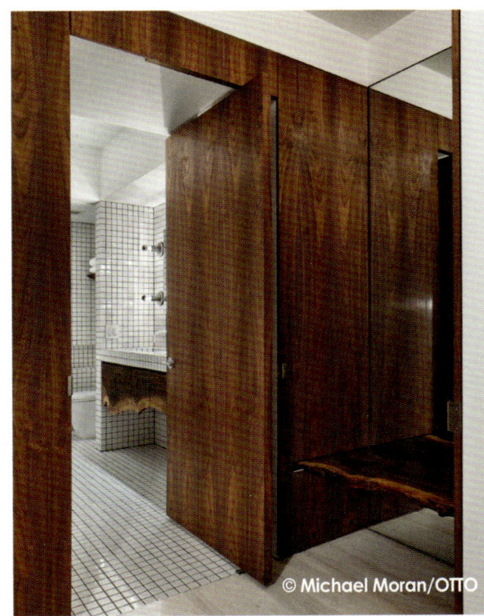

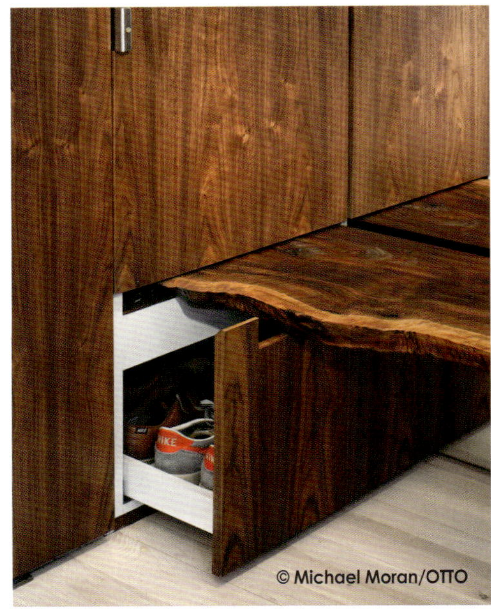

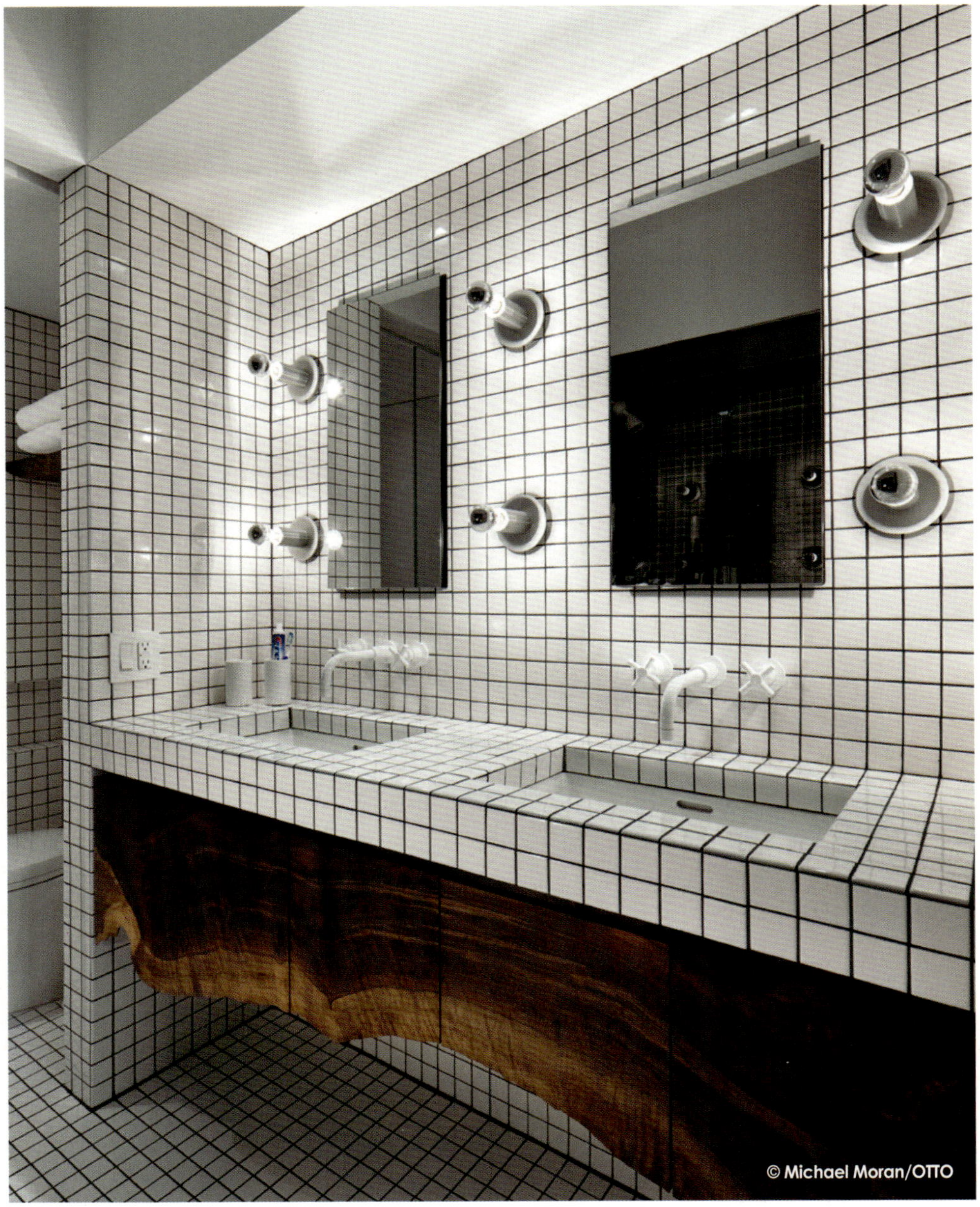
© Michael Moran/OTTO

'Aktipis' Flowershop, Patras, Greece

Point Supreme Architects

© Yiannis Drakoulidis

Axono

Facade

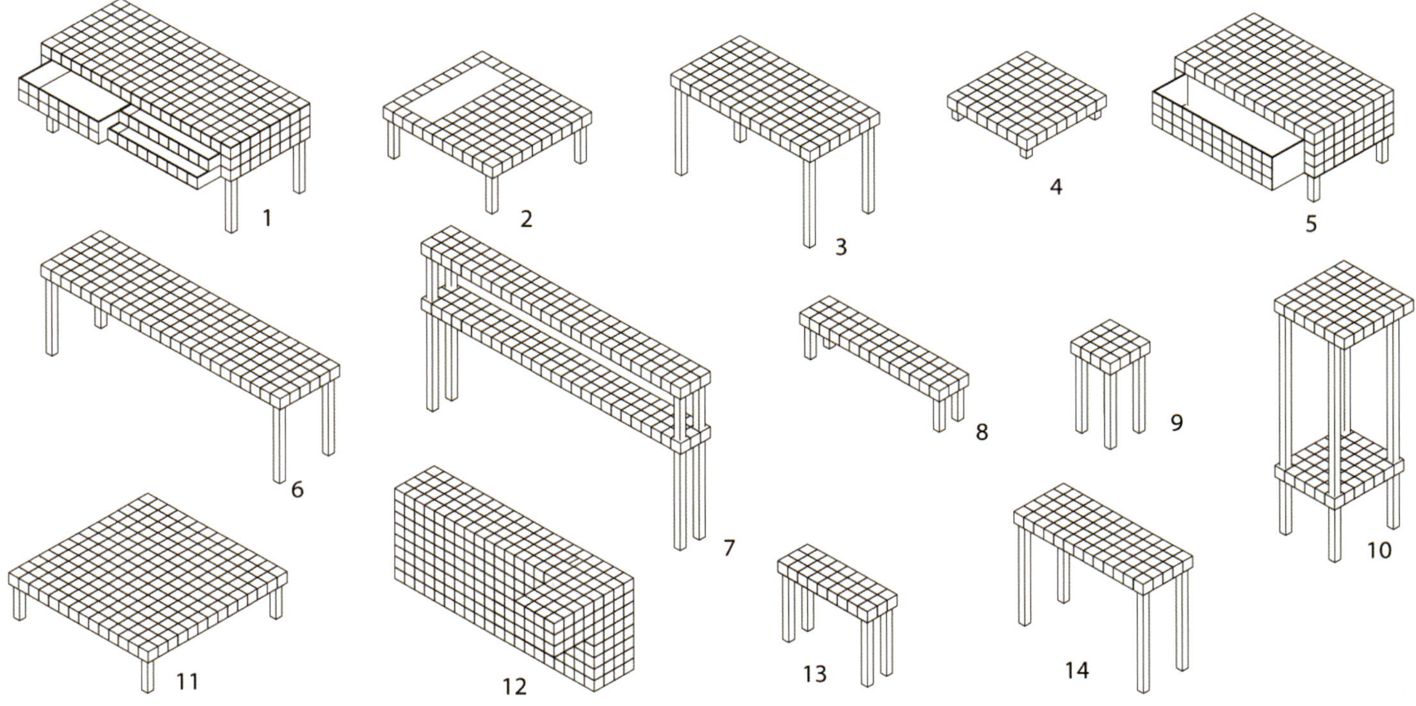

1.cashier/preparation table
2.low table/client seat
3.high table for low plants
4.very low table for heavy pots with big plants
5. table with big drawer for storing pots
6. high and long table for medium size plants
7. high, narrow and long table for small pots with flowers
8. vitrine table; fits in the 3dimensional facade void
9. table for special plant
10. table for plants with hanging branches
11. exhibition table
12. 'technical' table for computer, fax, phone
13. small table for special plants in the fridge
14. table in the fridge

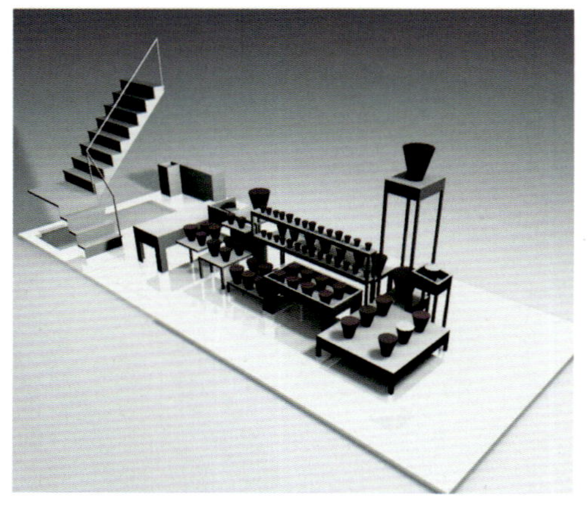

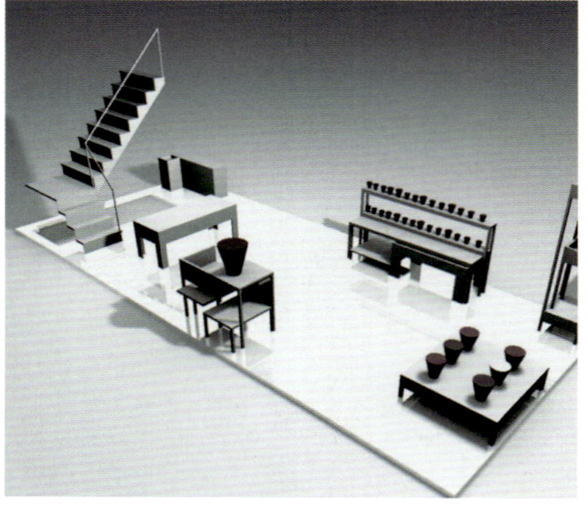

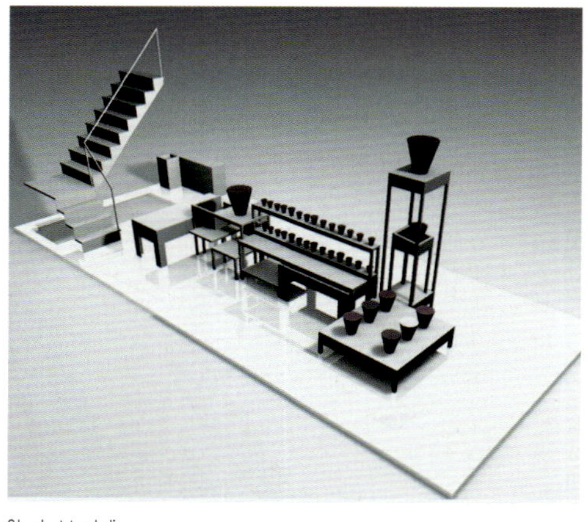

Study Modeling

Flowershop phases

© Yiannis Drakoulidis

© Yiannis Drakoulidis

Amasion, Amsterdam, the Netherlands

Concrete Architectural Associates

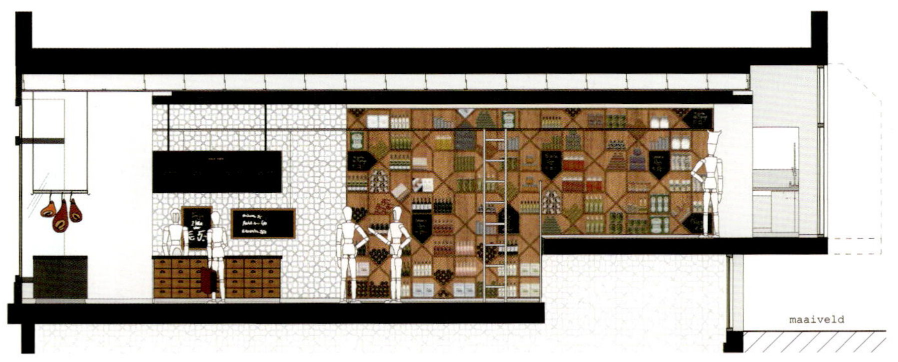

Section B-B

Section C-C

Kalavrita Café, Kalavrita, Greece
Point Supreme Architects

© Yiannis Drakoulidis

VIVID

Minto Midtown Penthouse, Toronto, Canada

II BY IV Design Associates Inc

© David Whittaker

© David Whittaker

© David Whittaker

NINETTE HOUSE, Madrid, Spain

Arenas Basabe Palacios Arquitectos

Ground Floor

First Floor

© Imagen Subliminal

© Imagen Subliminal

© Imagen Subliminal

© Imagen Subliminal

No Visitor House KOSID, Seoul, South Korea

jay is working

90 C H
11 0 2
33 22 49
100 CH 13
63 97 03

Construction Process

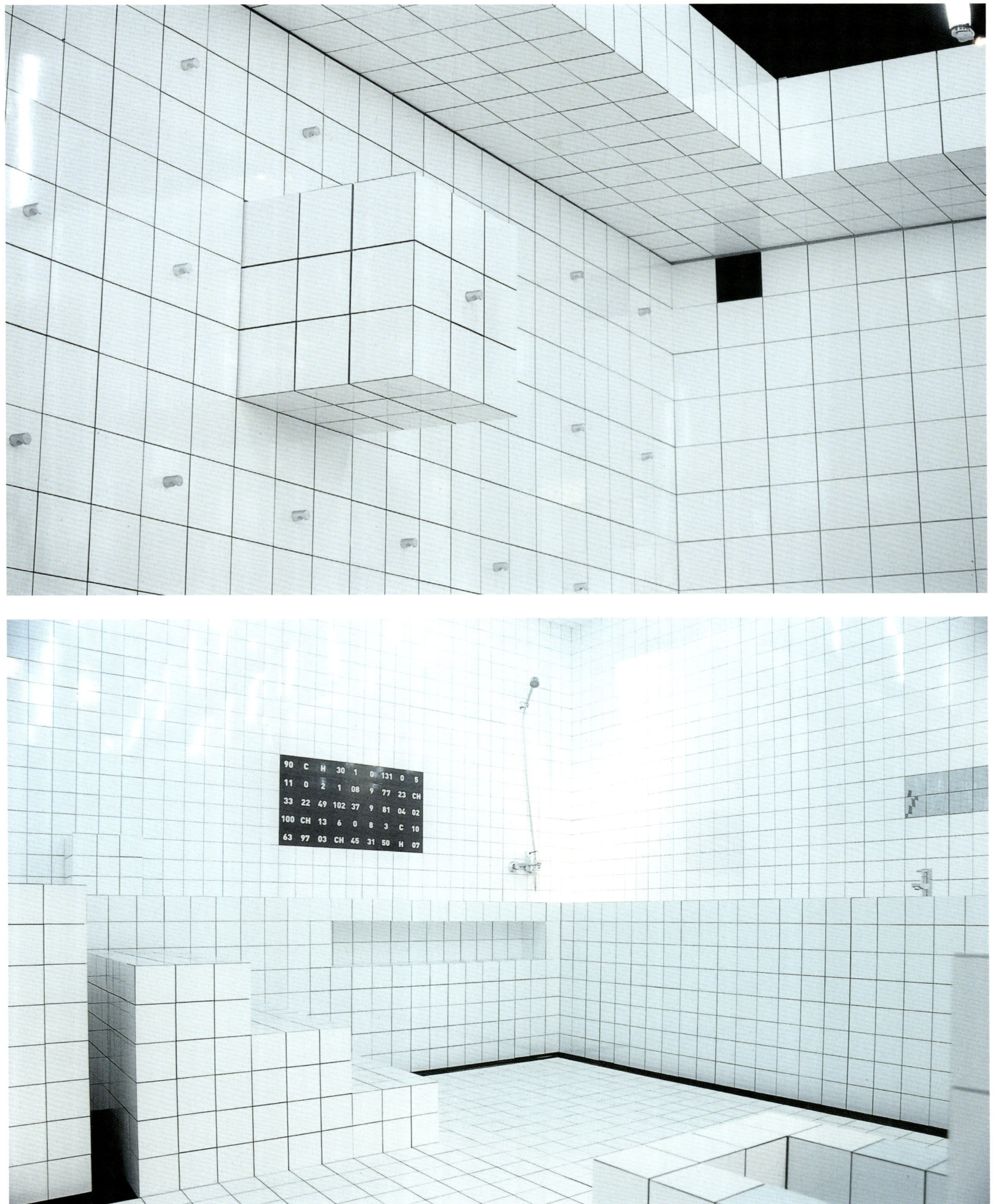
90 C H 30 1 0 131 0 5
11 0 2 1 08 9 77 23 CH
33 22 49 102 37 9 81 04 02
100 CH 13 6 0 8 3 C 10
63 97 03 CH 45 31 50 H 07

Office and Hotel Room, Rotterdam, Netherlands

BOARD

© Sonia Arrepia

Concept

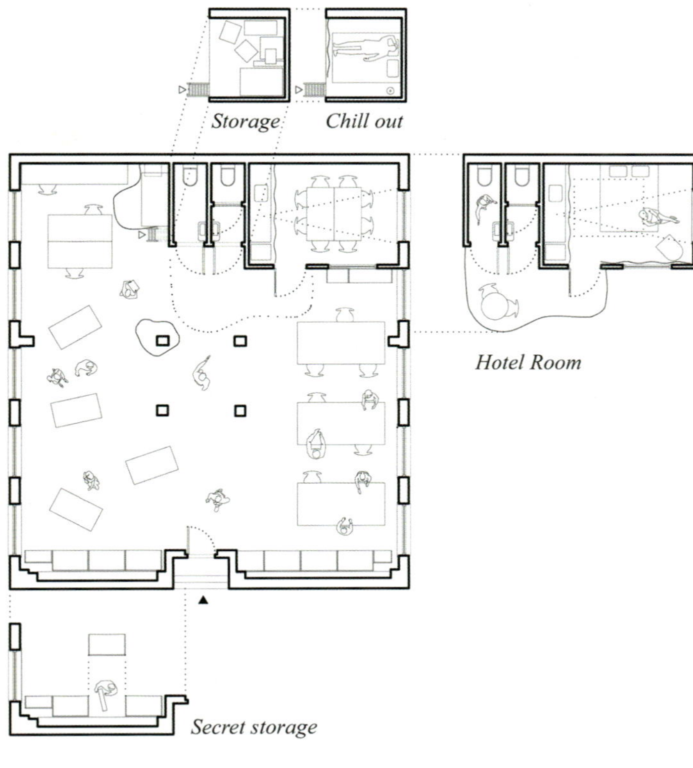

Floor Plan

Study Modeling

Tiled entrance door closed and under construction by artisan Francisco Javier Ramo Soriano

Interior of the lavatory that contains the hand-made mosaic depicting a palm tree

© Sonia Arrepia

TILESCAPE, Hamburg, Germany

BudCud

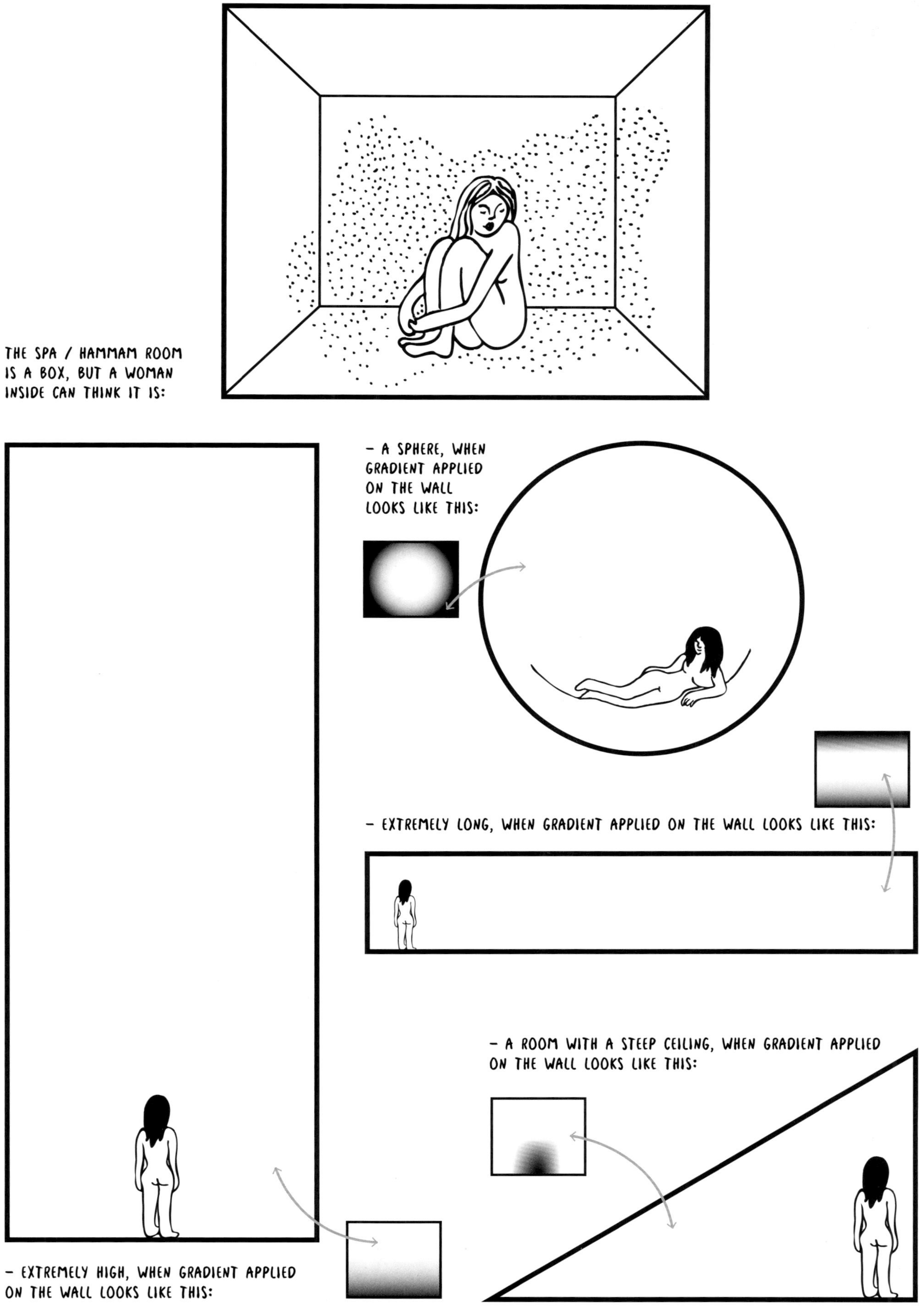

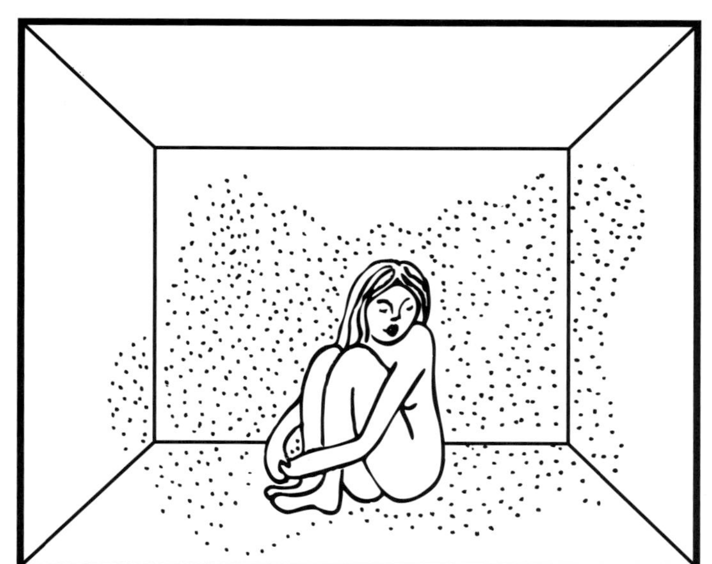

THE SPA / HAMMAM ROOM IS A BOX, BUT A WOMAN INSIDE CAN THINK IT IS:

– A SPHERE, WHEN GRADIENT APPLIED ON THE WALL LOOKS LIKE THIS:

– EXTREMELY LONG, WHEN GRADIENT APPLIED ON THE WALL LOOKS LIKE THIS:

– A ROOM WITH A STEEP CEILING, WHEN GRADIENT APPLIED ON THE WALL LOOKS LIKE THIS:

– EXTREMELY HIGH, WHEN GRADIENT APPLIED ON THE WALL LOOKS LIKE THIS:

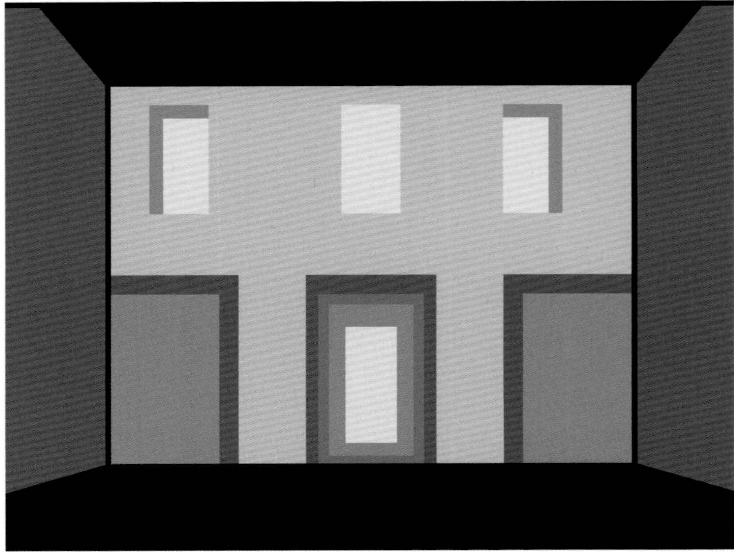

A. House of Siricius, Pompeii

B. The Light Inside. James Turrell

Perspective is used for building the image depth, performing three-dimensional space onto a flat surface of the painting. In house of Siricius in Pompeii optical tricks with perspective built narration of additional chambers, created the notion of infinite 'dream' space (picture A). These means are also used nowadays. Optical acrobatics is visible in contemporary art of James Turrell. The artist - 'illusionist' provides spatial distortions, using surfaces of light and color panels (picture B).

spa space as it is

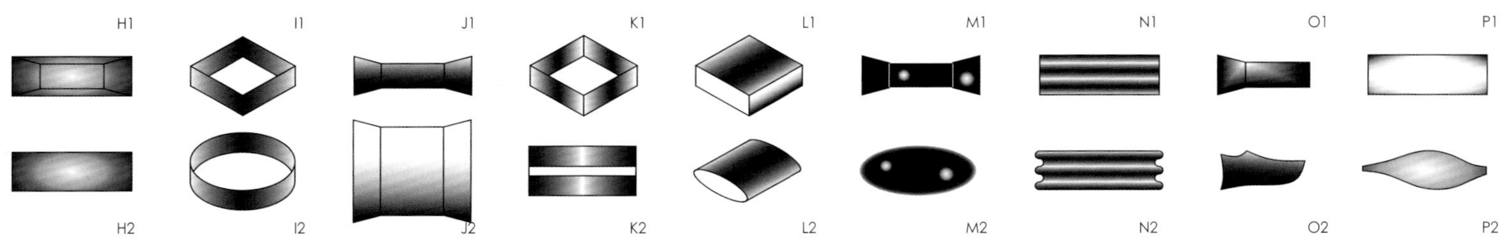

spa space as it seems

A specific usage of color gradients can distort the regular proportion of a spa room, putting the visitor into more abstract space. The diagram shows how the distortion of the spatial volume can be done with different application of monochrome combination.

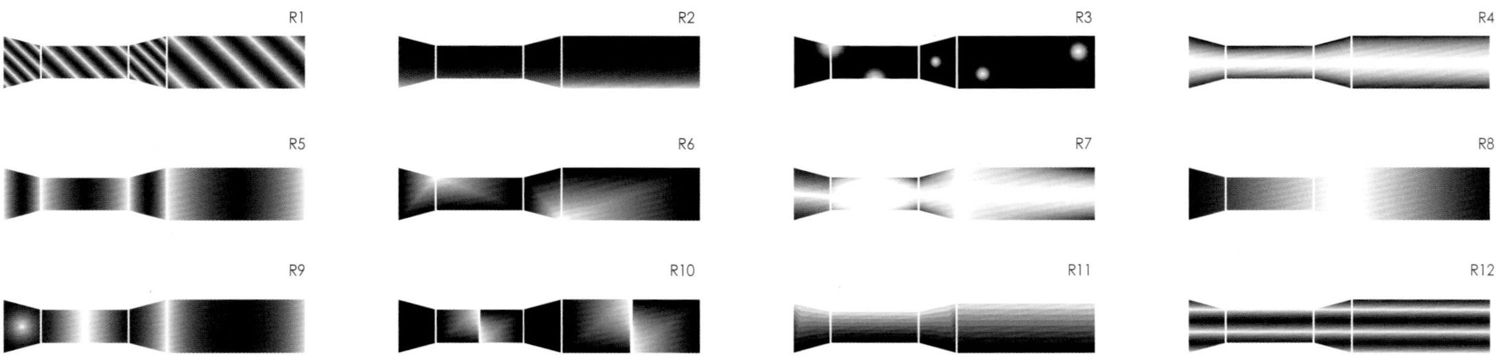

Types of optical distortion that could be used in the project.

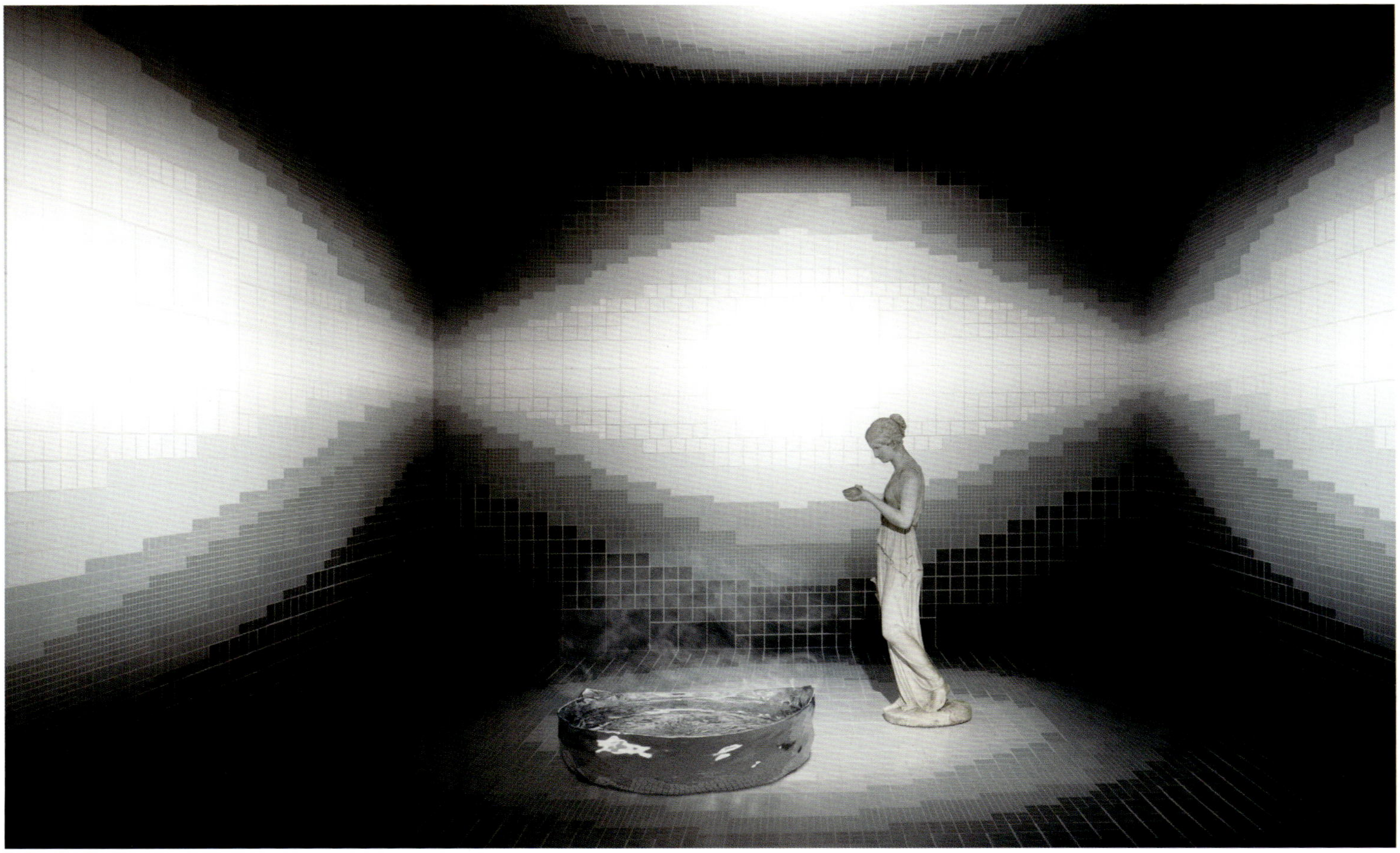

Witloof, Maastricht, the Netherlands

Maurice Mentjens Design

© Noa Kalina

CD House, Miane, Italy

Donner Sorcinelli Architecture

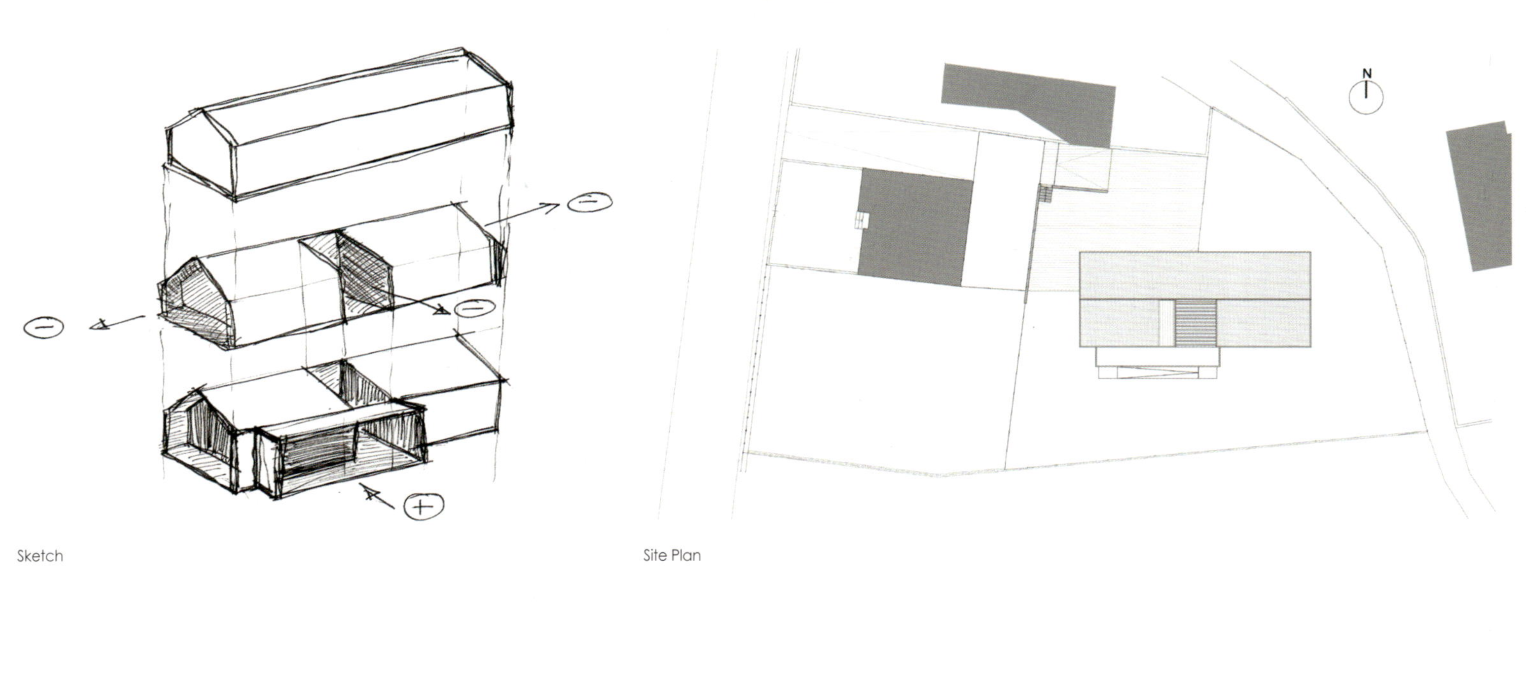

Sketch

Site Plan

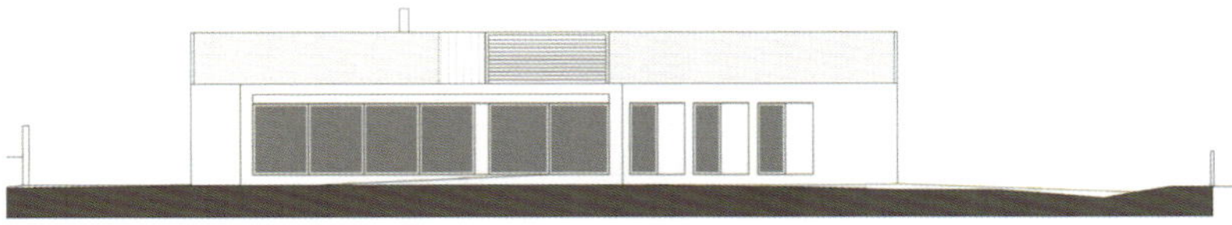

South Elevation

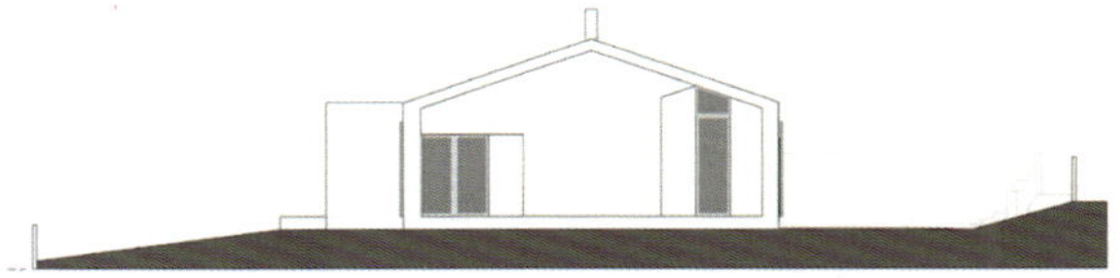

East Elevation

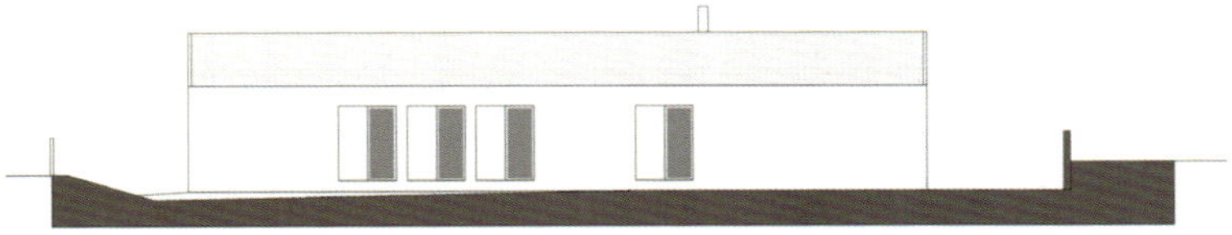

North Elevation

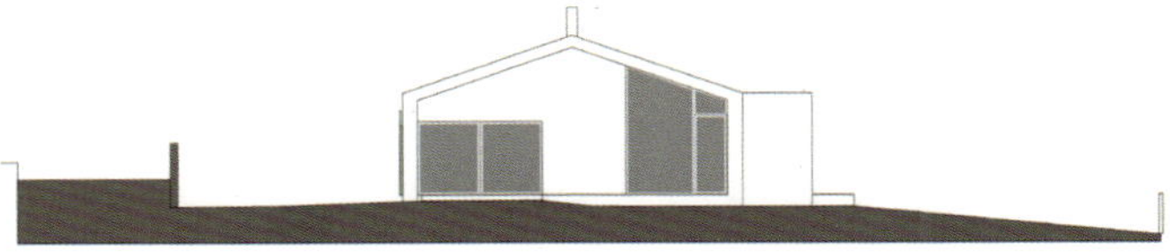

West Elevation

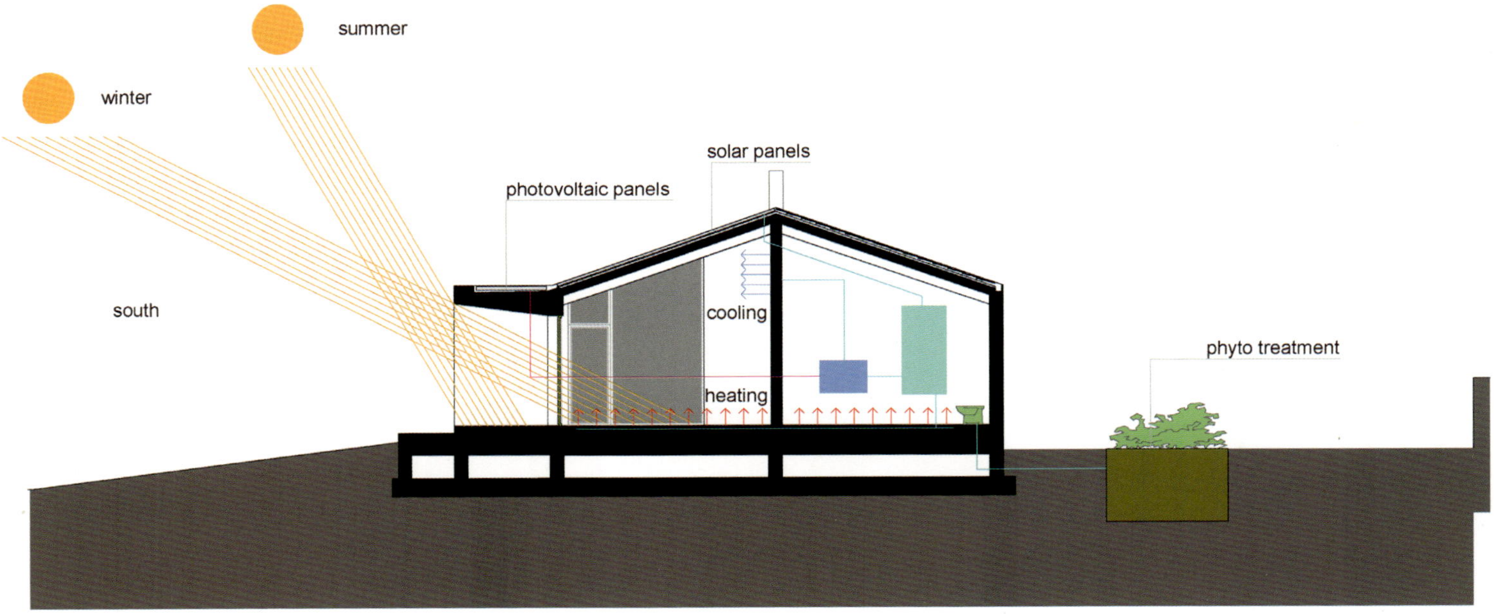

summer

winter

solar panels

photovoltaic panels

south

cooling

heating

phyto treatment

Sustainability

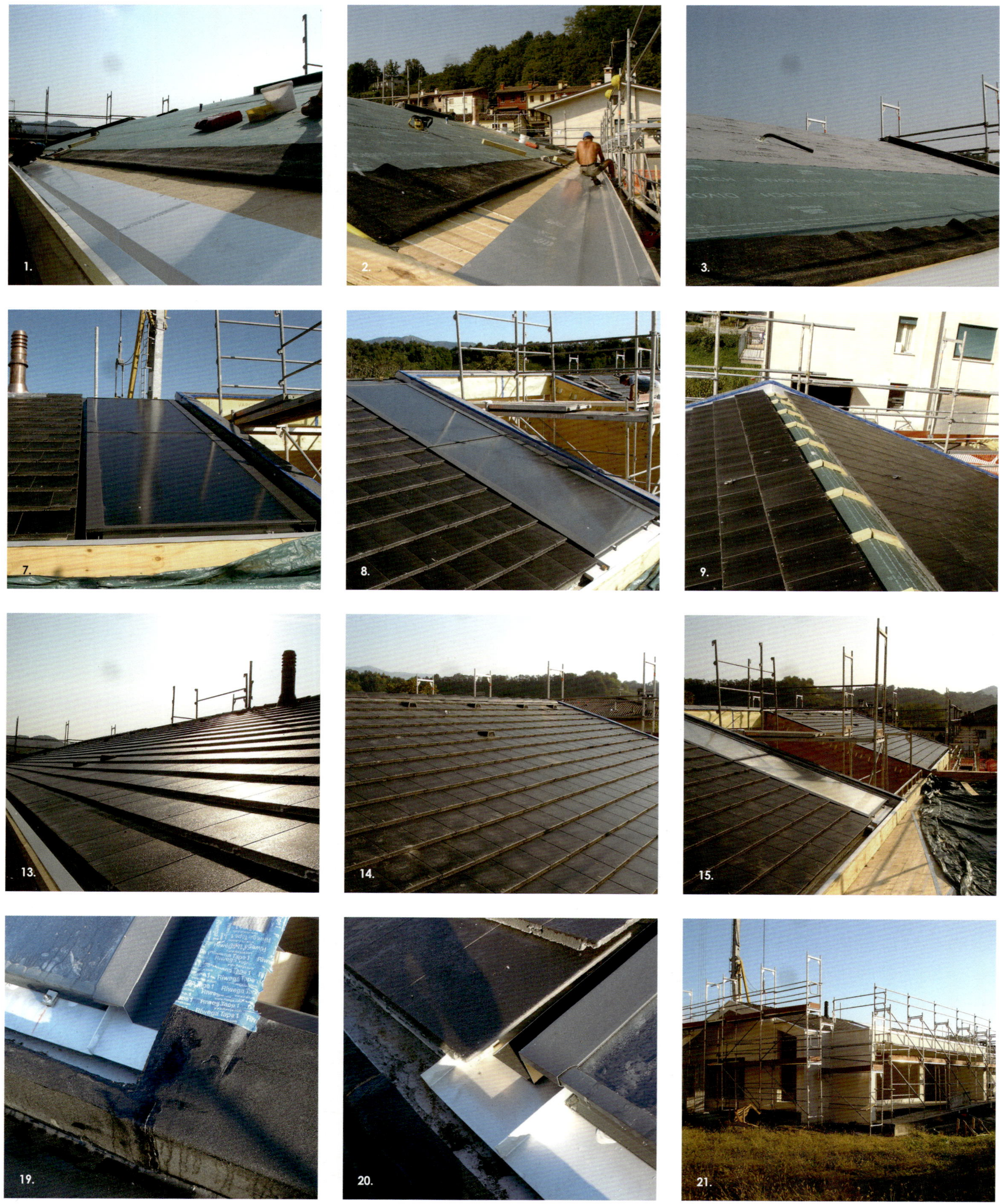

Progress of roof tiles

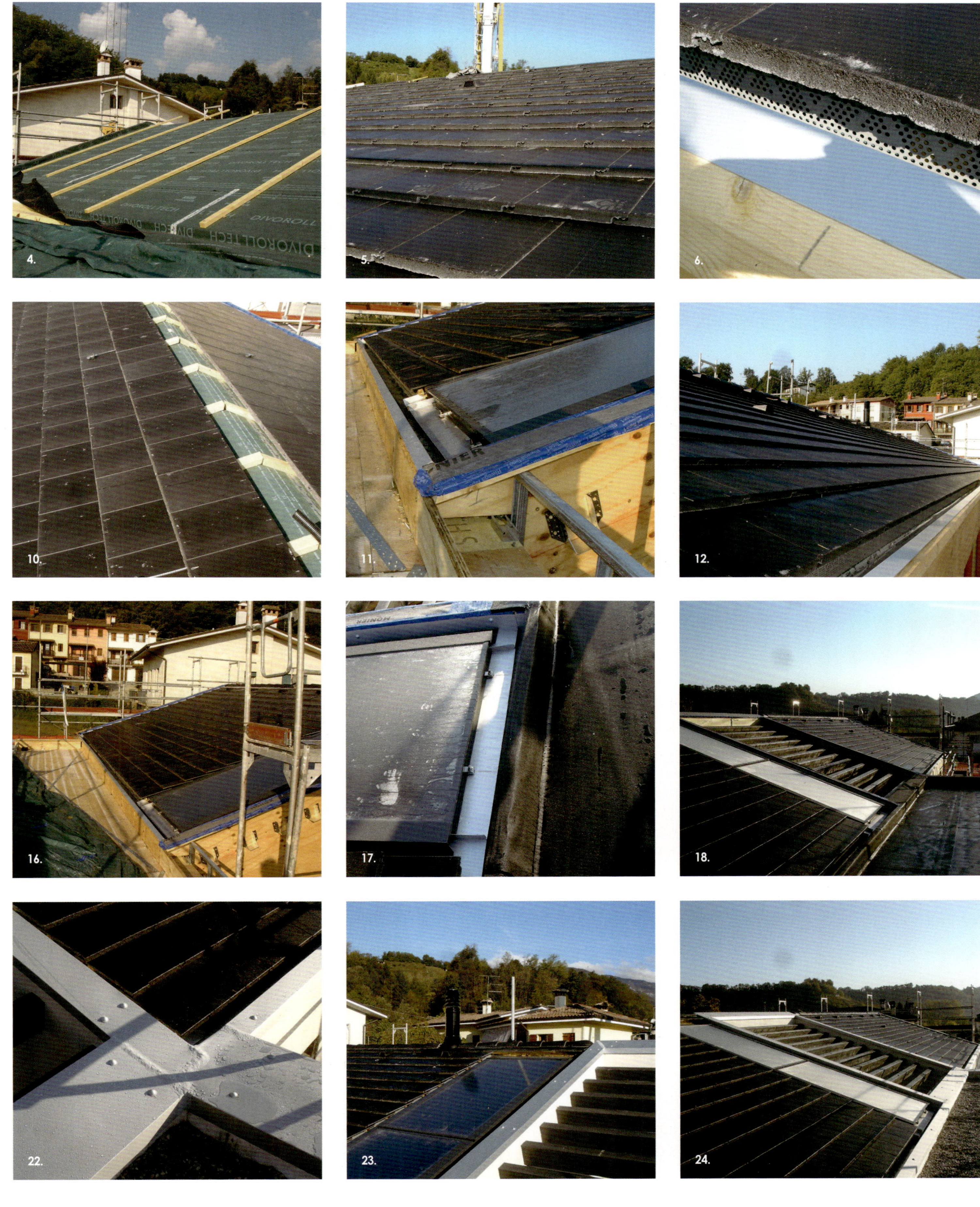
4.
DIVOROLL TECH
5.
6.
10.
11.
12.
16.
17.
18.
22.
23.
24.

Pété Mane, Gundlupet, India

architecture paradigm

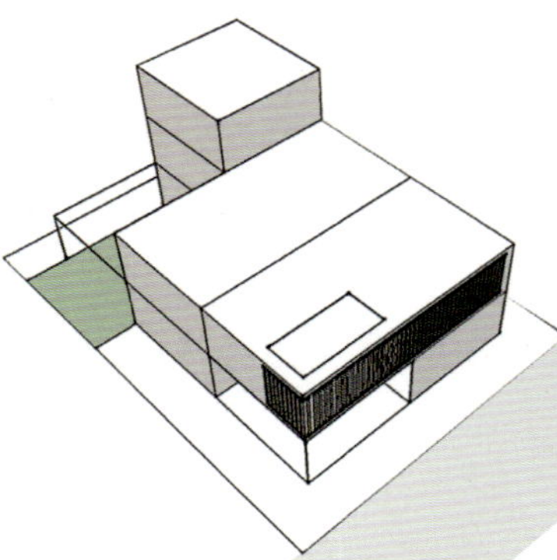

The screens are designed along the roadside offering privacy to the residence. This aspect adds to the idea which is a result of orienting the spaces towards the interiors rather than the street or outside and critically acknowledges changing structure of the town and the loss of vibrancy generated in the earlier tightly knit streets.

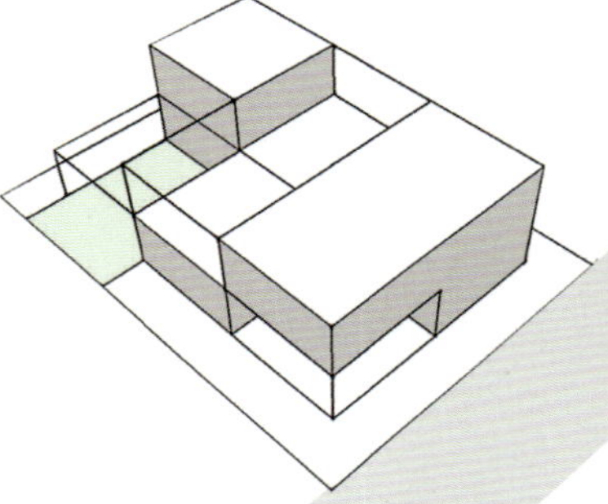

The Central bay is seen as double height where the upper level spaces look into it furthering the idea of focusing internally

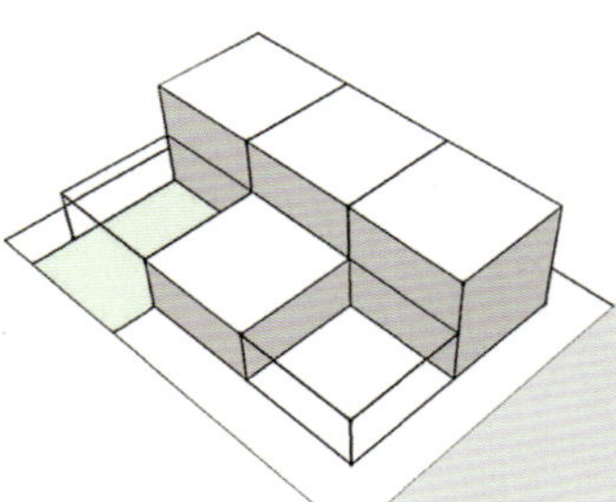

Two volumes are subtracted from this and is moved to the upper level. One of the open space is seen as multipurpose space / car port. The other is a open to sky court yard located towards the rear, this space is seen as an anchor around which the organization works.

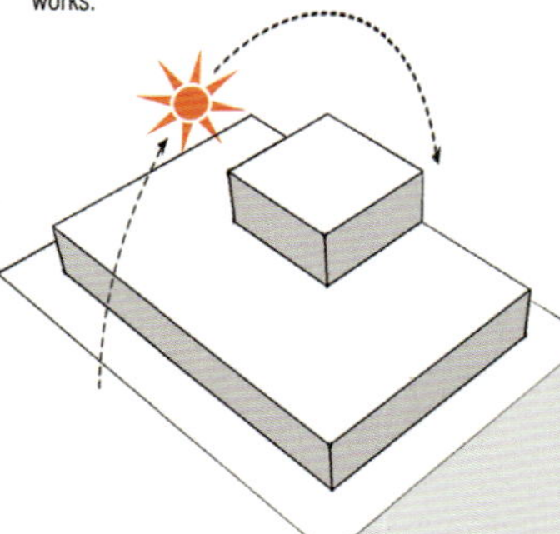

The program when blocked occupies the entire lower level and also a part of the upper level.

Properly breathing house, Hanoi, Vietnam

H&P Architects

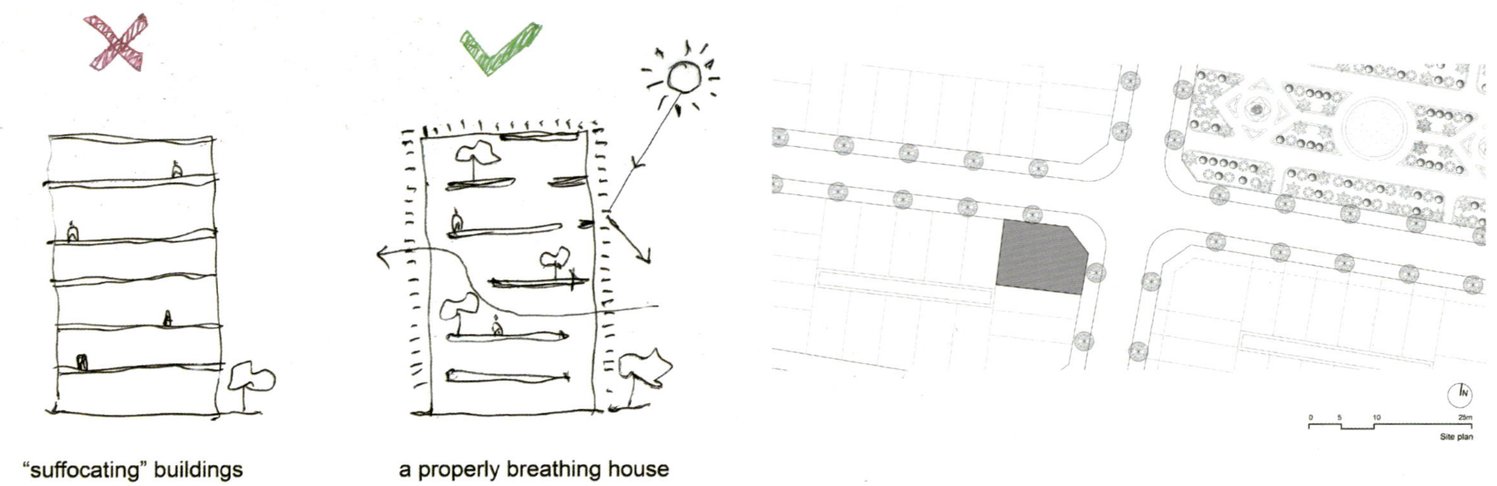

Sketch

Site Plan

Detail

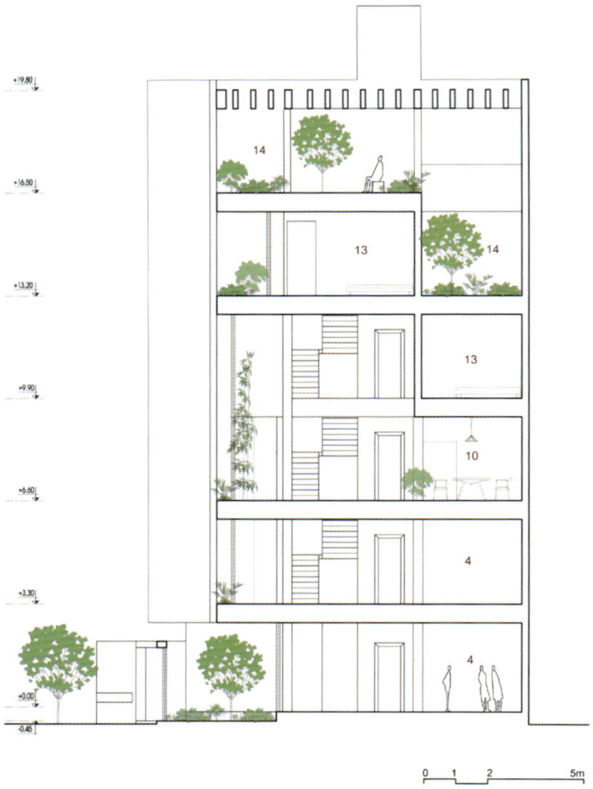
14
13
14
13
10
4
4

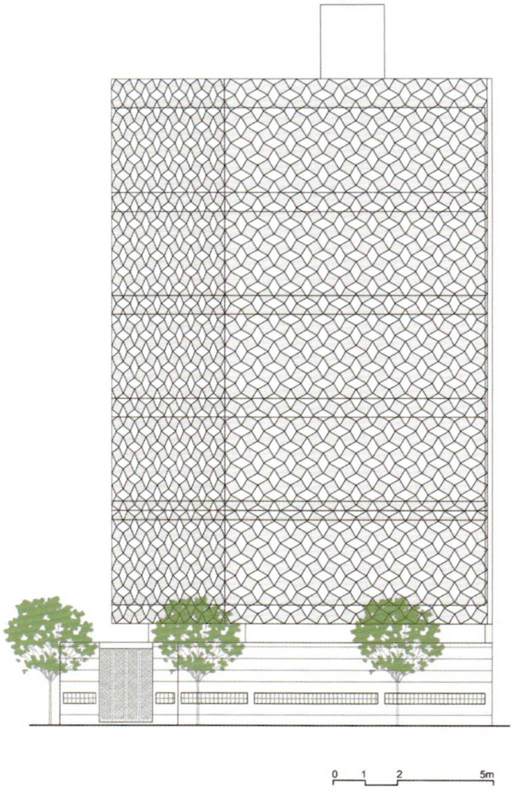

Section

Elevation

1. Entrance
2. Garden
3. Garage
4. Working Space
5. Kitchen
6. Elevator
7. Toilet
8. Workship
9. Living Room
10. Dinning
11. Void
12. Multifunction space
13. Bedroom
14. Terrace
15. Library
16. Washing

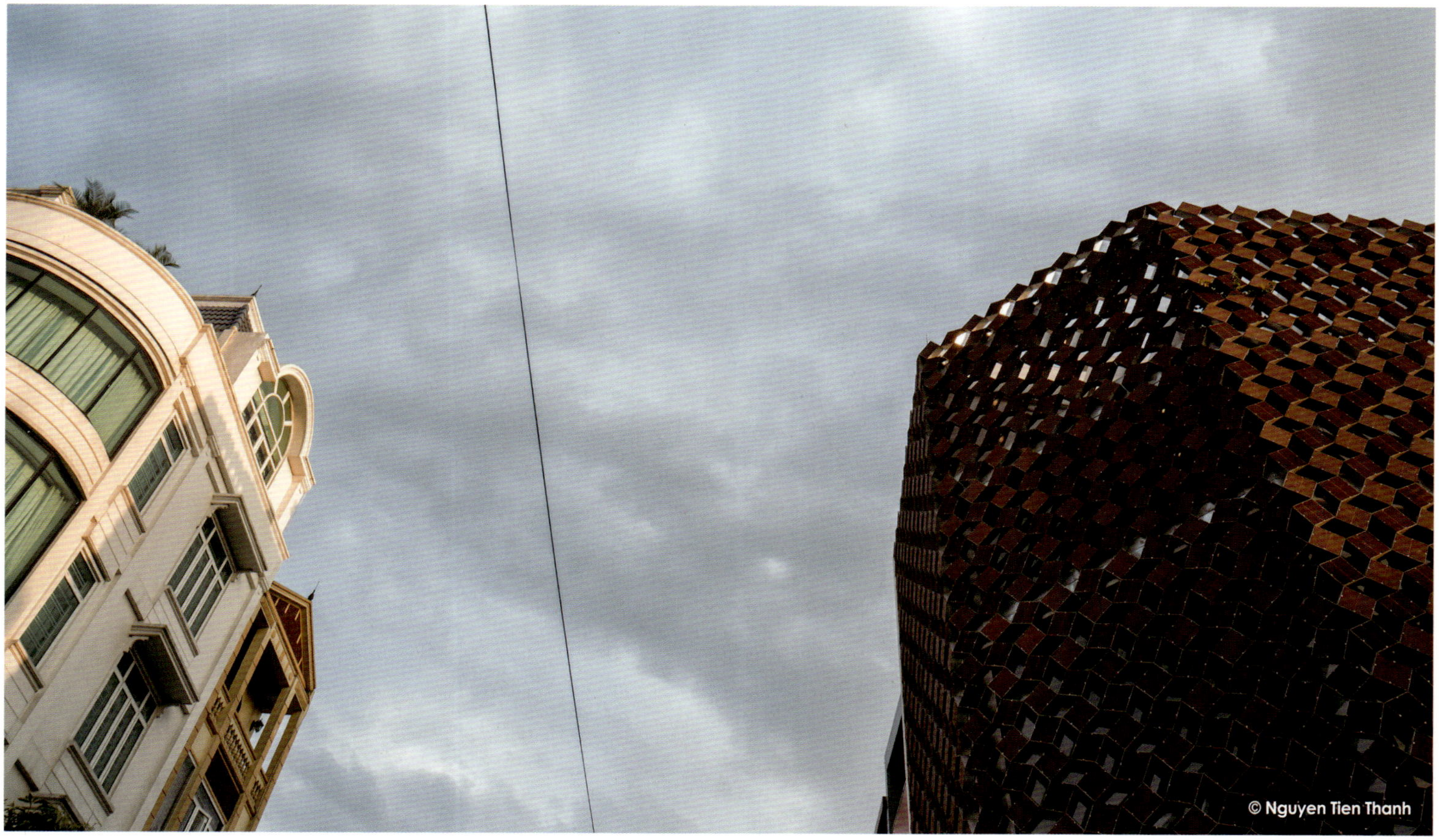
© Nguyen Tien Thanh

© Nguyen Tien Thanh

© Nguyen Tien Thanh

© Nguyen Tien Thanh

© Nguyen Tien Thanh

© Nguyen Tien Thanh

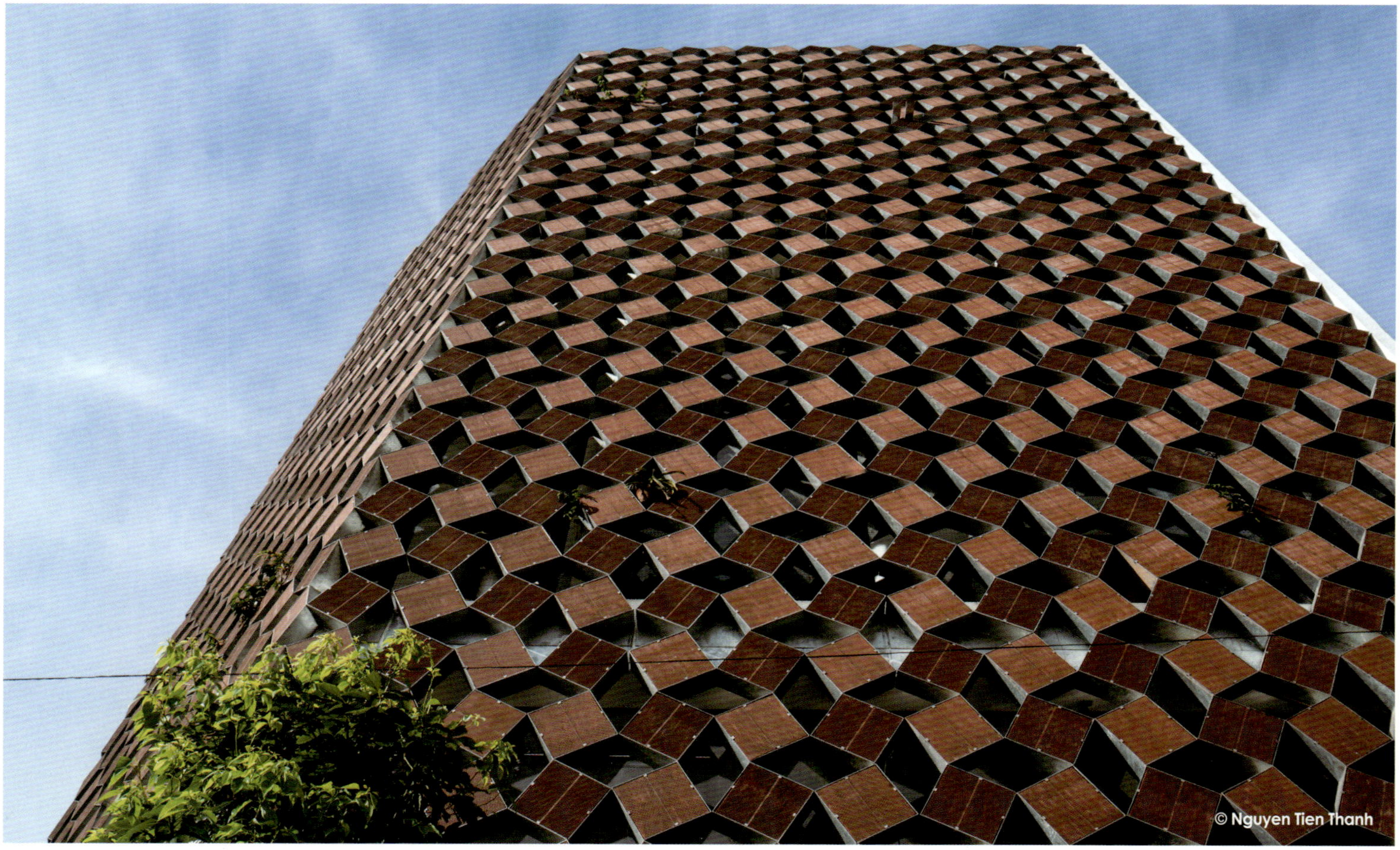

© Nguyen Tien Thanh

VM Houses, Copenhagen, Denmark
BIG

© Stuart Mcintyre

© Tobias Toyberg

57C

Doosan Art Sqaure, Seoul, South Korea
jay is working

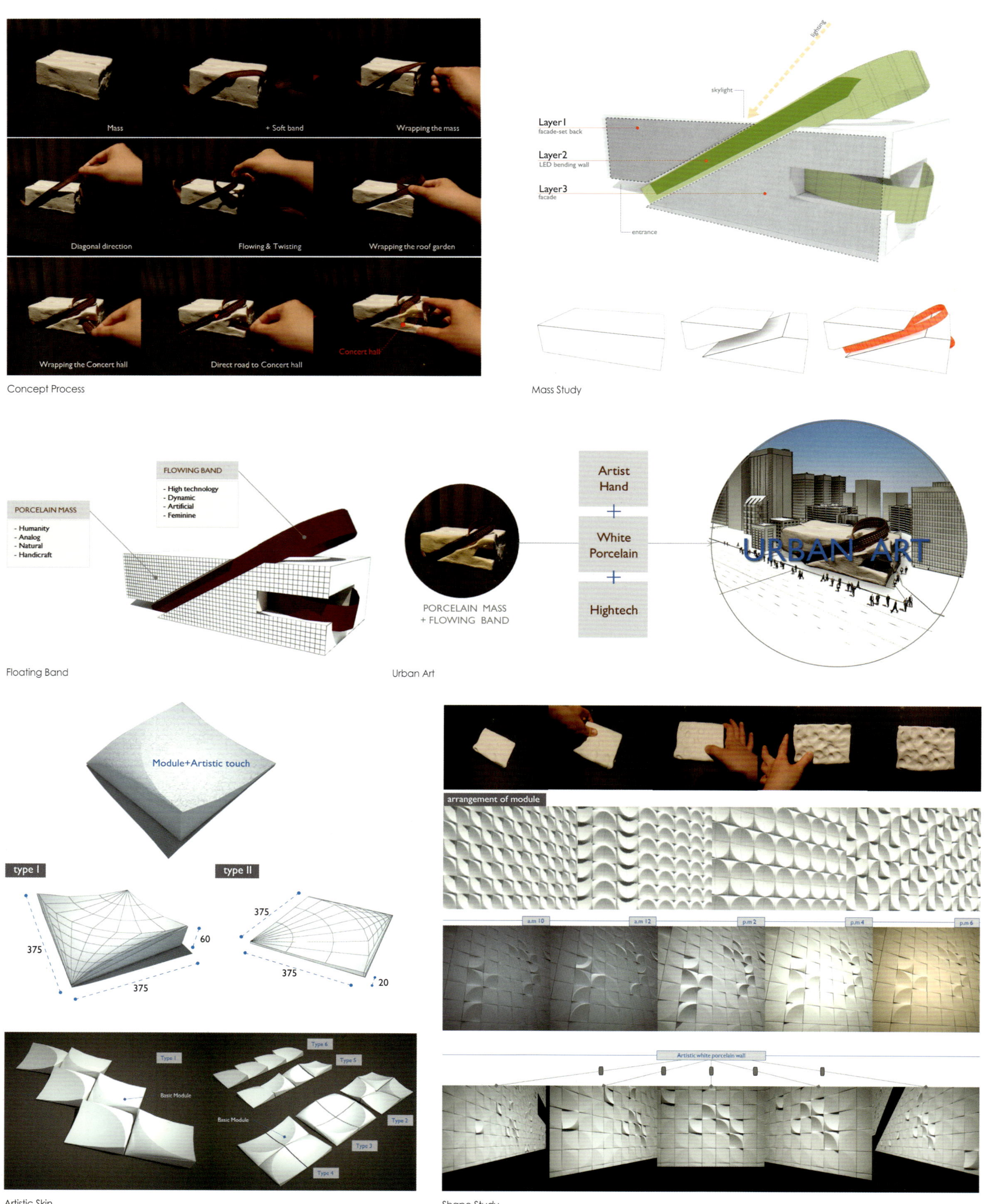

Concept Process

Mass Study

Floating Band

Urban Art

Artistic Skin

Shape Study

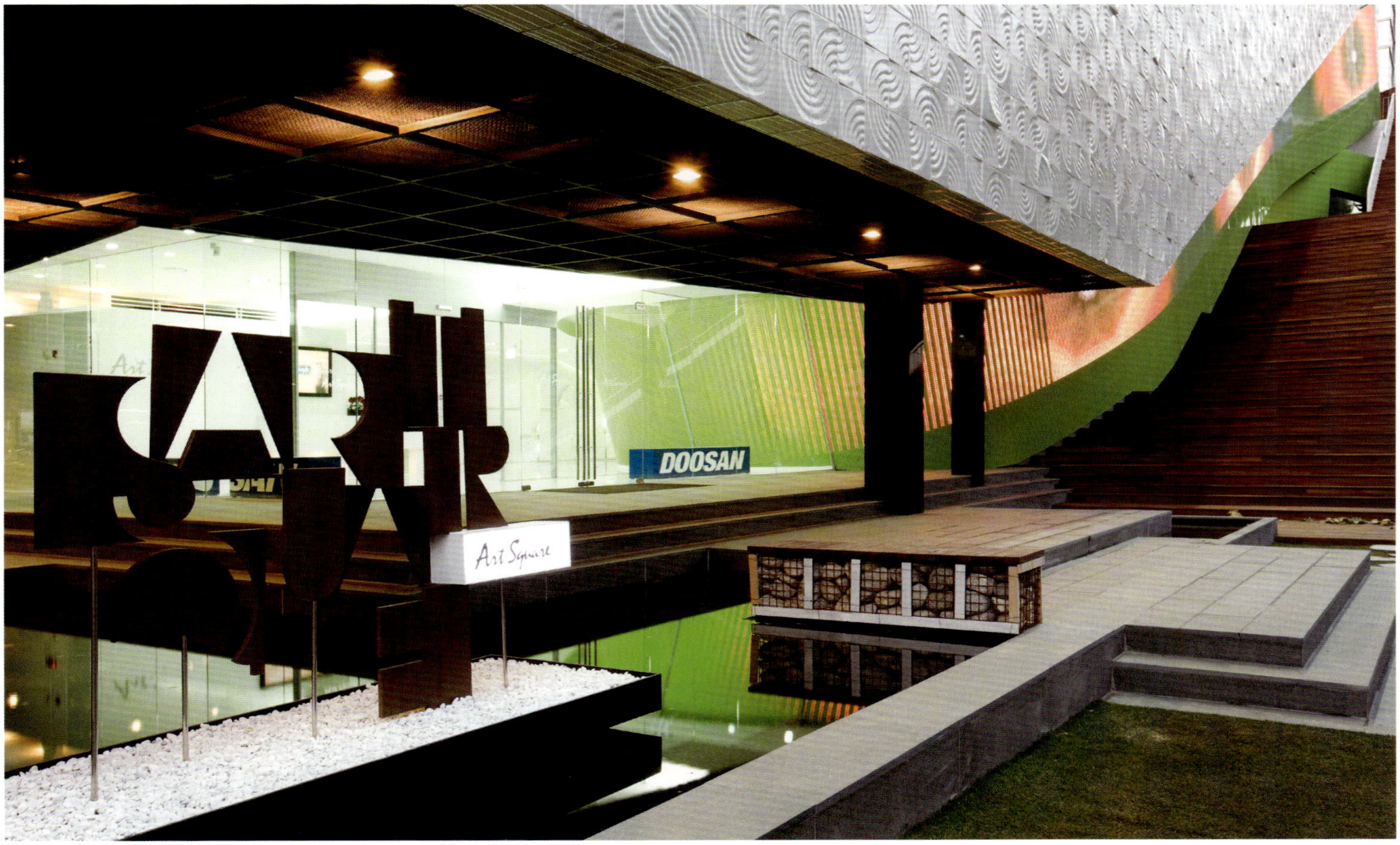
DOOSAN
Art Square

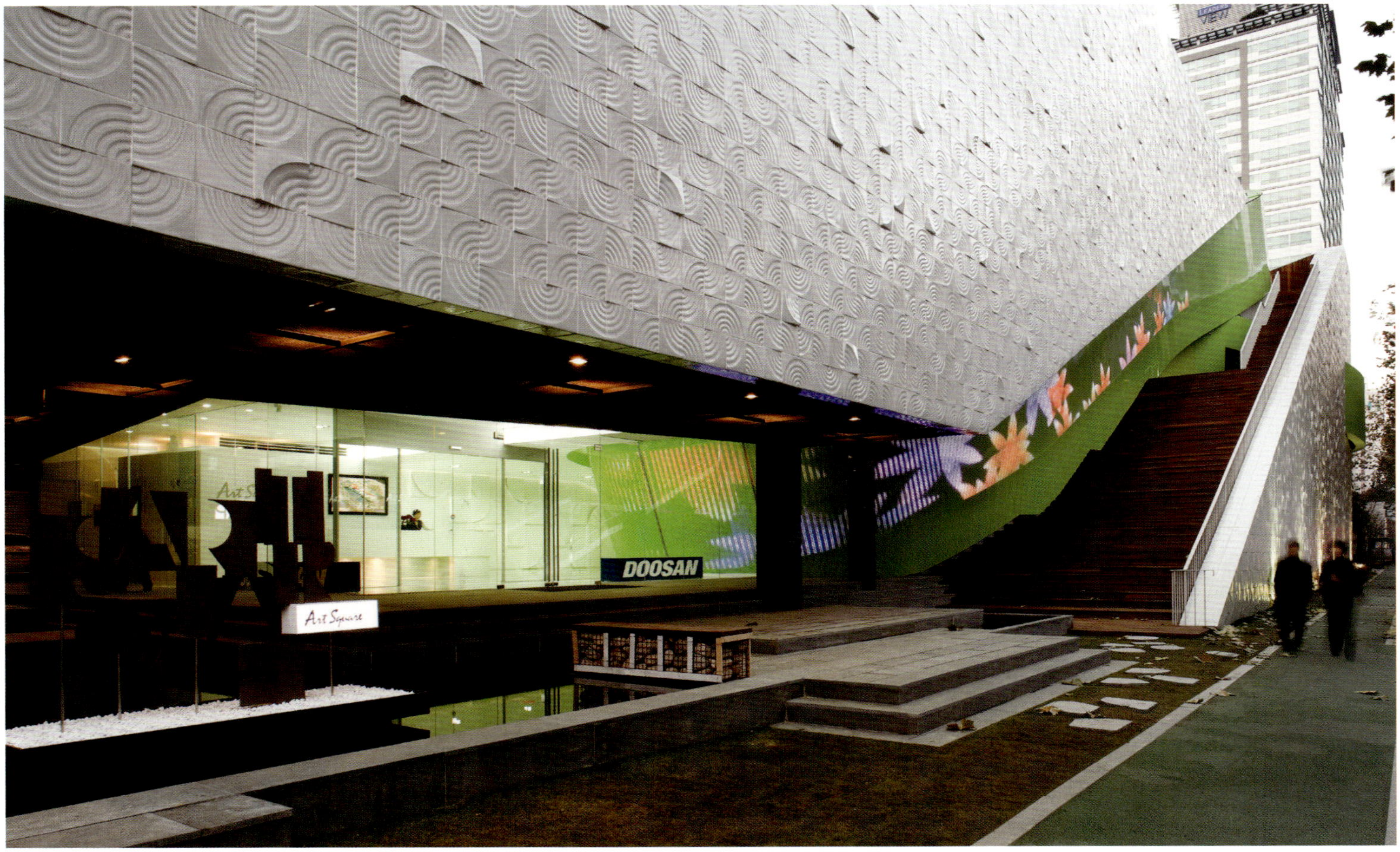
DOOSAN
Art Square

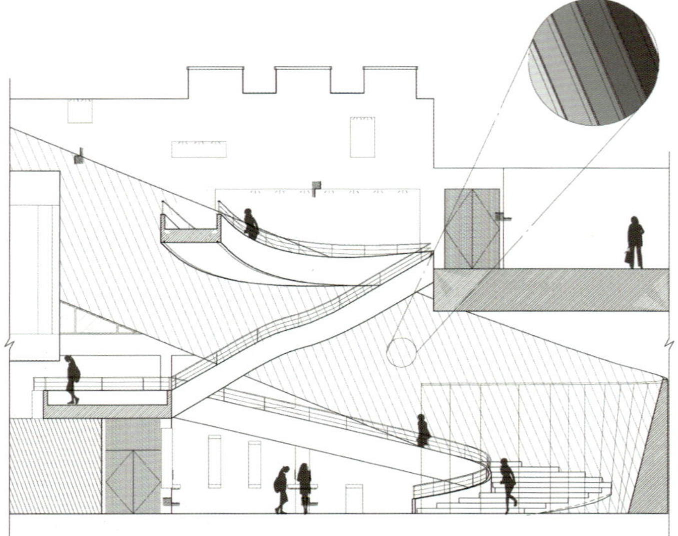

Section

3d Modeling

Construction Process

Art Square
SAMSUNG
1 삼성
삼성
다원

포스코트
POSCOURT

LEGO House, Billund, Denmark

BIG

© Iwan Baan

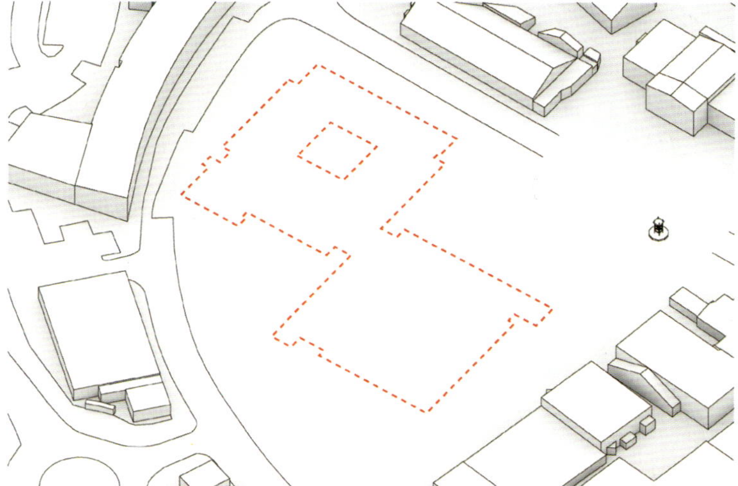

THE CENTER OF THE CAPITAL OF CHILDREN
Since the LEGO House will be at the city center of the Capital of Children, we thought why not design it like a city center—or rather a town square?

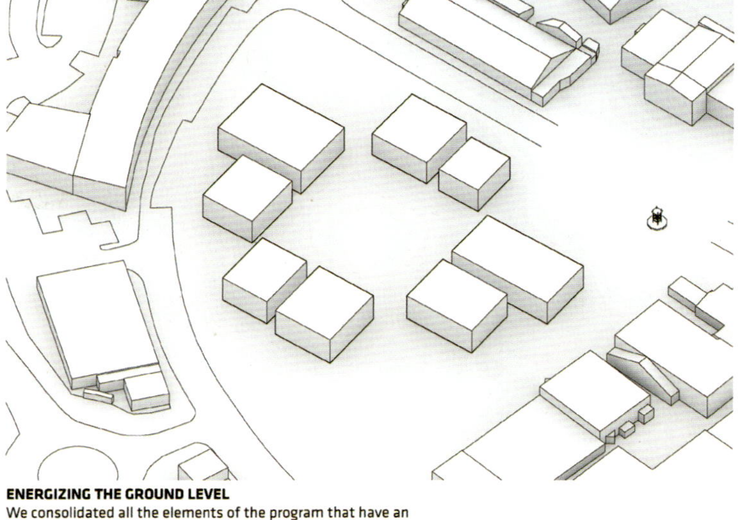

ENERGIZING THE GROUND LEVEL
We consolidated all the elements of the program that have an outward-oriented, everyday-like urban character around a central space: café, forum, LEGO store, ticket offices, wardrobe and restrooms, offices and loading.

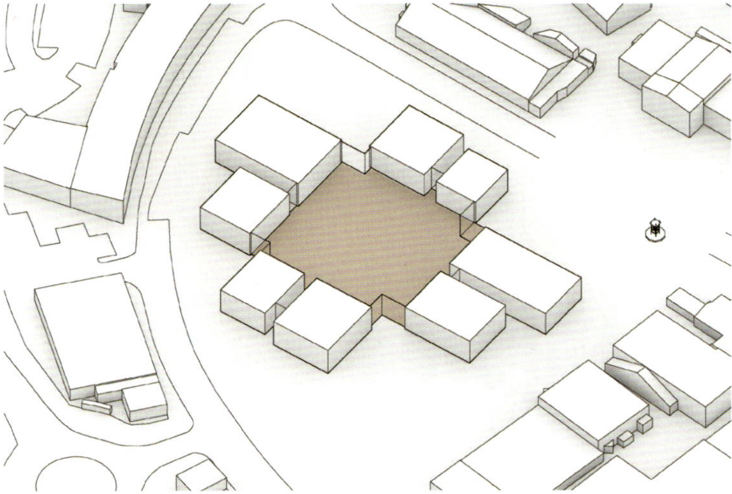

CHILDREN'S SQUARE
Placed like individual buildings framing a square, they allow daylight and views to pass between them while letting people enter from multiple directions and allowing shortcuts through the building—like crossing a plaza.

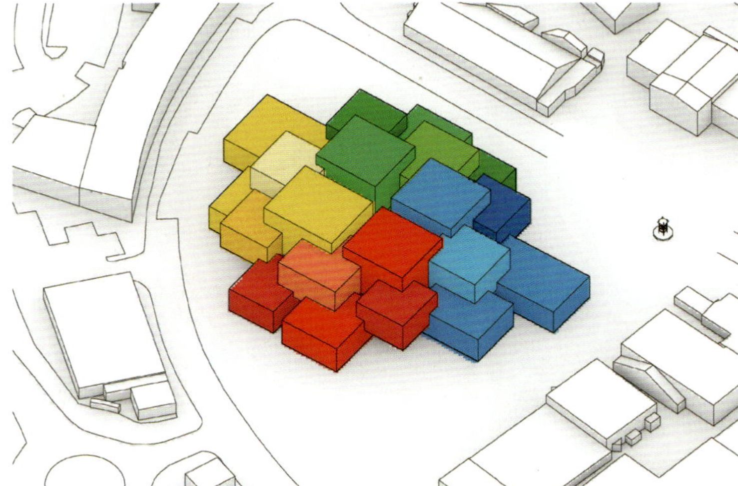

GALLERIES
Above the square, a cluster of galleries overlap to create a continuous sequence of exhibitions.

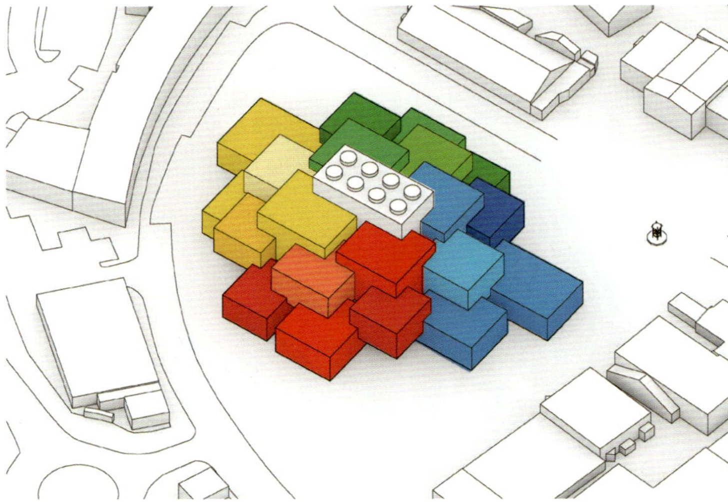

MASTERPIECE GALLERY
At the top of the pile of bricks the Masterpiece Gallery provides a bridge between all the corners of the exhibition, and serves as a sky-lit gallery for LEGO as an art form.

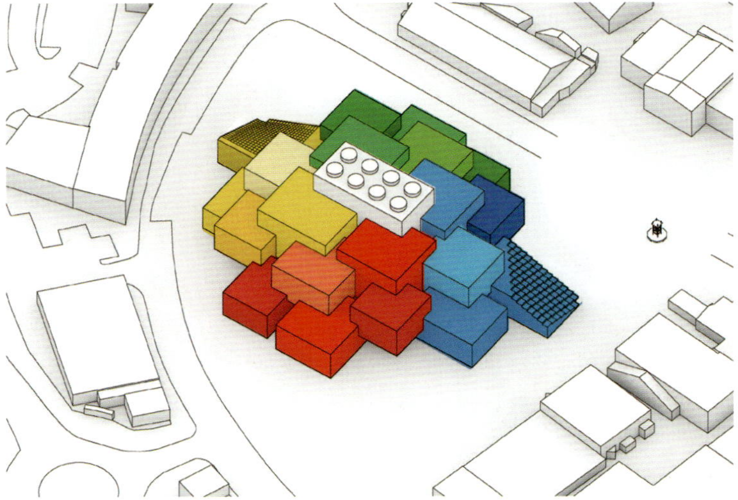

MELT
Two of the volumes seem to melt in a pixelated way to form informal auditoria for people watching or public performances.

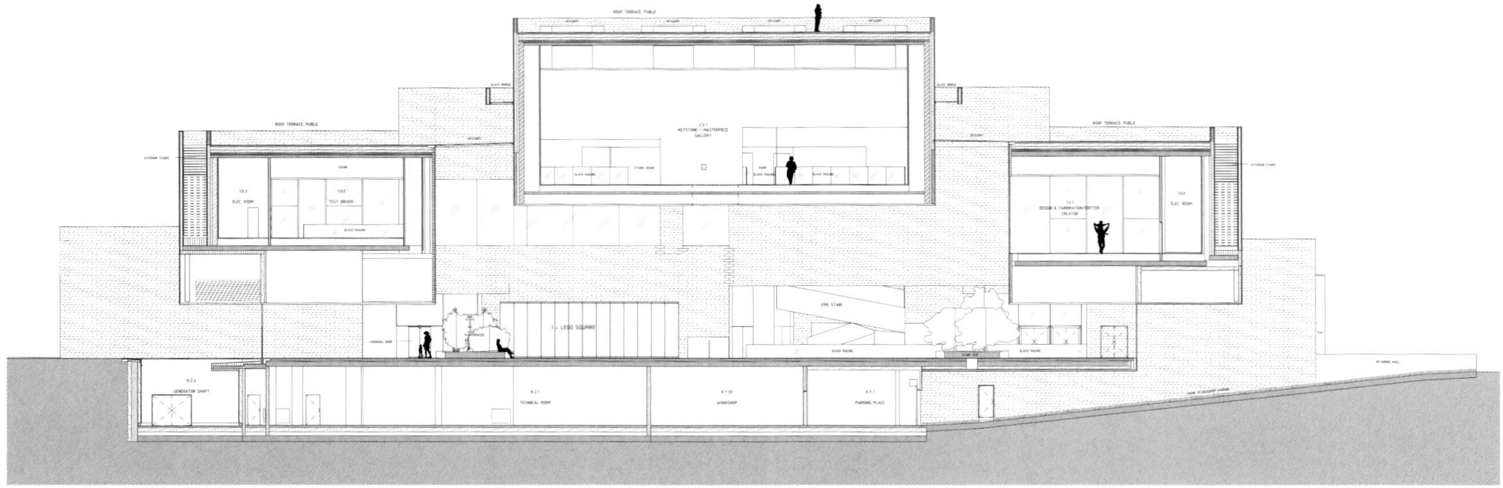

Section A-A

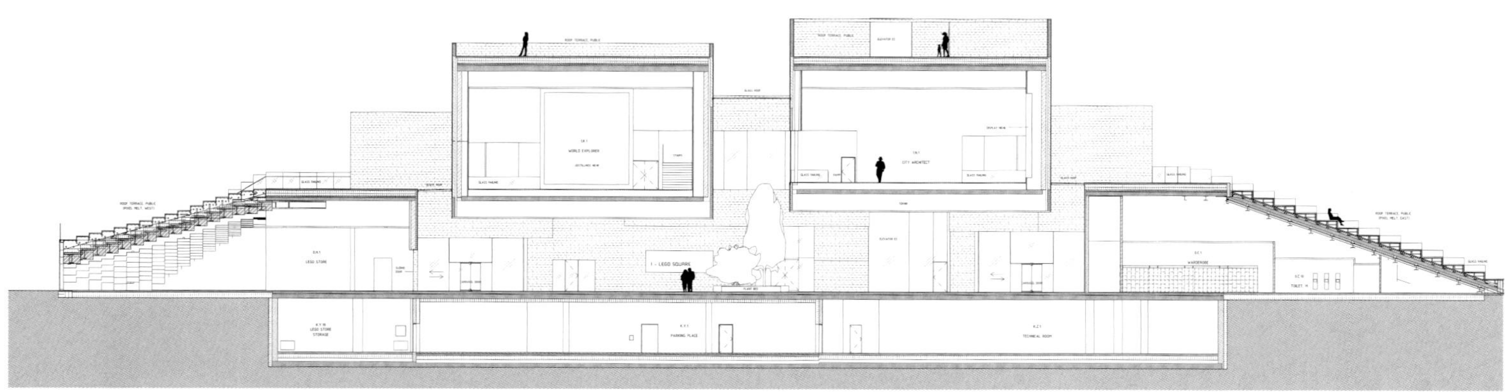

Section B-B

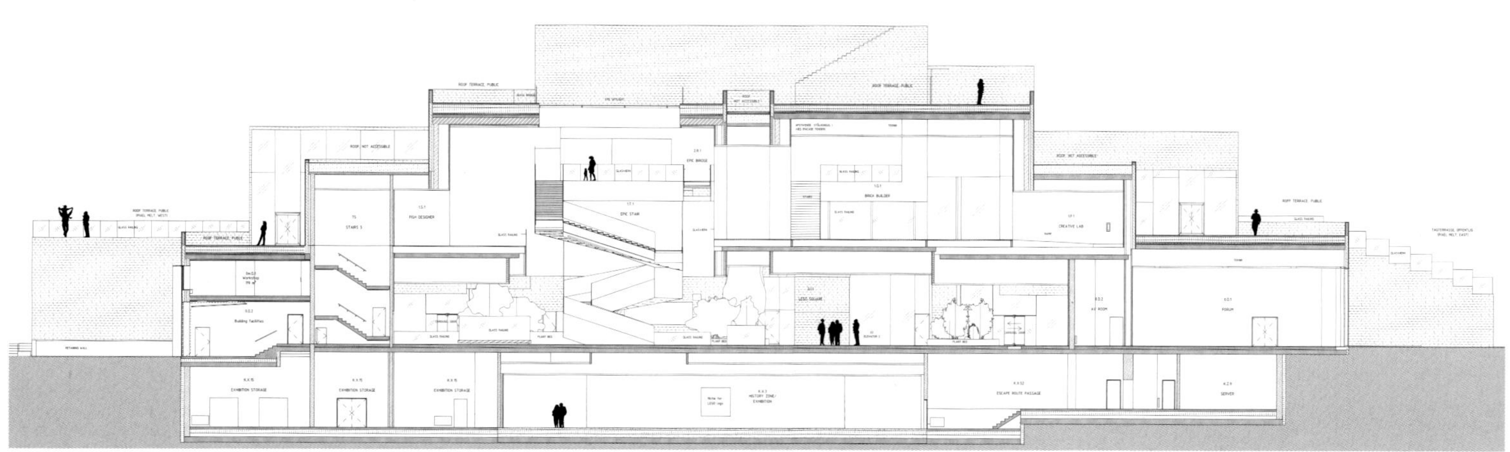

Section H-H

© Iwan Baan

© Iwan Baan

© Iwan Baan

Toilets
© Iwan Baan

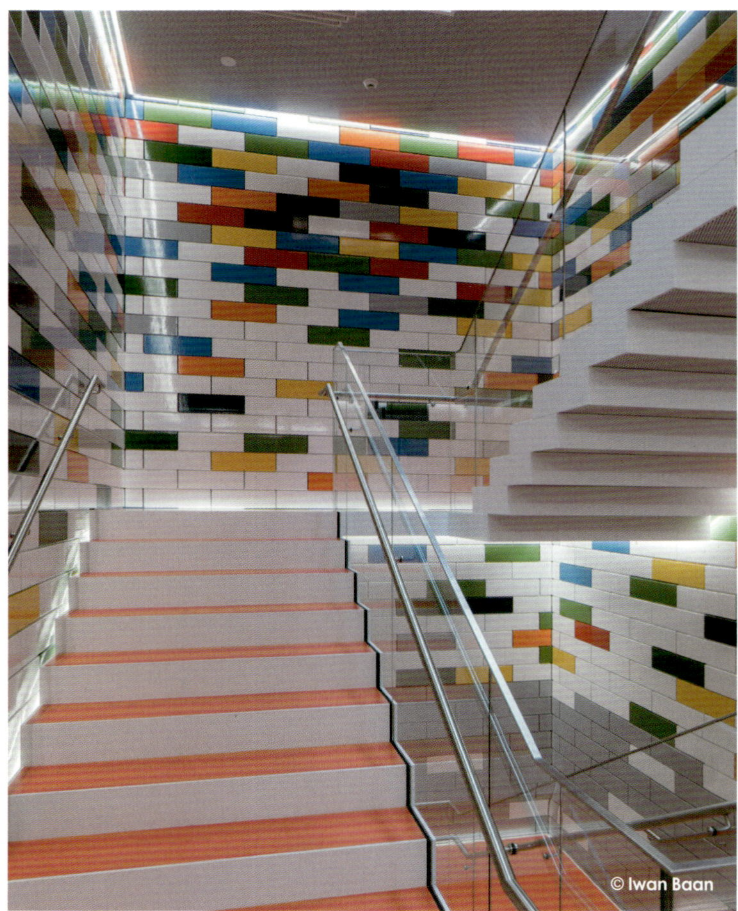

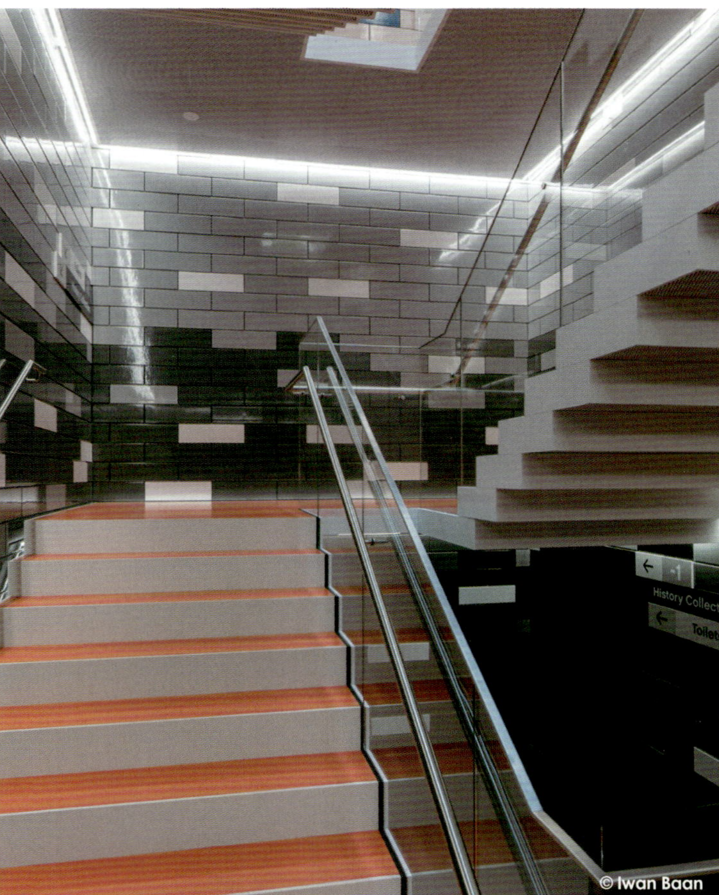

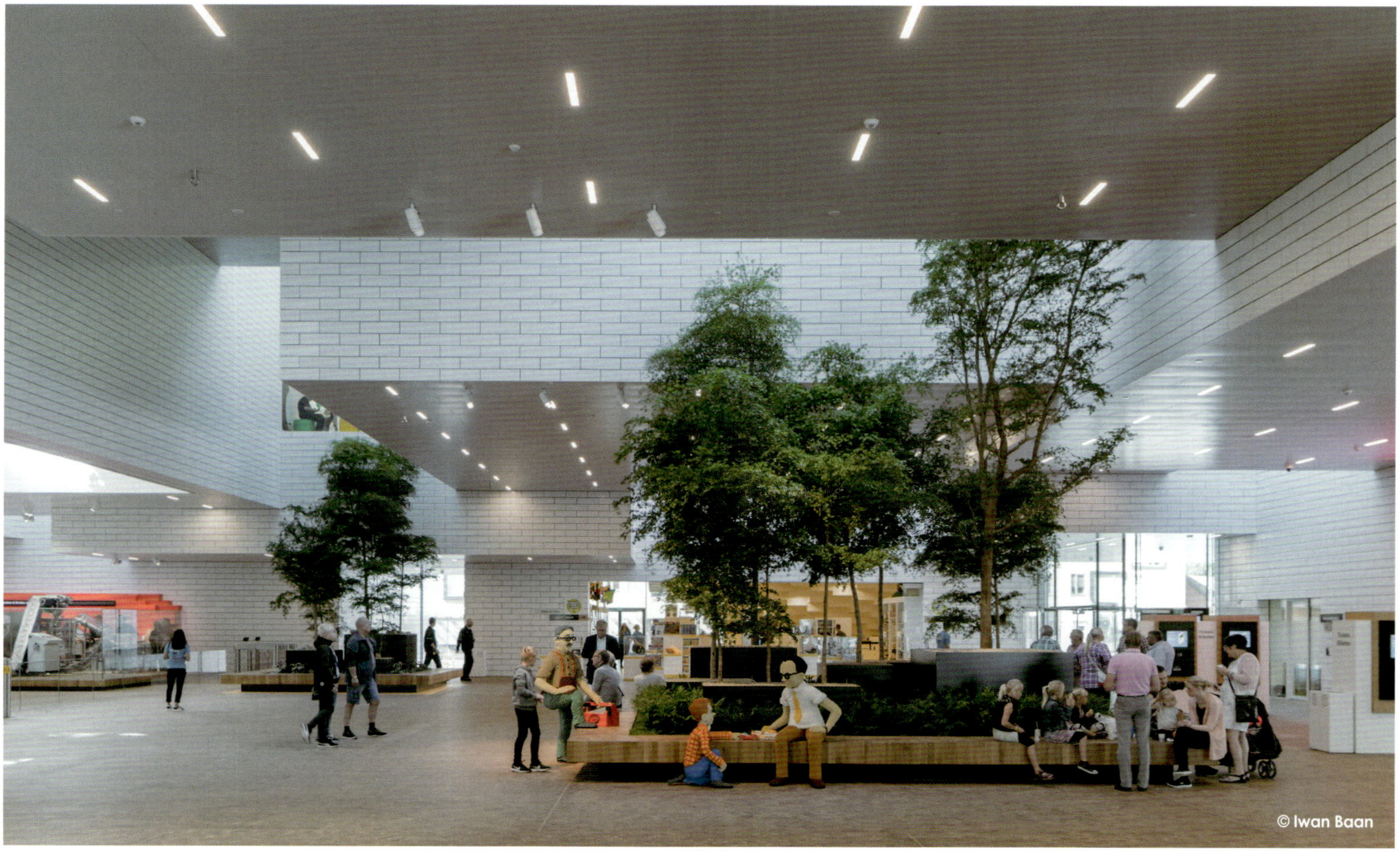
© Iwan Baan

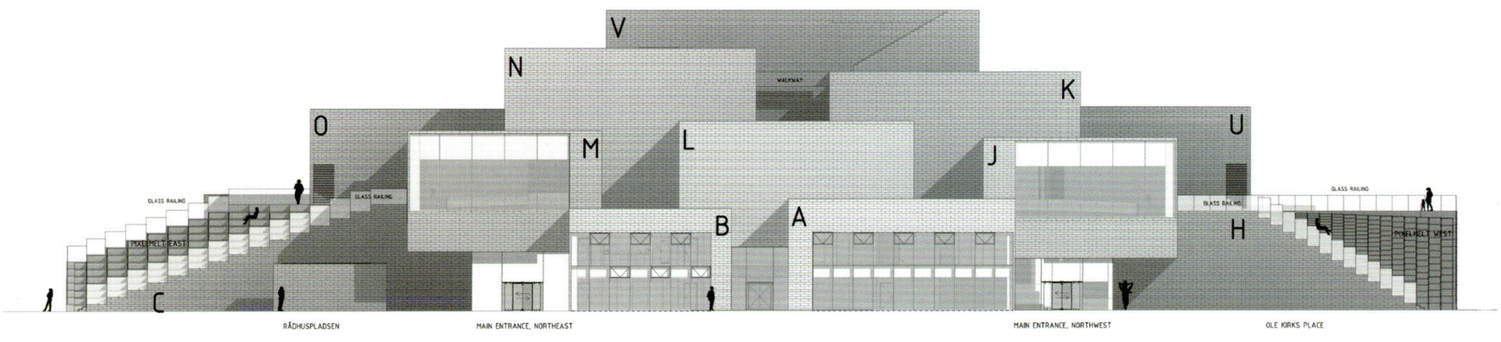

North Elevation

© Iwan Baan

Warehouse 8B, Madrid, Spain

Arturo Franco

© Carlos Fernandez Piñar

© Carlos Fernandez Piñar

© Carlos Fernandez Piñar

© Carlos Fernandez Piñar

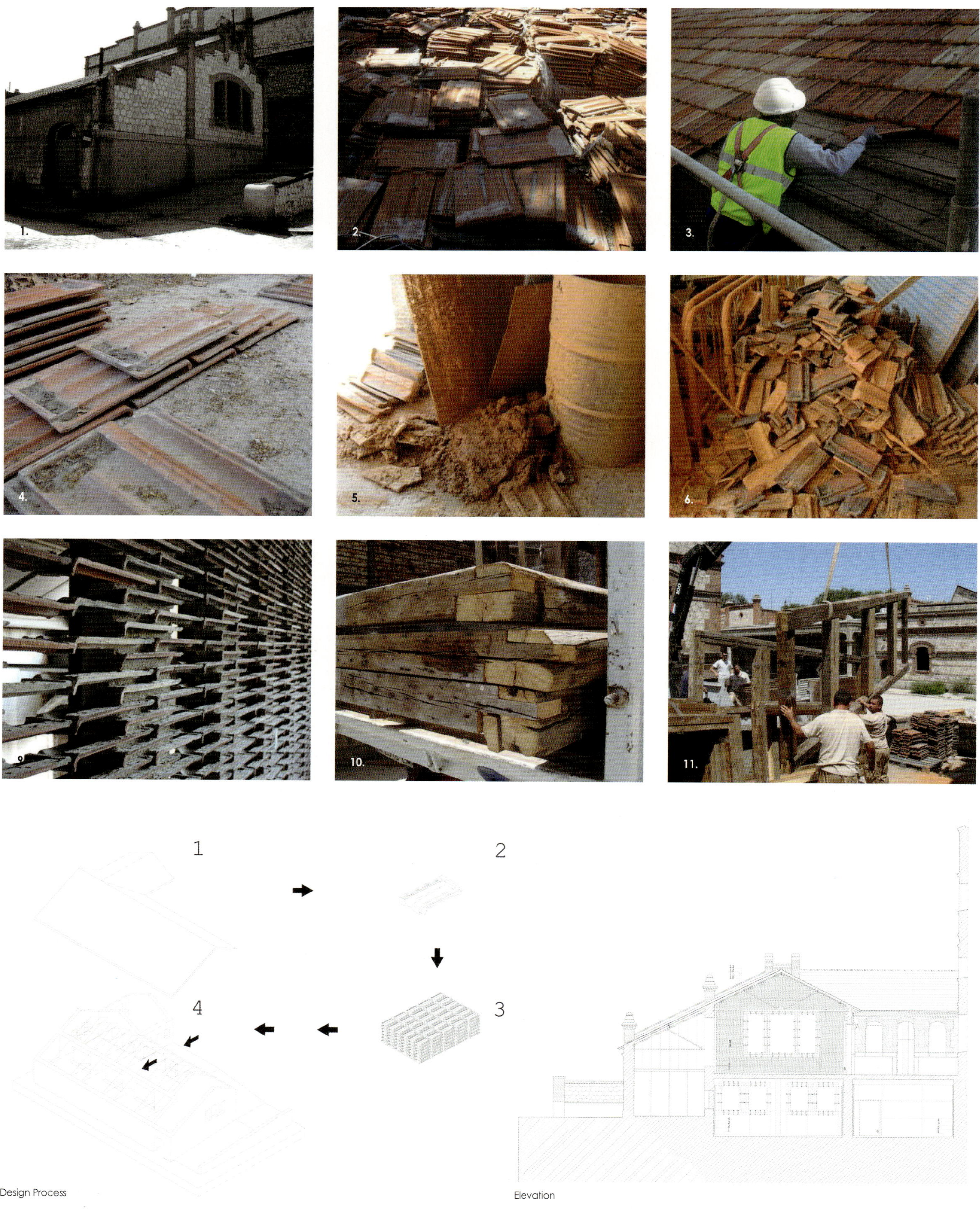

Design Process

Elevation

4.

7.

12.

Construction Process

1. Warehouse 8B in its original state
2. Tips and flat tile heaps near the warehouse from other interventions were discovered.
3. It was decided to remove the flat tiles of the roof one by one, stacking the tiles for use in the interior.
4. Interior of the Warehouse 8B top floor after the removing of the roof
5. We studied possible tile bonds to reuse it and to de-contextualize it in the interior.
6,7. Image of the workshop installed at building site to adapt the ceramic parts.
8. Occasional lattice of stacked tiles at building site
9. Internal lattice that appears removing a tile inside the bond used.
10. Woodwork proceeding from big one-piece old wood chunks, 25x25 section, with simple knots
11. Installation of large wood frames
12. The glass stops are fixed with carpenter's jacks with permanent compression. This eliminates nails and bolts.

© Carlos Fernandez Piñar

© Carlos Fernandez Piñar

wasl Tower, Dubai, UAE

UNStudio

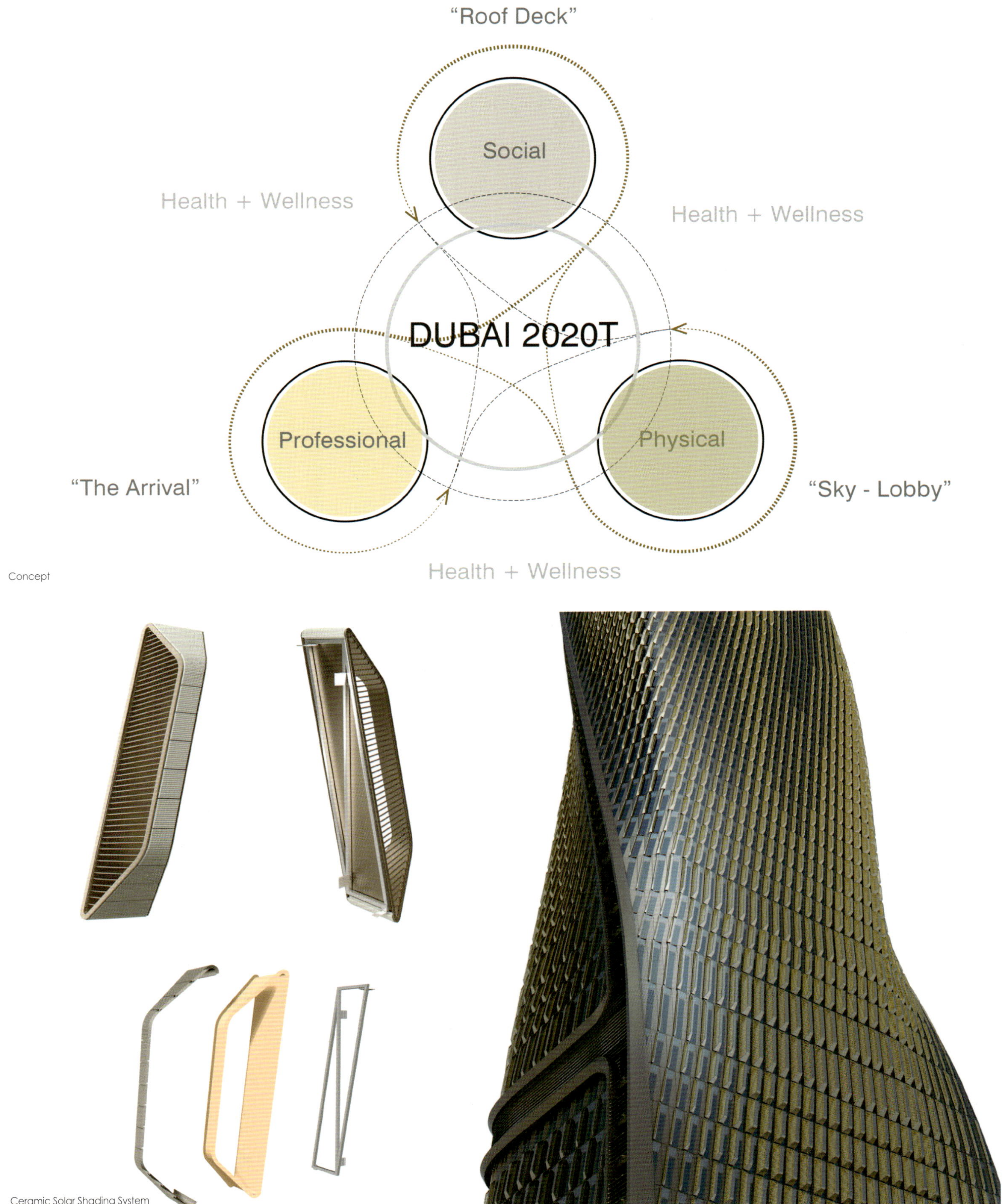

Concept

Ceramic Solar Shading System

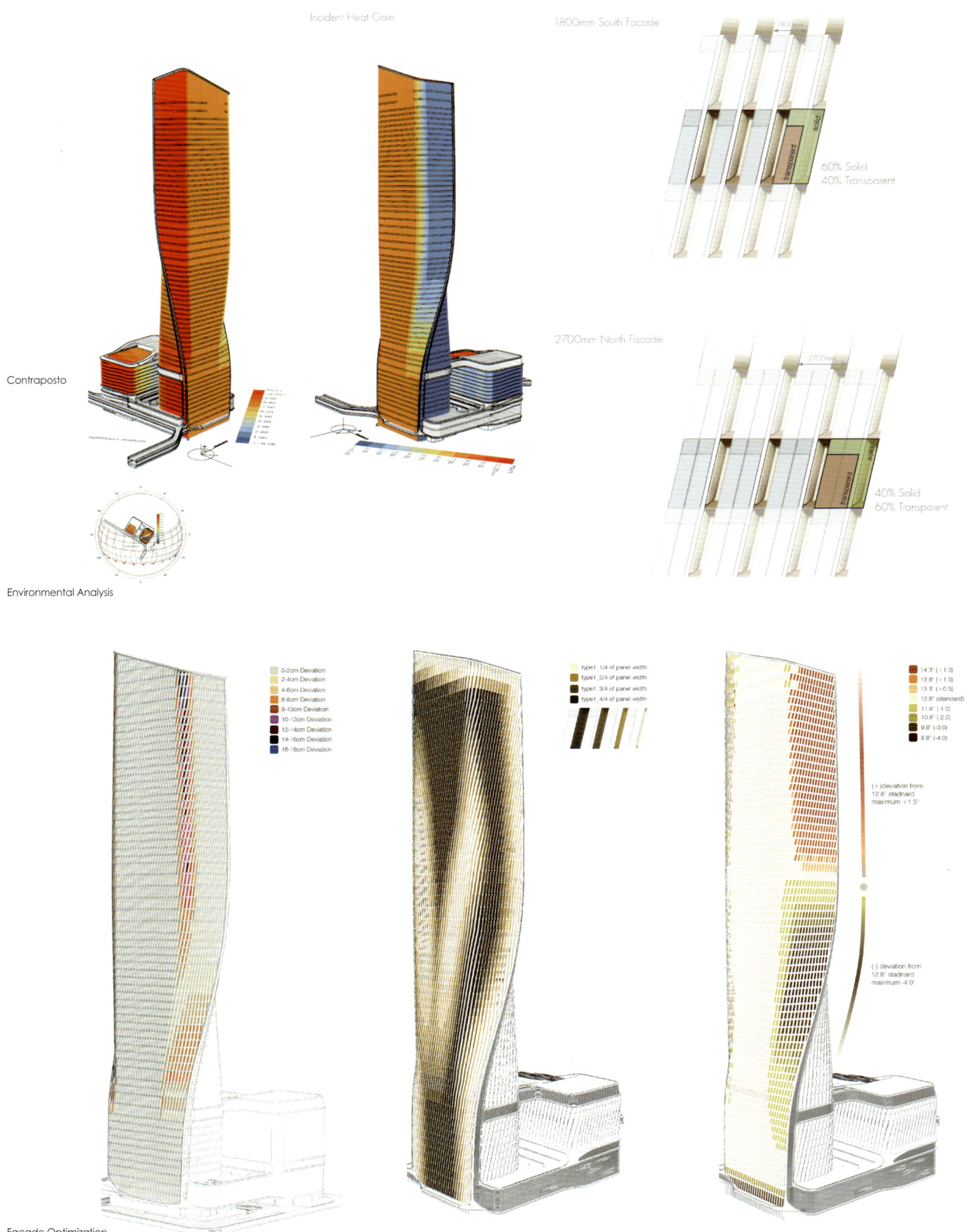

Contraposto

Environmental Analysis

Facade Optimization

Mountain, Barcelona, Spain

TAKK Architecture

Model

Distribution of Tiles on the Different Rows

Tests for the Tiles

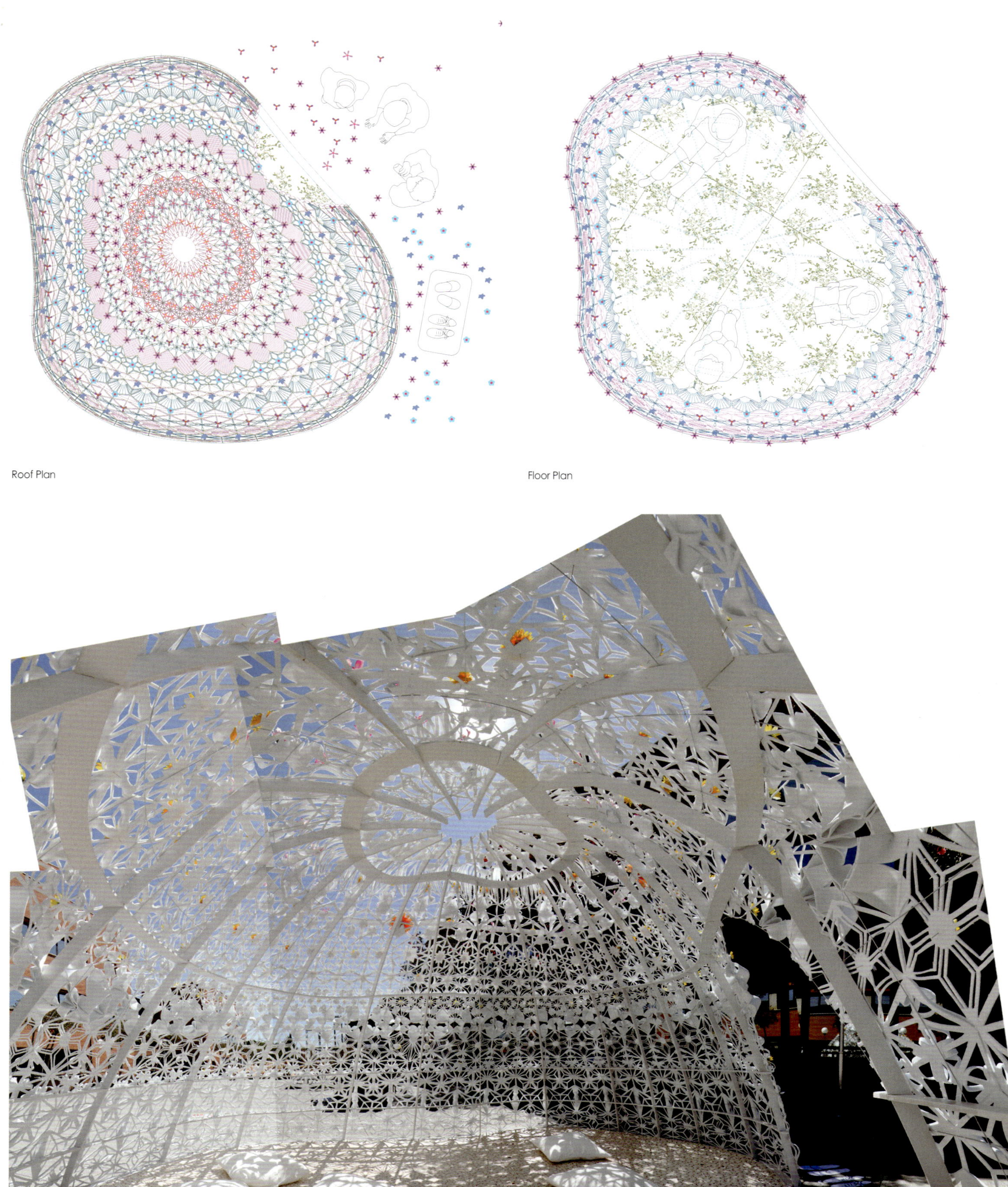

Roof Plan

Floor Plan

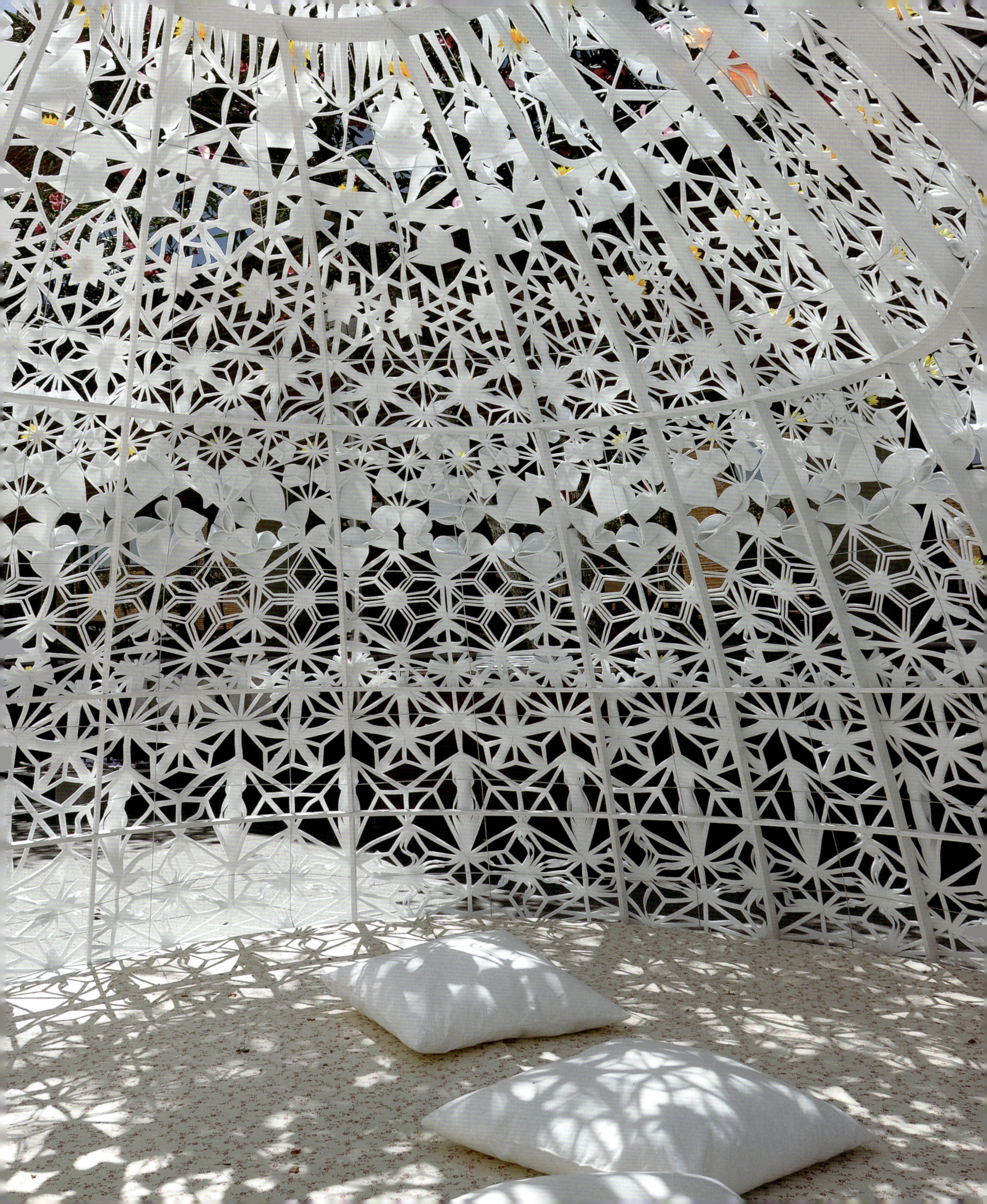

Jinzhou, Jinzhou, China
Casanova+Hernandez Architects

© Casanova + Hernandez architects Ben McMillan

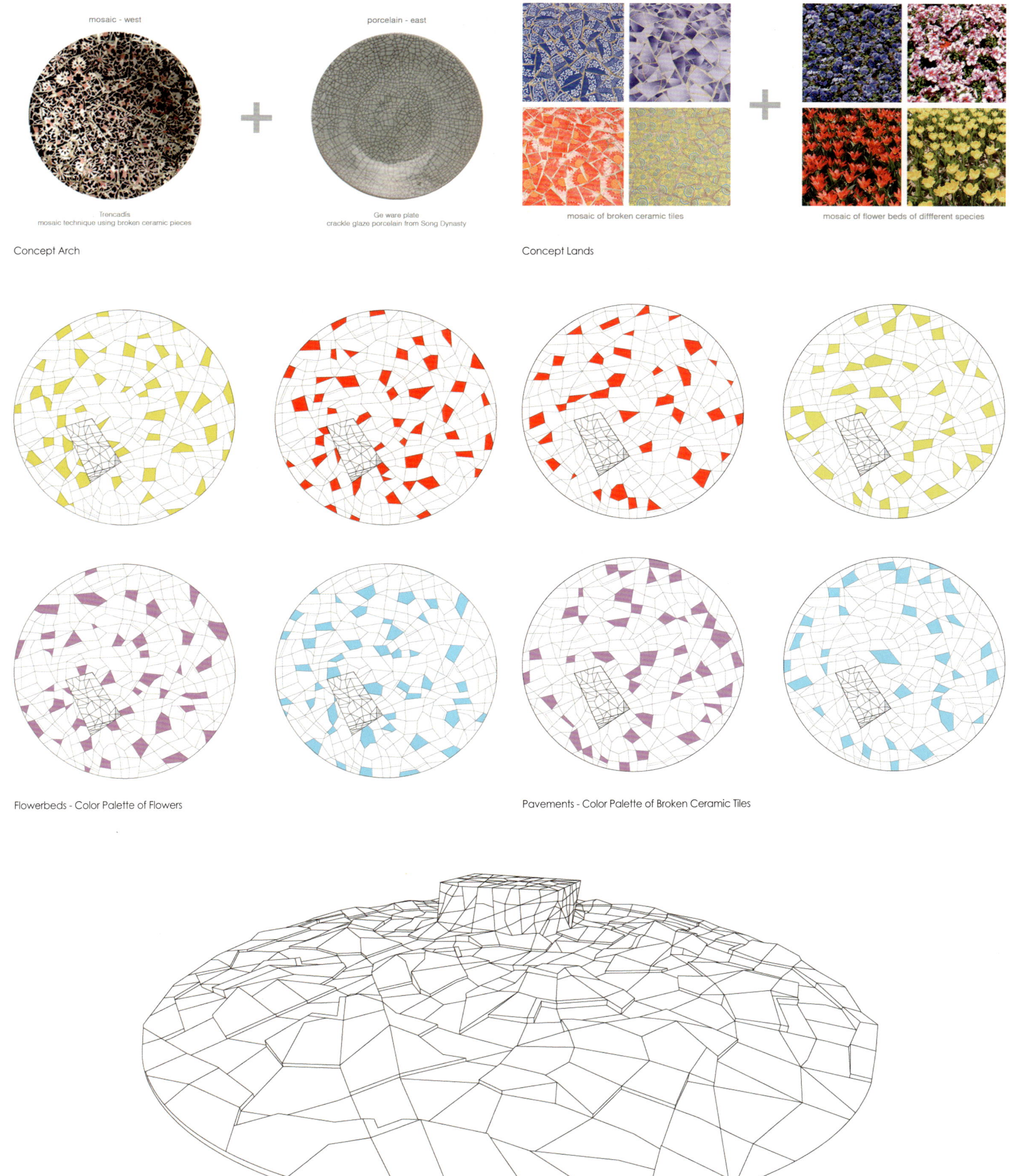

Concept Arch

Concept Lands

Flowerbeds - Color Palette of Flowers

Pavements - Color Palette of Broken Ceramic Tiles

Axonometry white

Study Modeling

Construction Process

Section

unfolded facades

© Casanova + Hernandez architects Ben McMillan

© Casanova + Hernandez architects Ben McMillan

© Casanova + Hernandez architects Ben McMillan

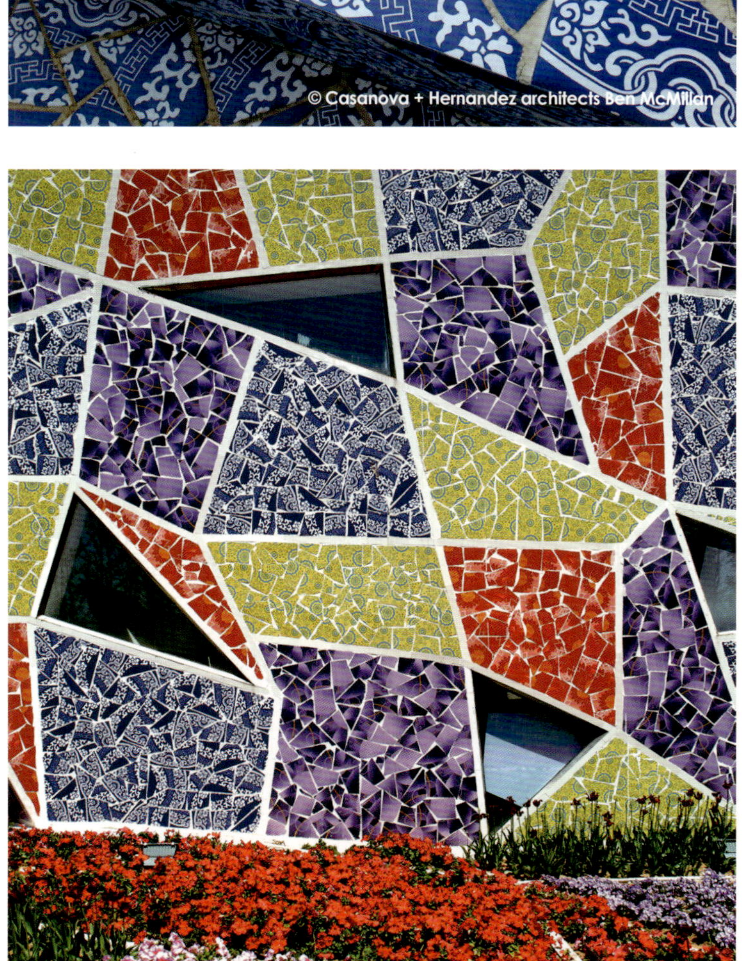
© Casanova + Hernandez architects Ben McMillan

© Casanova + Hernandez architects Ben McMillan

Library Creativity from Knowledge,

ma0

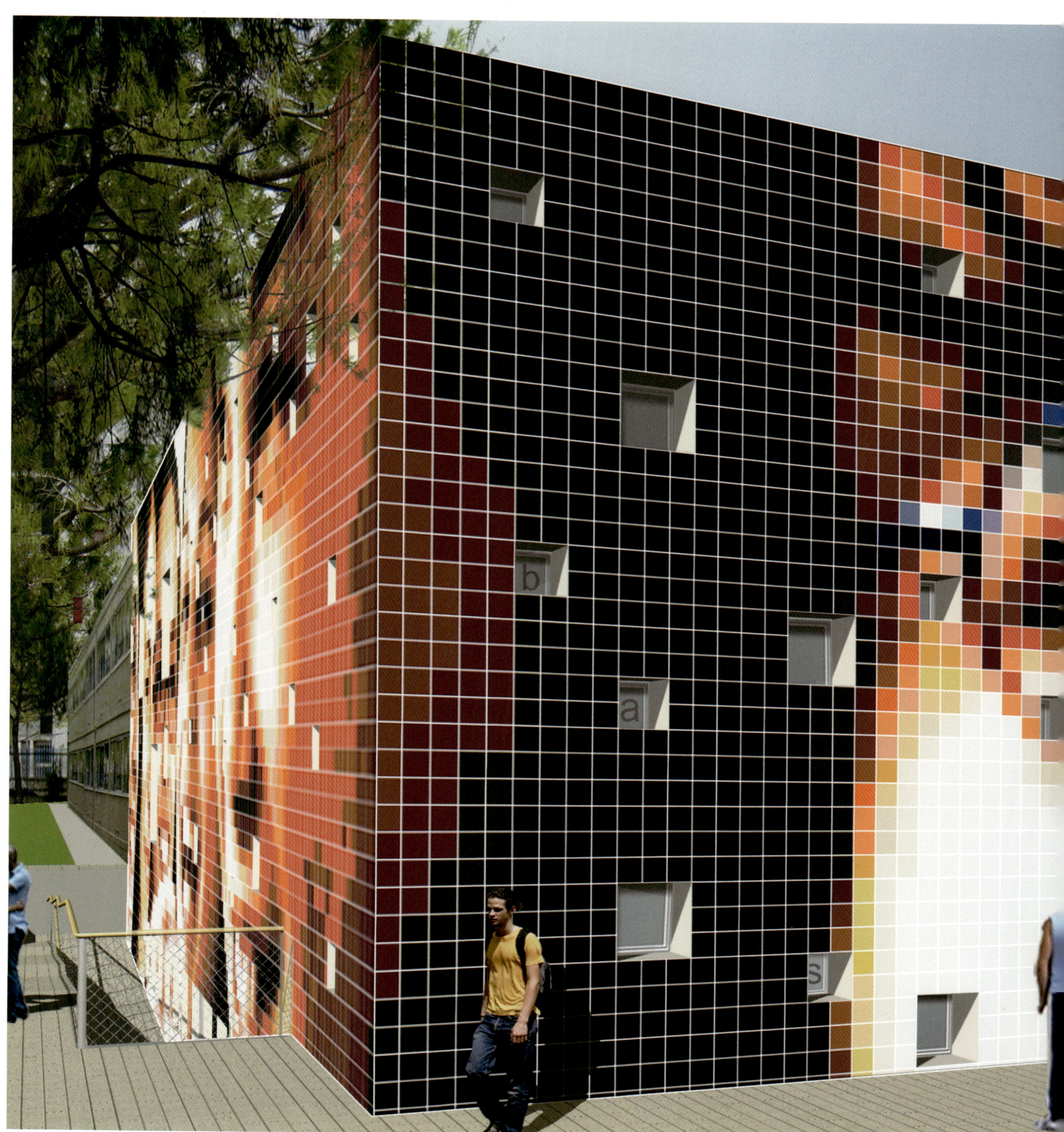

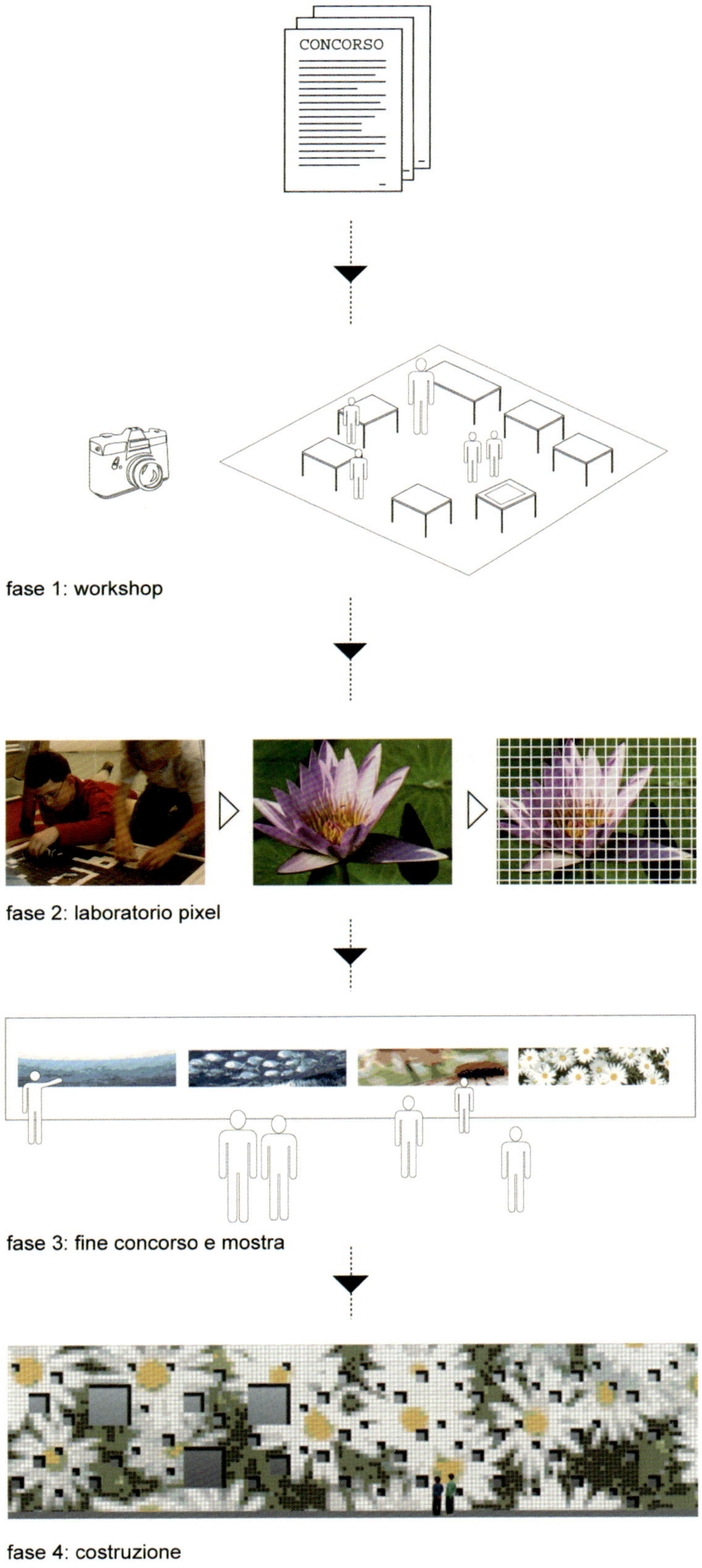

CONCORSO

fase 1: workshop

fase 2: laboratorio pixel

fase 3: fine concorso e mostra

fase 4: costruzione

The Participative Process

Concept 01 / Every Book is a Window on the World

Concept 02 / Creativity comes from Knowledge

Construction Process

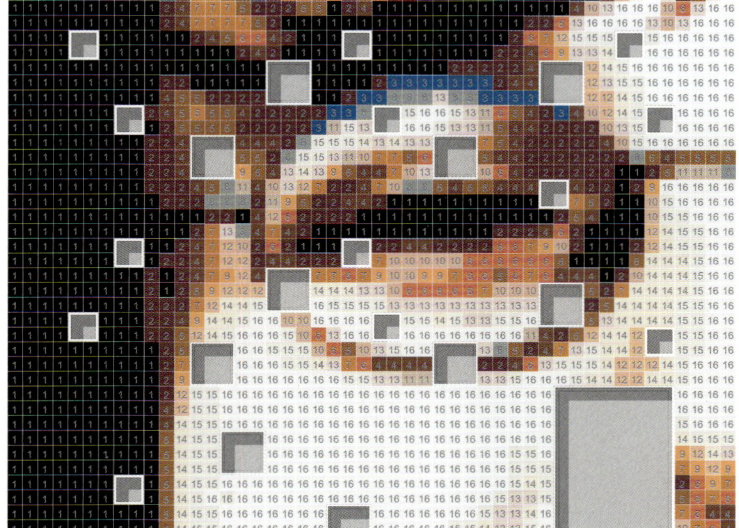

Tiles Numeration

Plaza del Salvador Rehabilitation, Talavera de la Reina, Spain

OOIIO Architecture

© Eugenio H. Vegue y Francisco Sepúlveda.

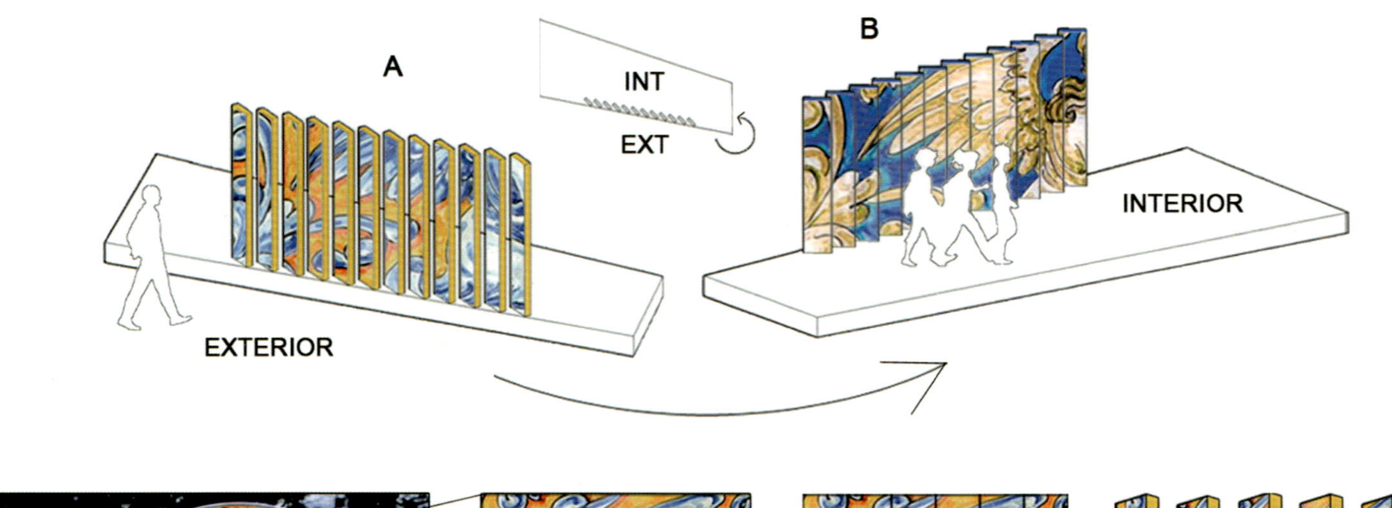

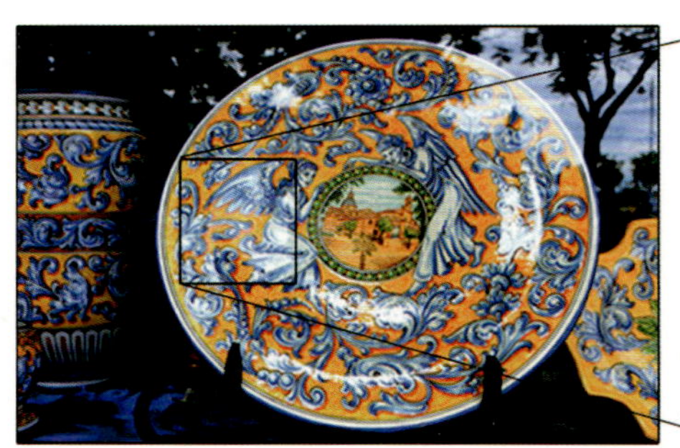

REINTERPRETATION OF TRADITIONAL LOCAL CERAMICS

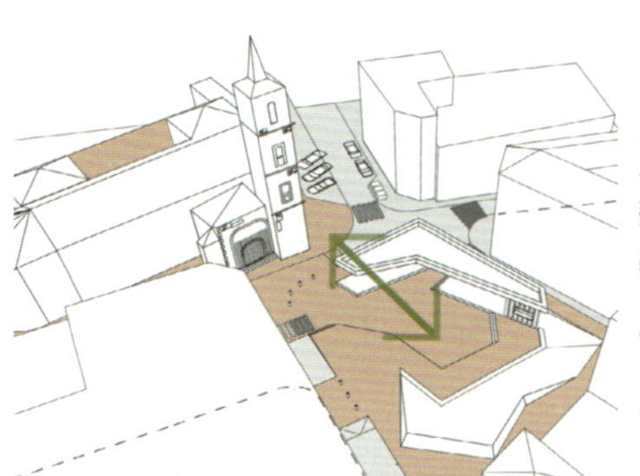

CONTINUIDAD ZONAS PUBLICAS FRENTE A LA IGLESIA DEL SALVADOR

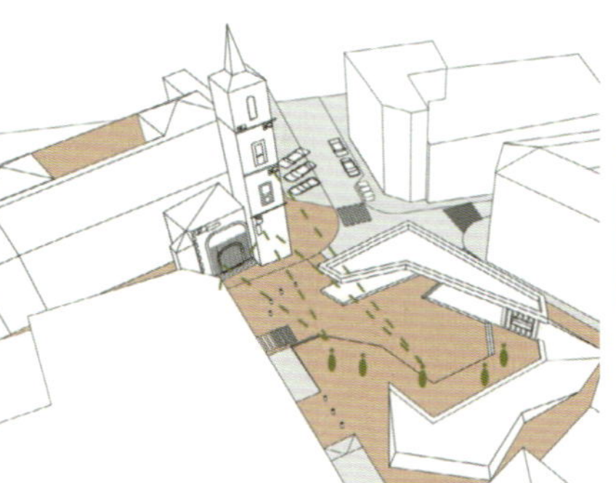

FOMENTAMOS LAS VISTAS A LA IGLESIA

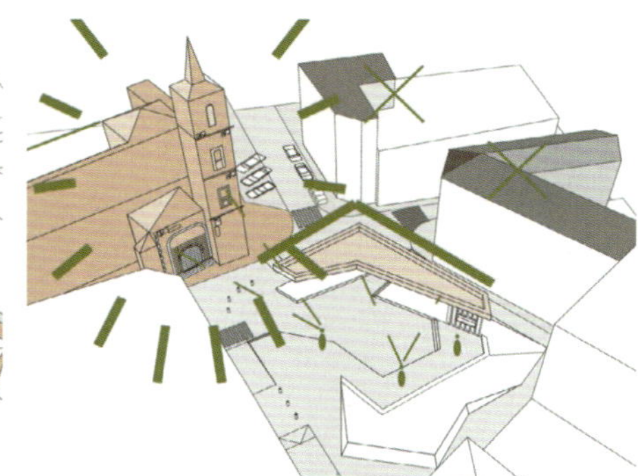

LOS NUEVOS ESCENARIOS BLOQUEAN LA VISTA DE LOS USUARIOS DE LA PLAZA A LOS EDIFICIOS

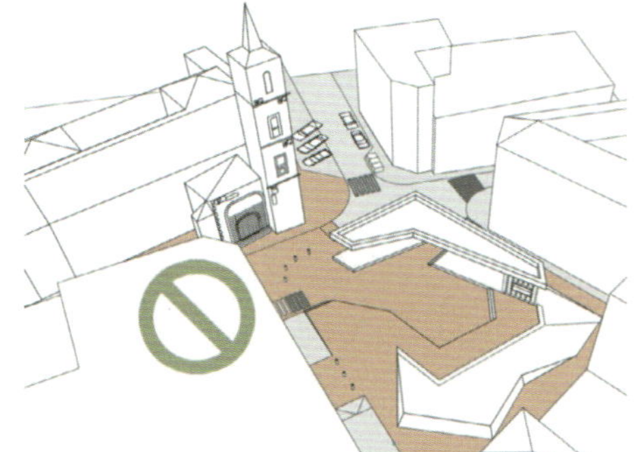

ELIMINAMOS EL PROBLEMA DEL PARKING FRENTE A LA FACHADA DE LA IGLESIA

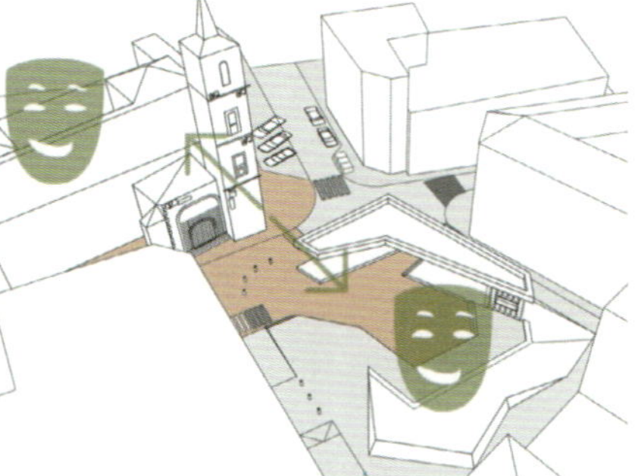

TEATRO & TEATRO MISMO PROGRAMA

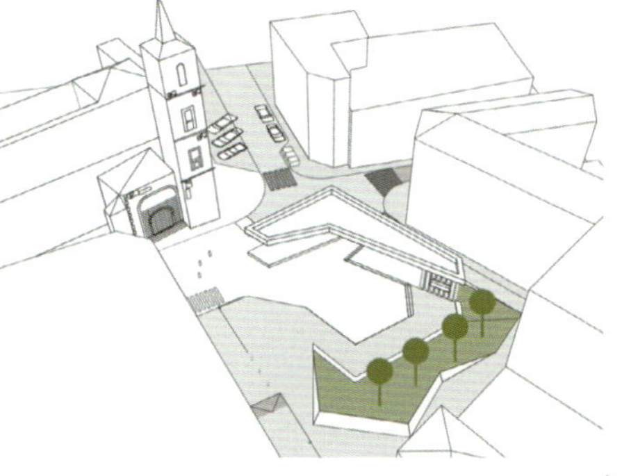

AUMENTO DE LAS ZONAS VERDES FRENTE AL MONUMENTO CERRANDO EL CONJUNTO

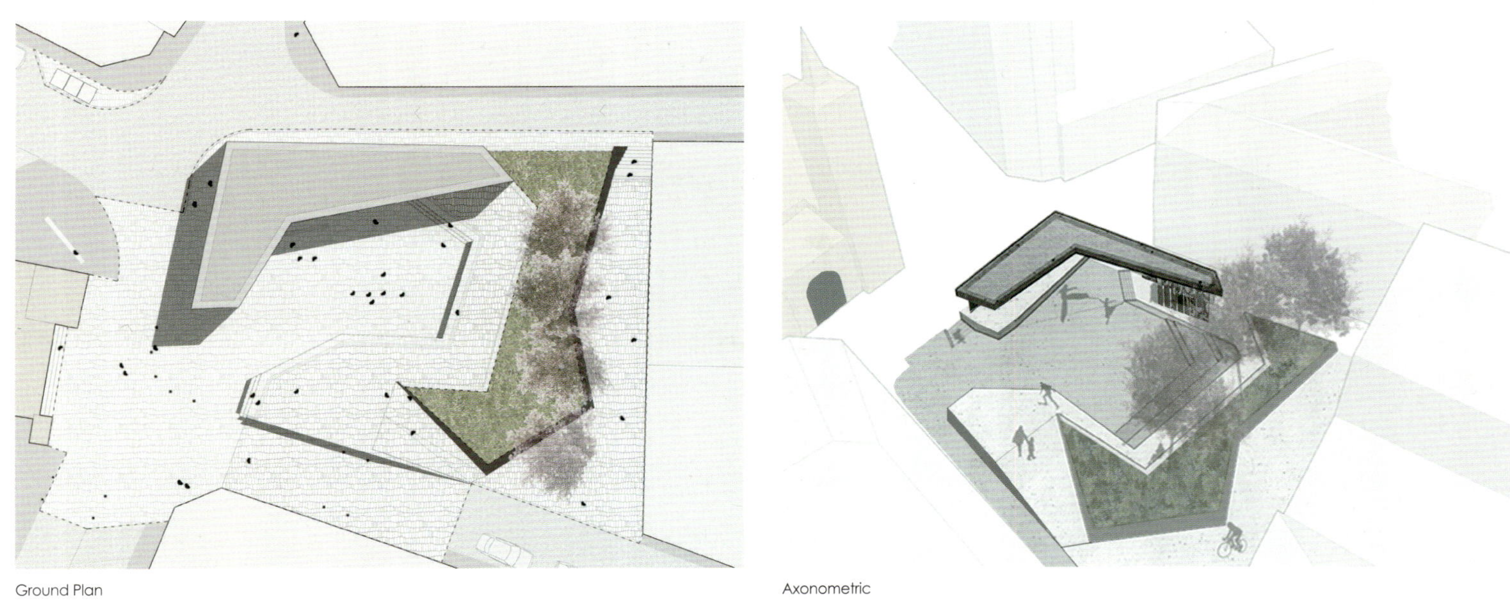

Ground Plan

Axonometric

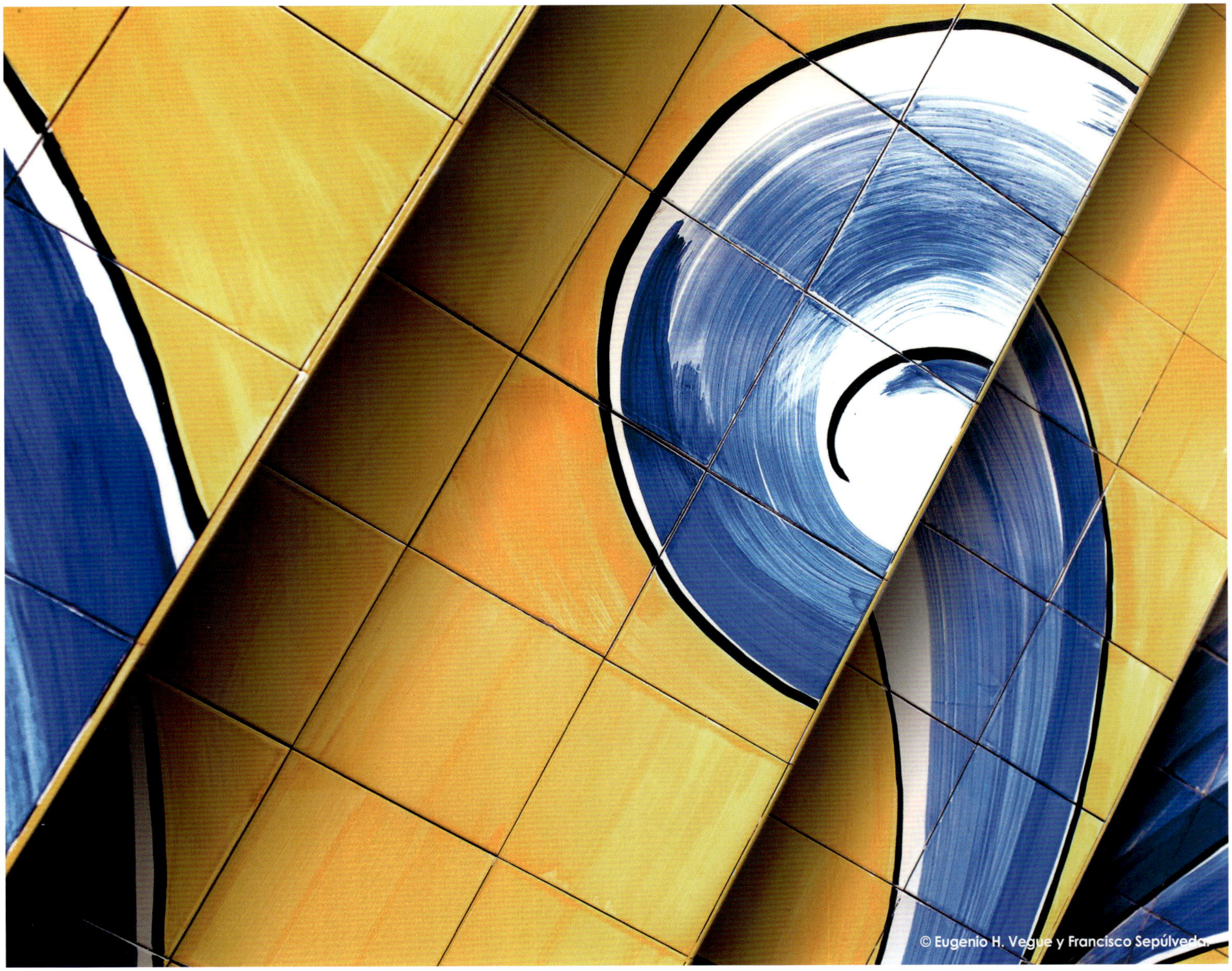

© Eugenio H. Vegue y Francisco Sepúlveda

Tile Detail

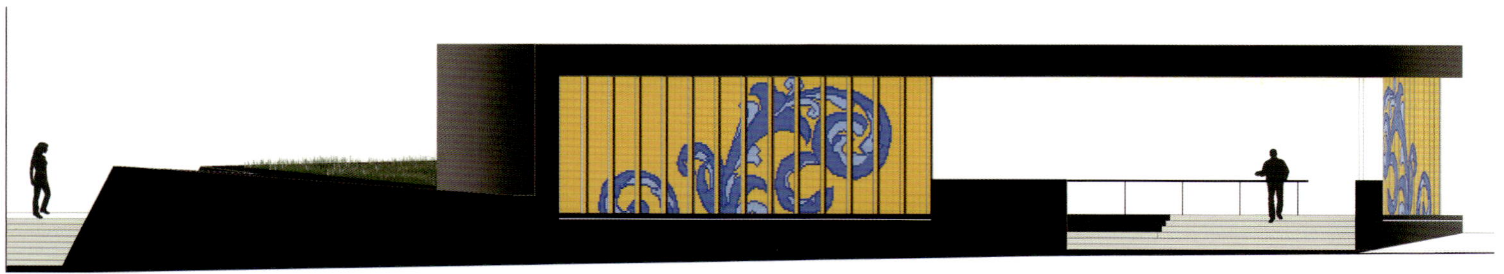

Elevation

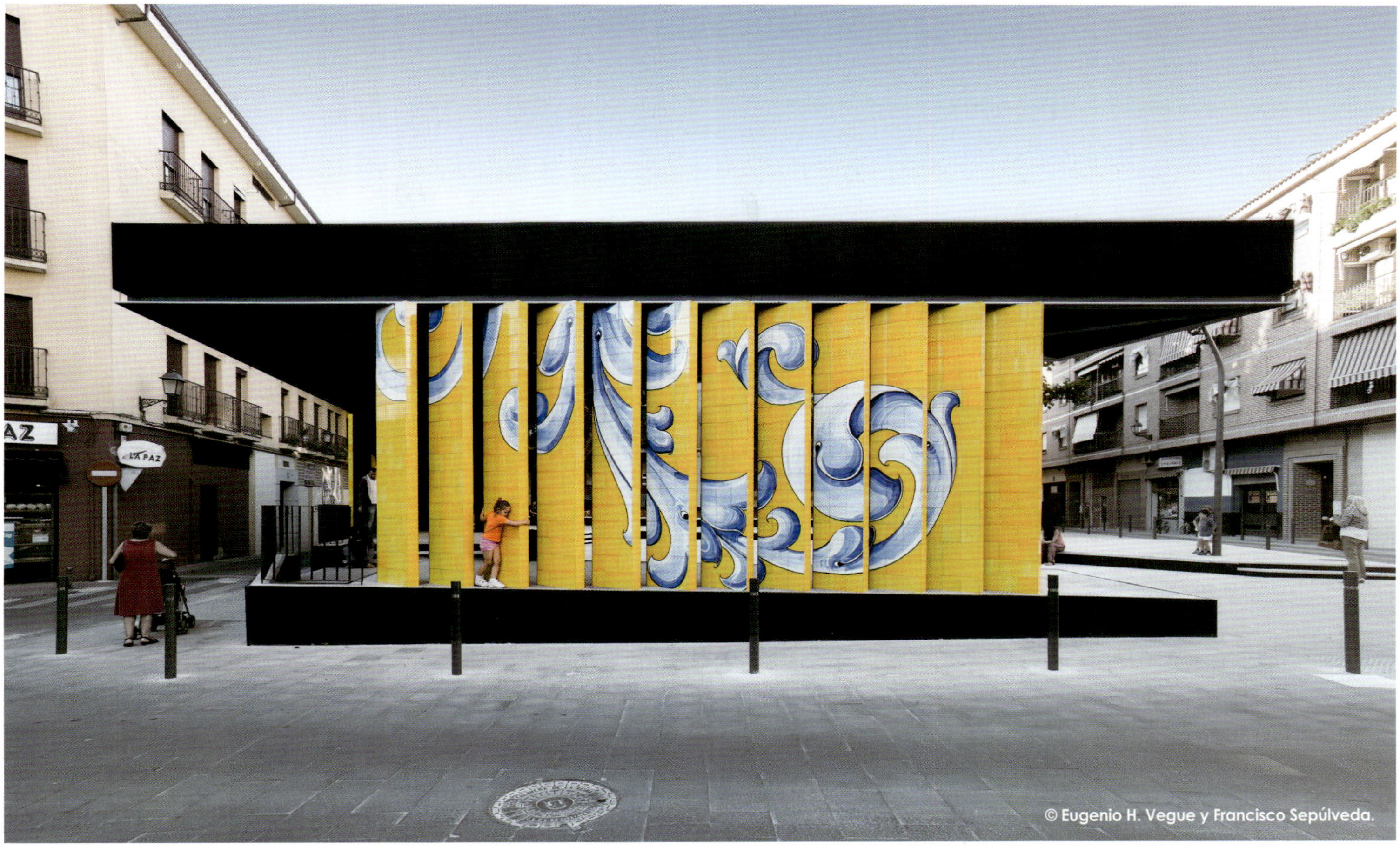

© Eugenio H. Vegue y Francisco Sepúlveda.

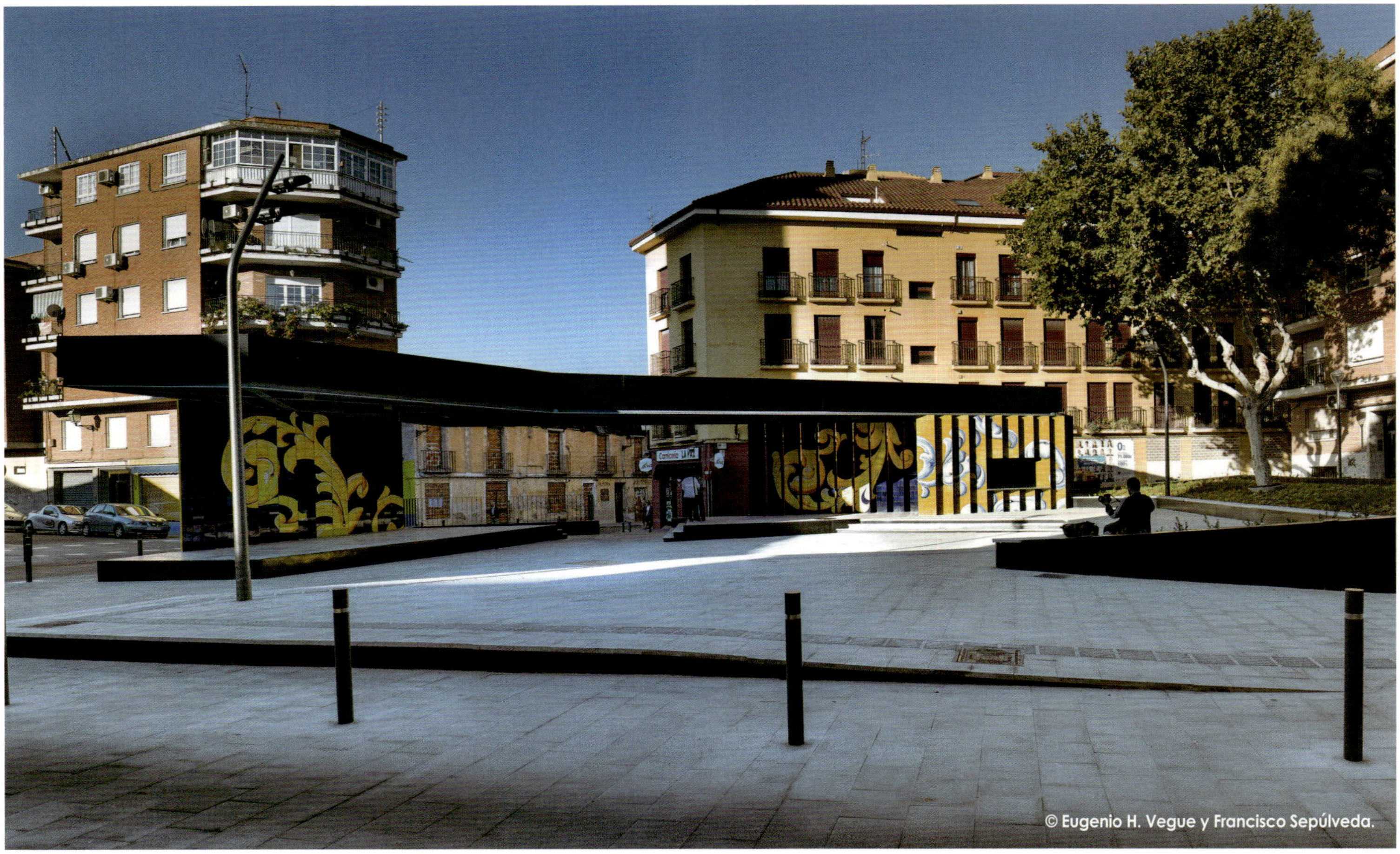

© Eugenio H. Vegue y Francisco Sepúlveda.

Bodybell

San Miguel Square Rehabilitation, Talavera de la Reina, Spain

OOIIO Architecture

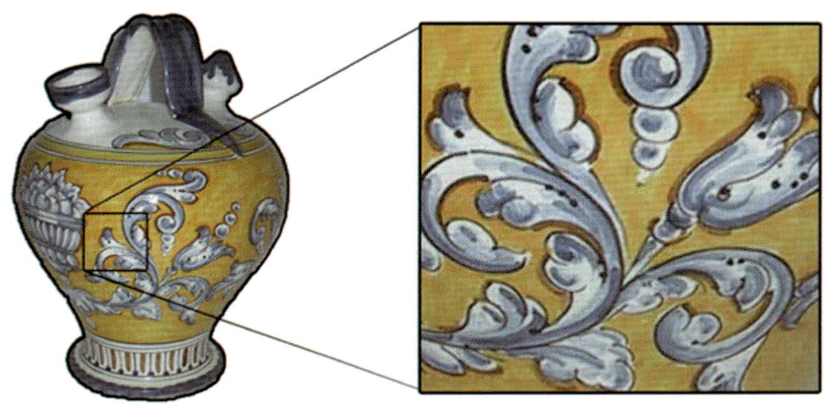

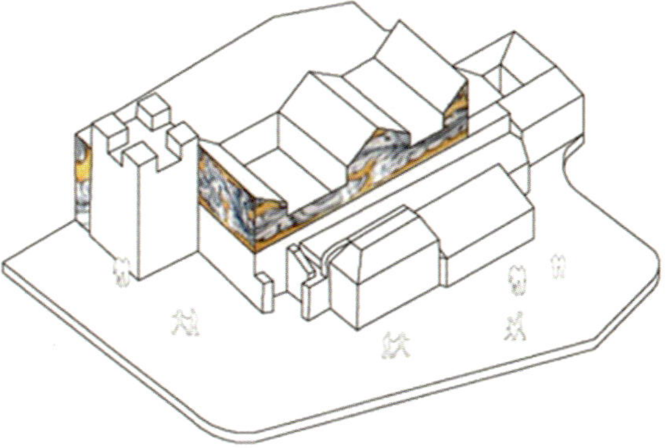

local traditional pottery

zoom into Renaissance motifs

make it grow to urban scale!

Concept

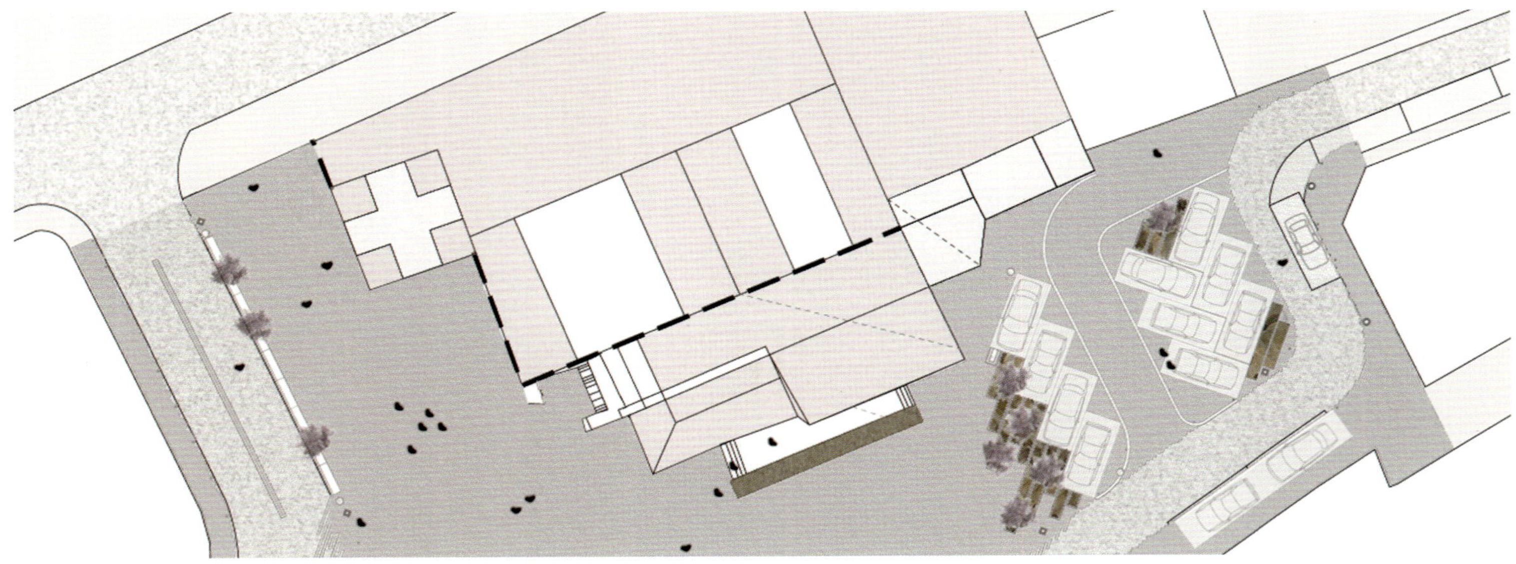

Site Plan

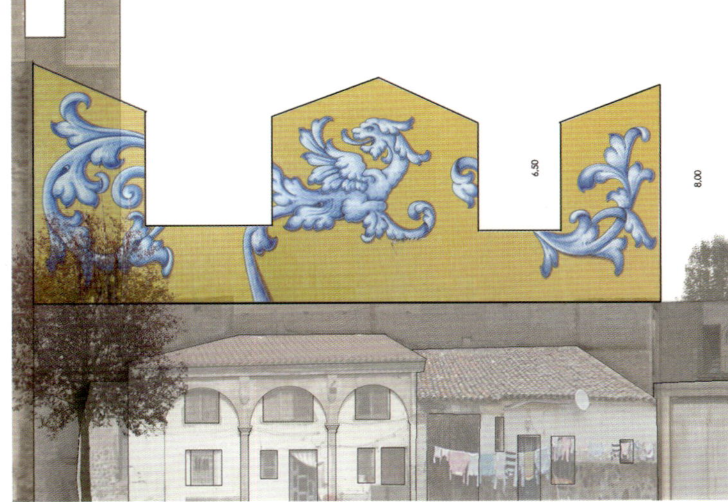

Elevation

Hand Craft Process of Tiles Mosaic Creation

1. Elegimos los motives inspirados en la Tradicion
2. Artesanalmente se prepara cada azulejos
3. Perfilando cada pieza
4. Seorganizan los lienzos mediatnte codigos
5. Los artesanos locales llinan de magia las piezas
6. Poco a poco el color inunda los motivos
7. Al horno para lacar
8. Comprobando que todas las piezas estan bien
9. Organizando cada detalle para llevar a obra
10. Se Transporta en cajas con mucho cuidado
11. Recepcion en obra
12. Pieza a pieza se van levantando los mosaicos

1.

2.

3.

4.

5.

Construction Process

© Eugenio H. Vegue y Francisco Sepúlveda.

© Eugenio H. Vegue y Francisco Sepúlveda.

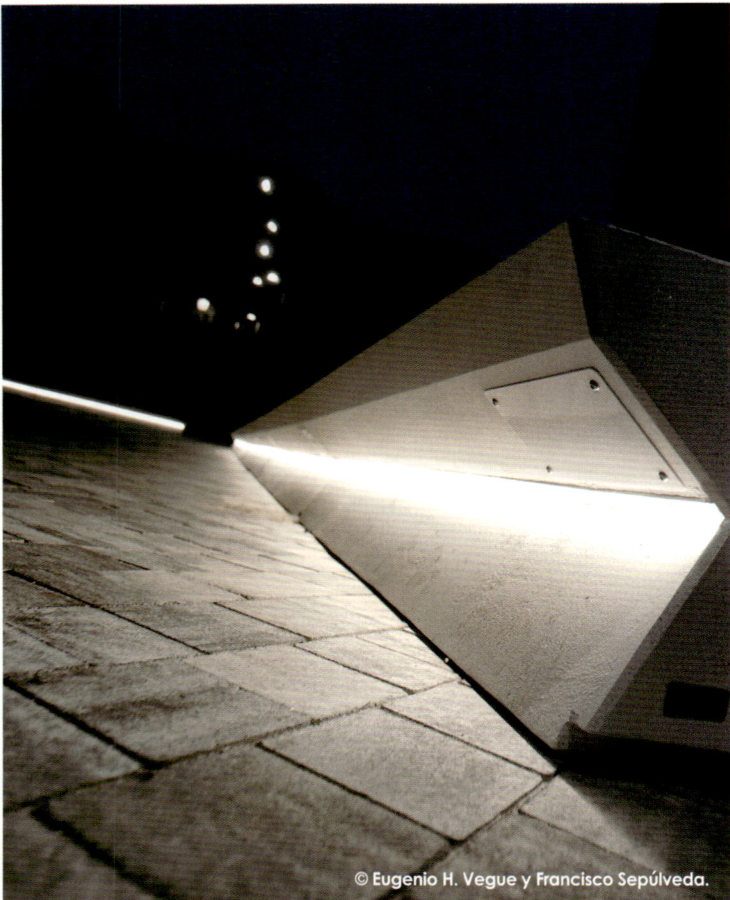

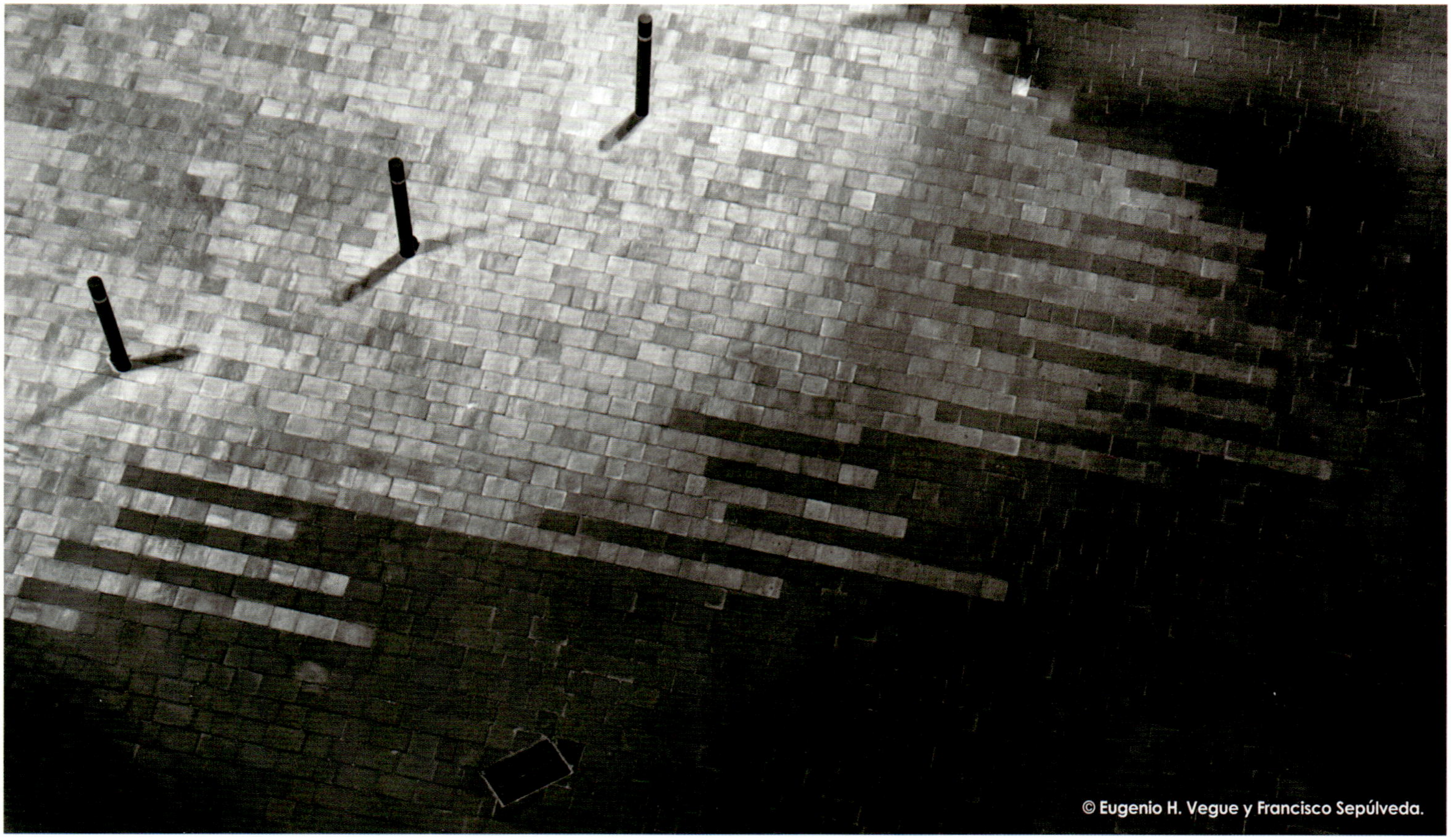
© Eugenio H. Vegue y Francisco Sepúlveda.

Tile Detail

Cafe Grumpy, New York, USA

Z-A studio& Cheng+Snyder

© Noa Kalina

Muy Bien, Seoul, South Korea
ogisadesign

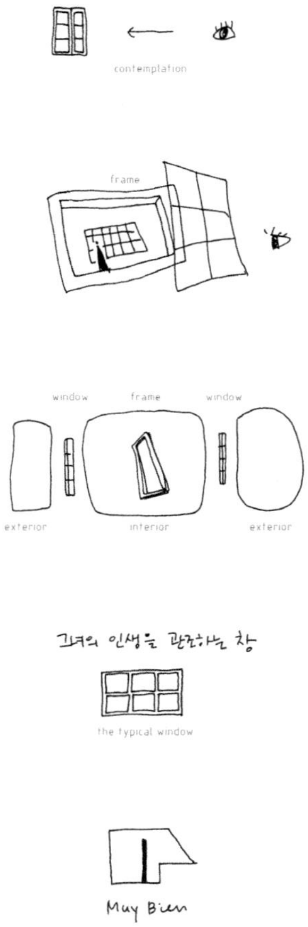

contemplation
frame
window
frame
window
exterior
interior
exterior
그녀의 인생을 관조하는 창
the typical window
Muy Bien

Concept Diagram

Muy Bien
그녀의 인생.

Sketch

FAN
2310
1950.

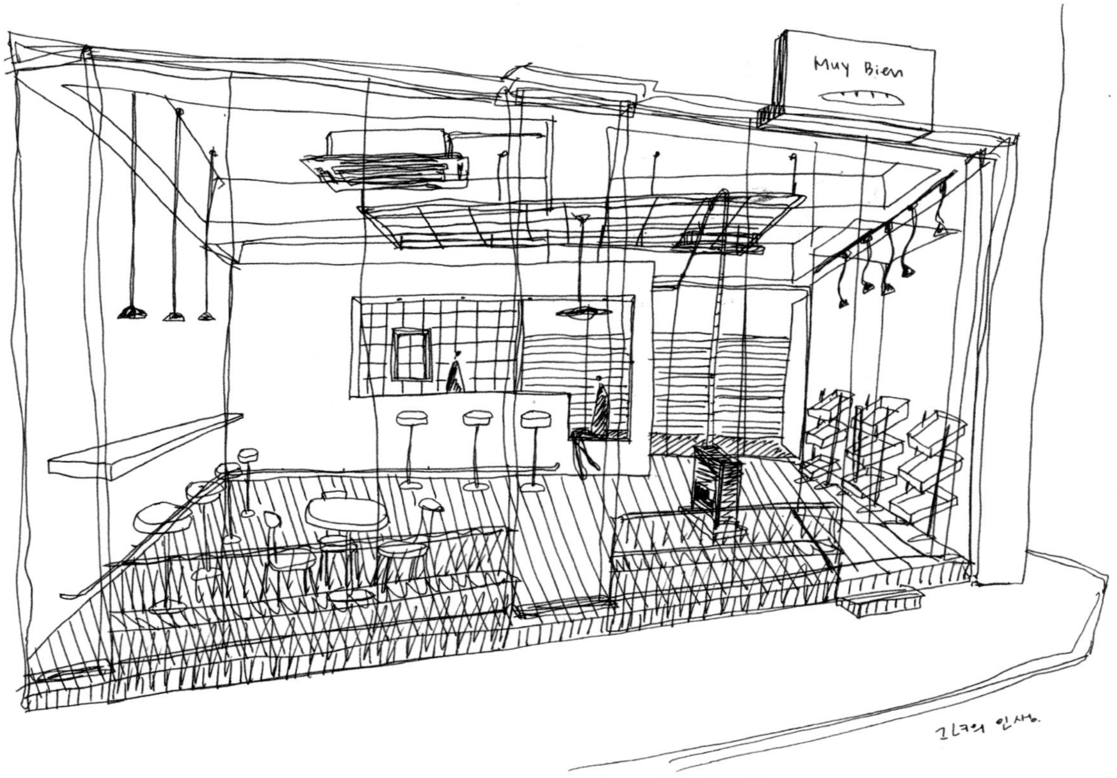

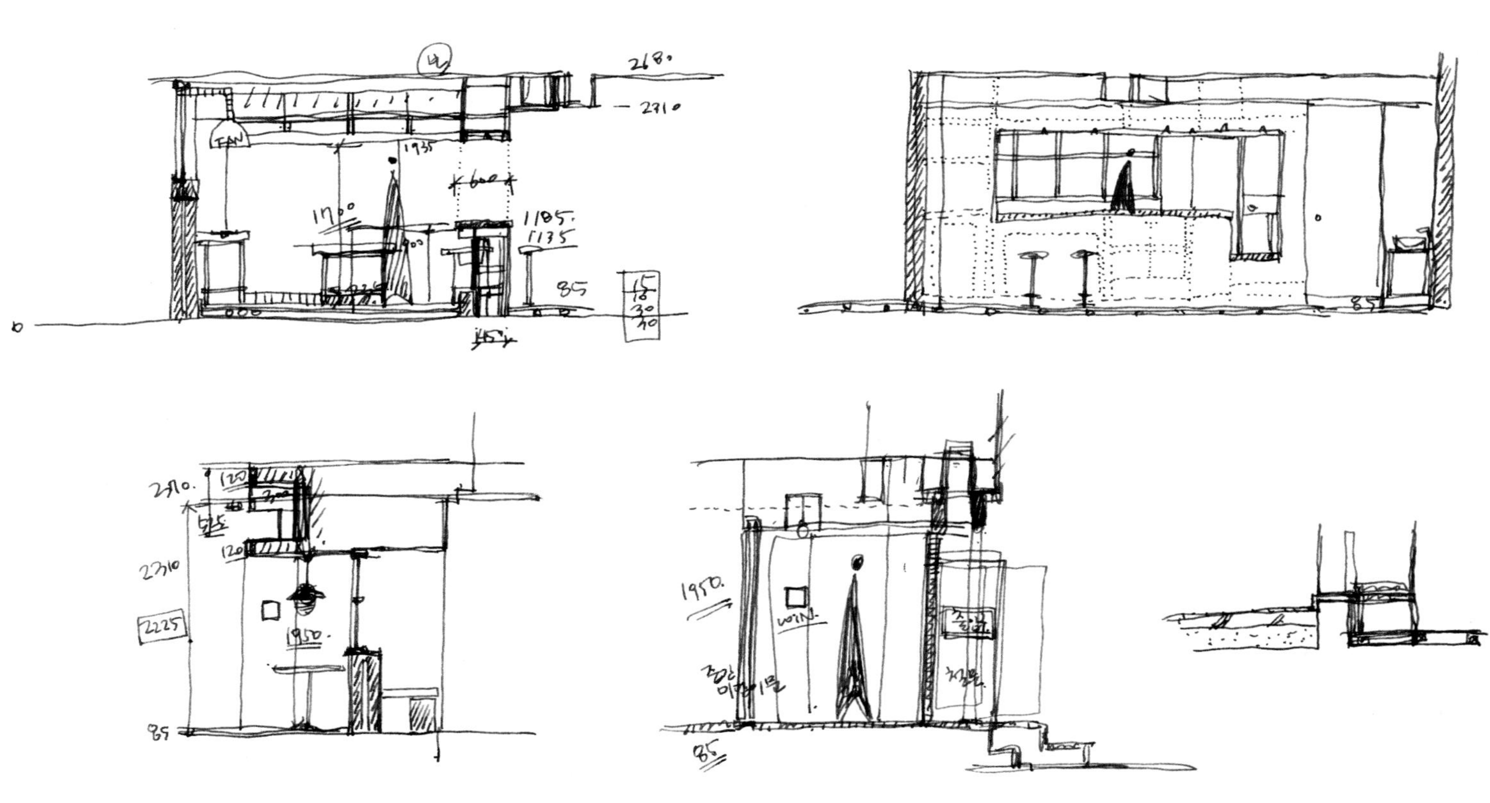

Detail Sketch

ADT

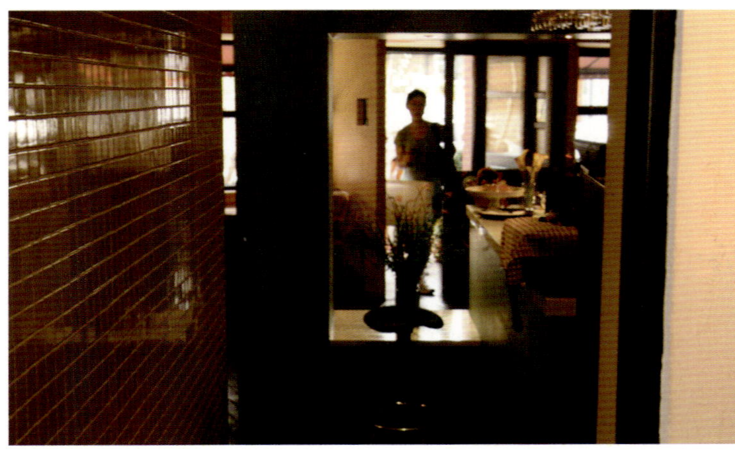

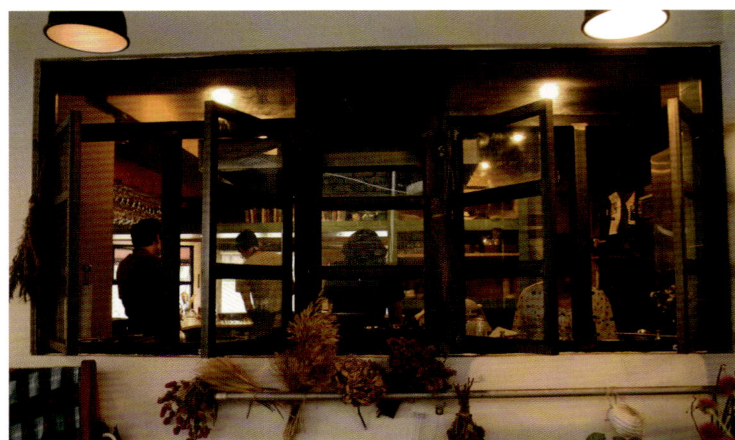

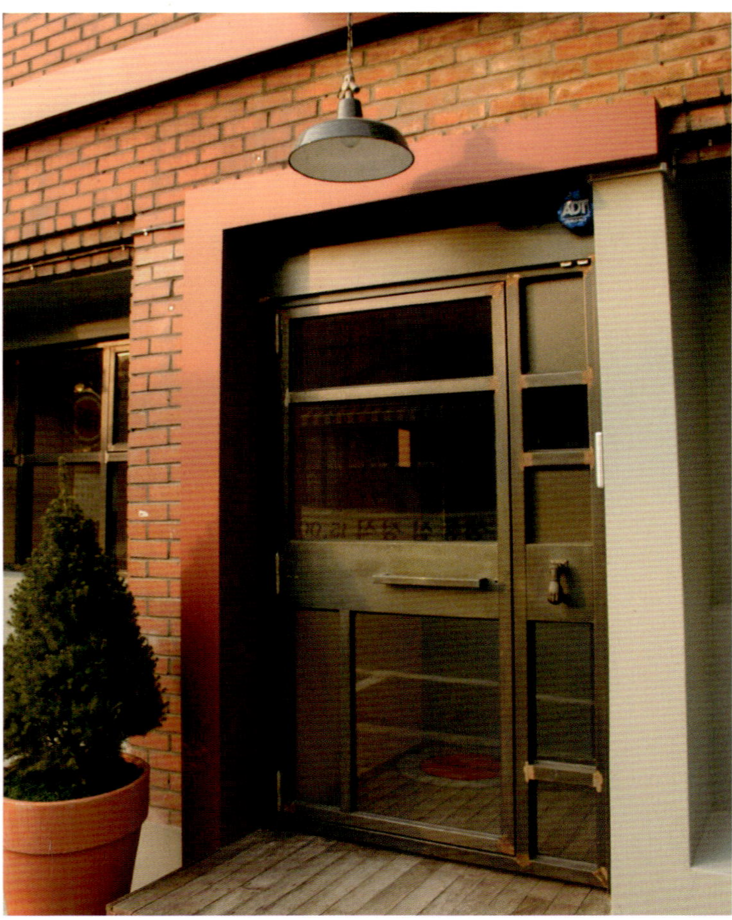

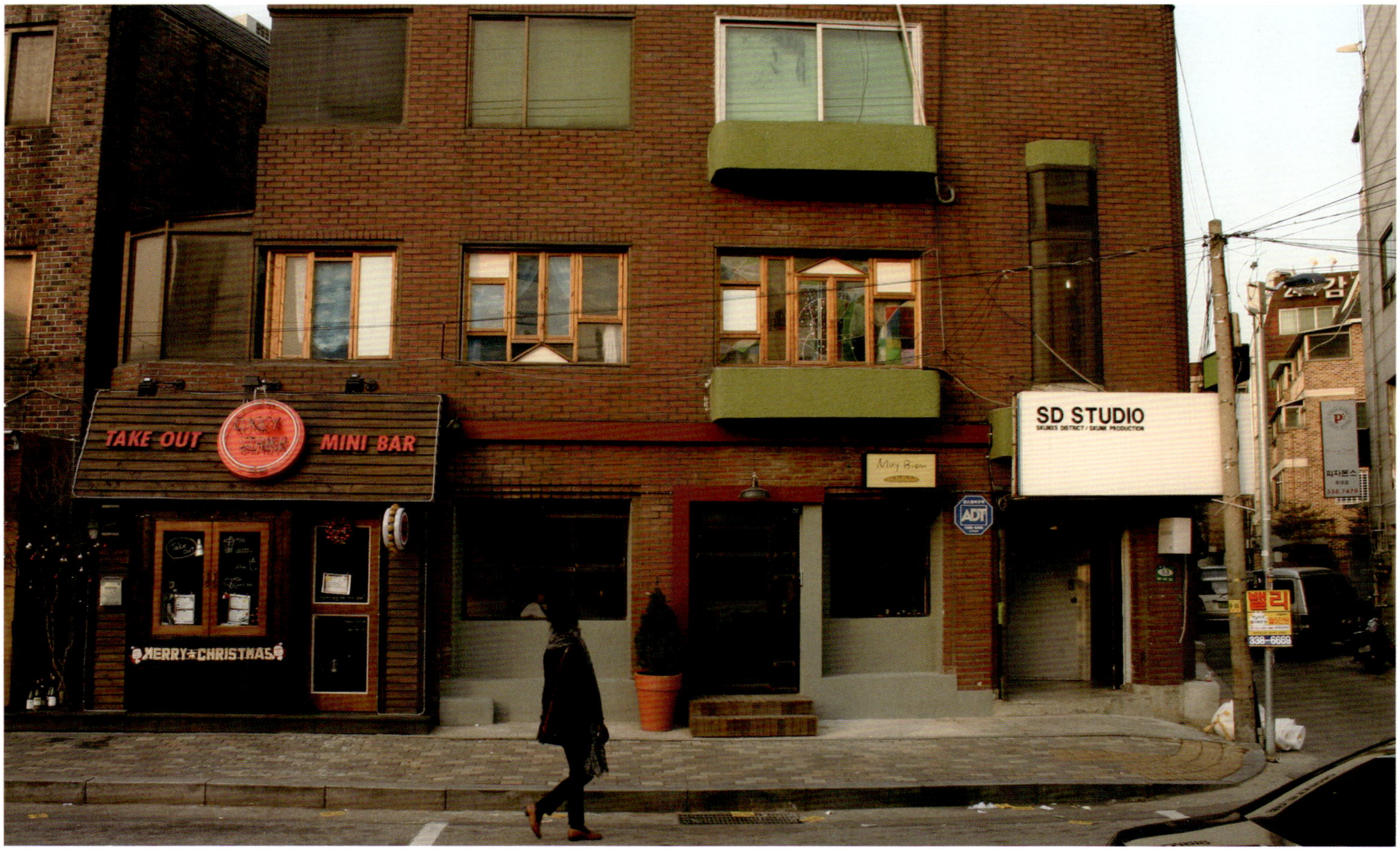
TAKE OUT
MINI BAR
MERRY★CHRISTMAS
ADT
SD STUDIO

coffee, wine
ADT

2 Houses with 6 Homes, Madrid, Spain

nodo17 Architects

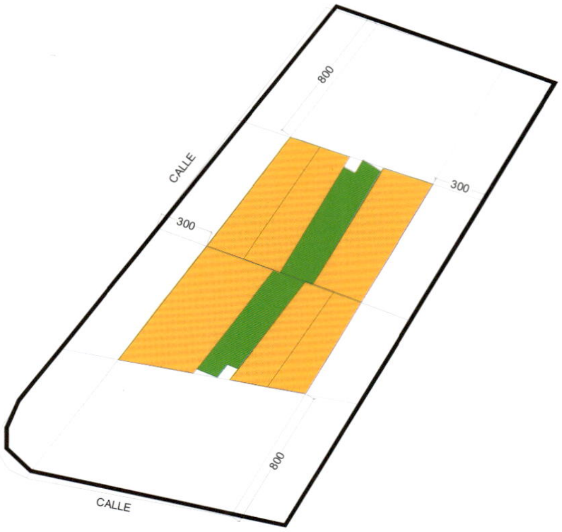

CALLE

800

300

300

800

CALLE

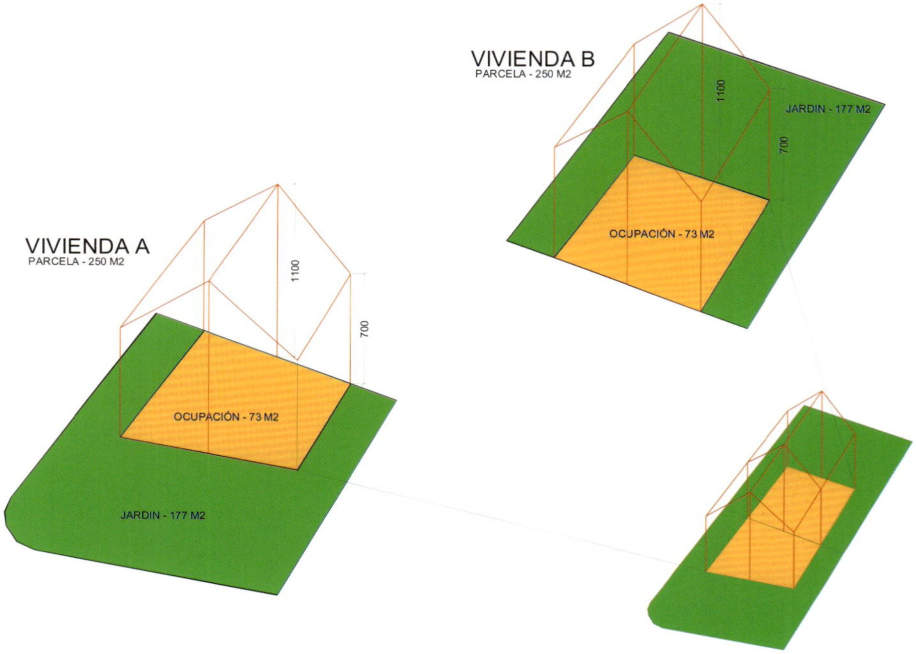

VIVIENDA B

PARCELA - 250 M2

1100

JARDIN - 177 M2

700

OCUPACIÓN - 73 M2

VIVIENDA A

PARCELA - 250 M2

1100

700

OCUPACIÓN - 73 M2

JARDIN - 177 M2

Architectural Diagram

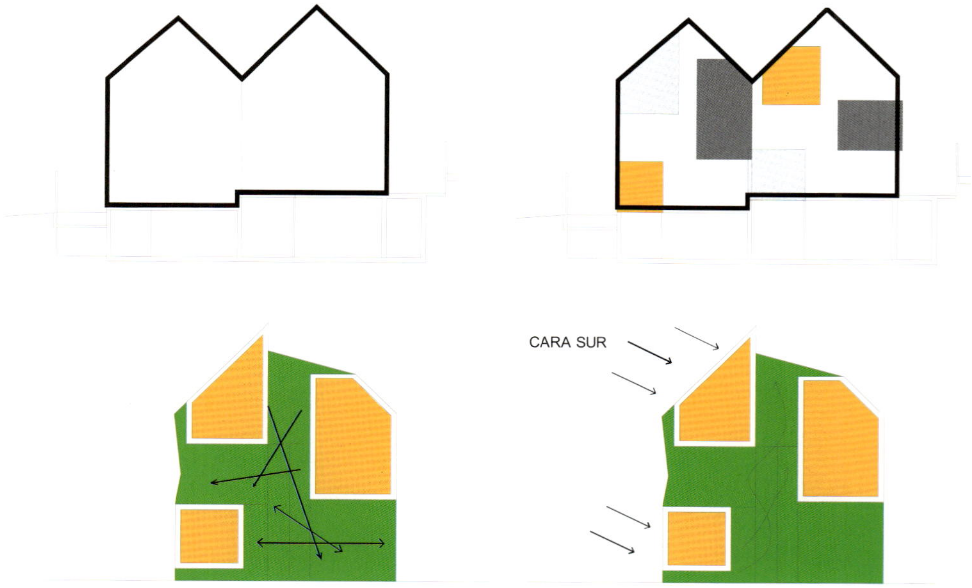

CARA SUR

LUZ, ESPACIO Y DISEÑO
DOS VIVE

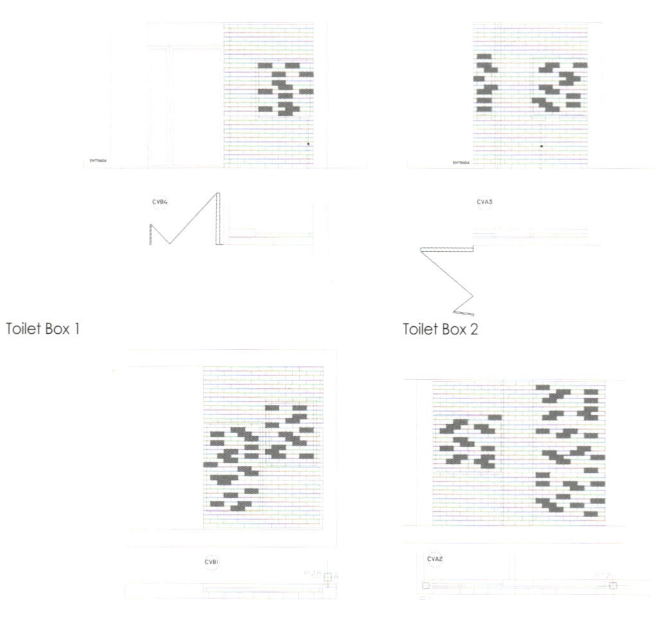

Toilet Box 1

Toilet Box 2

Shower Box 1

Shower Box 2

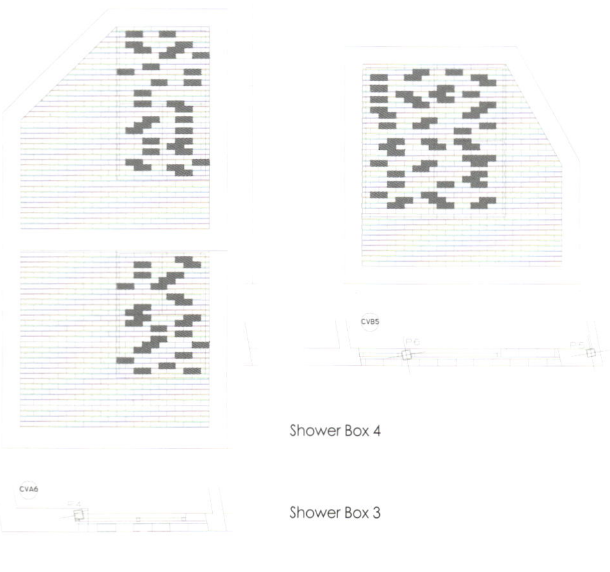

Shower Box 4

Shower Box 3

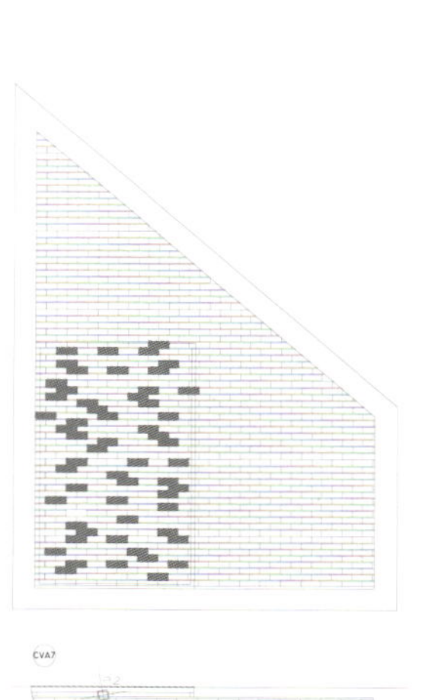

Shower Box 5

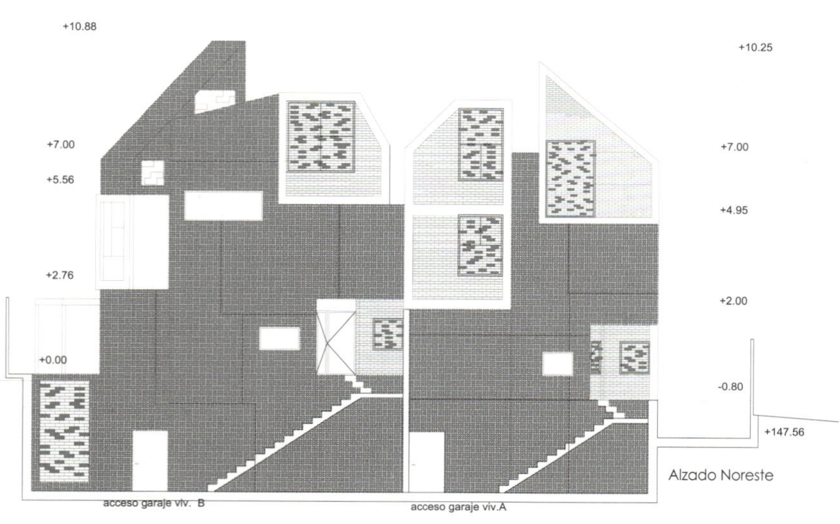

Alzado Noreste

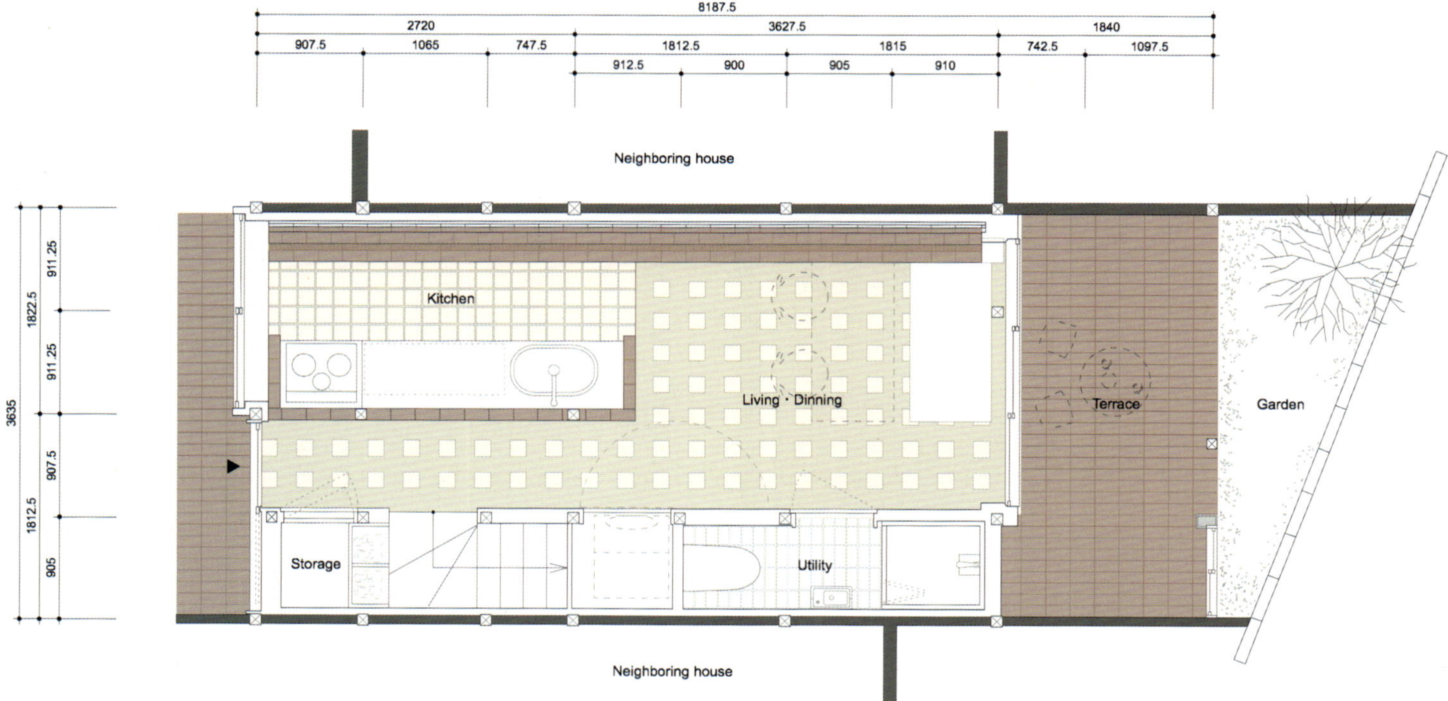

First Floor

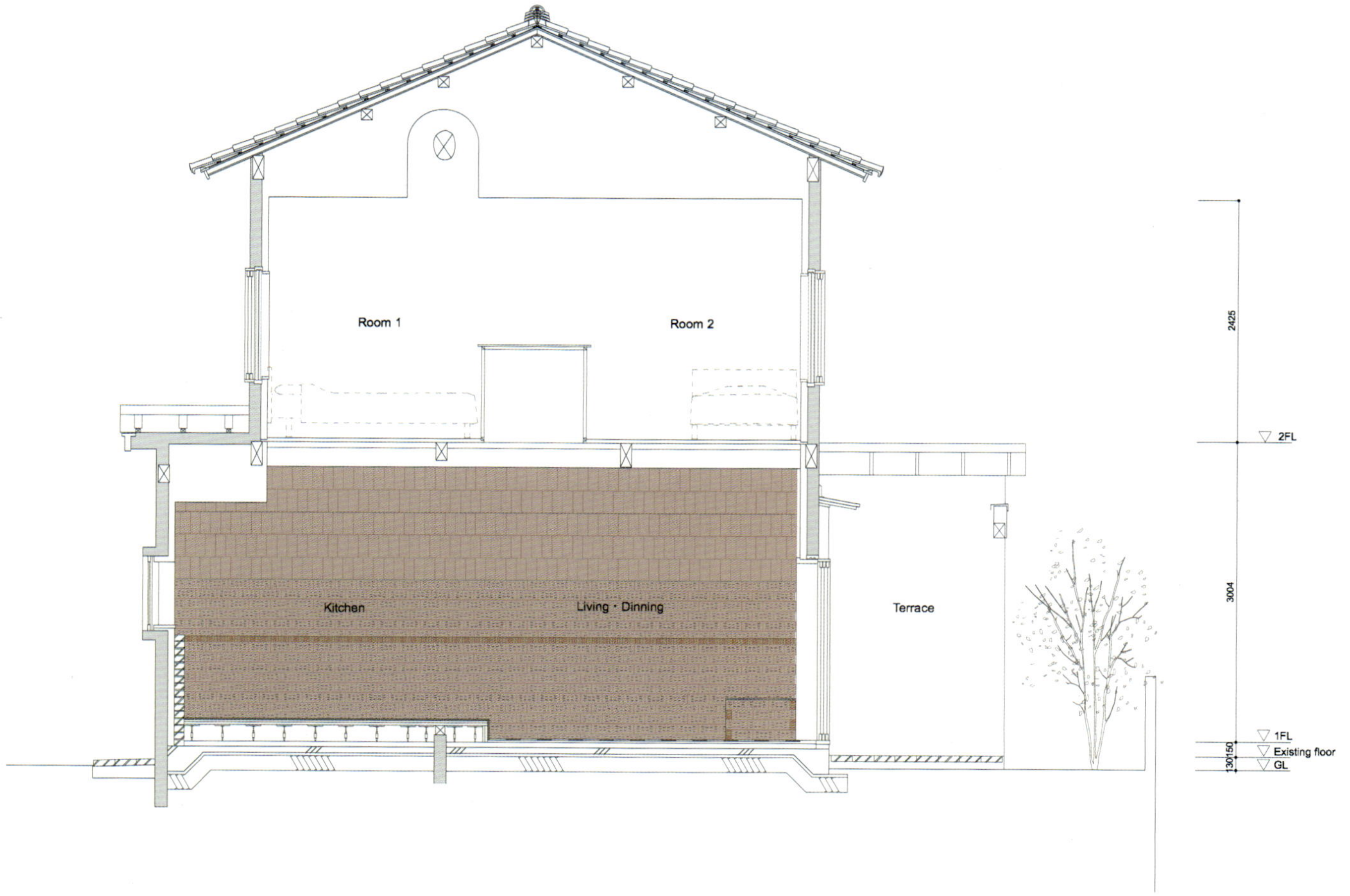

Room 1
Room 2
Kitchen
Living · Dinning
Terrace
2425
3004
2FL
1FL
Existing floor
GL

Section

Before Renovation

29 Jan 2012

A long-vacant house.

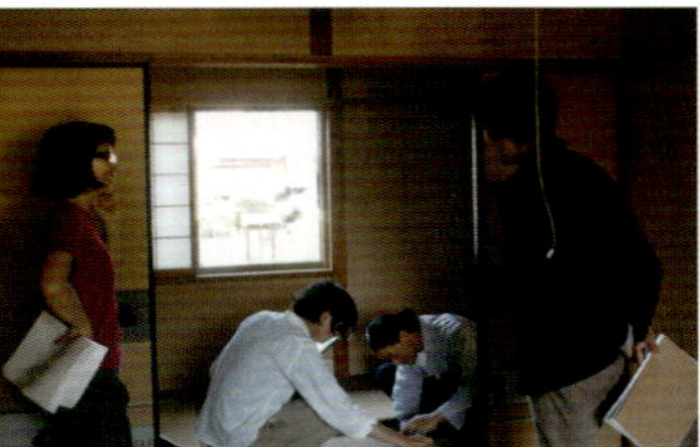

28 May 2014

Architectural survey.

20 July 2014

An architecture model.

1 Aug 2014

Visit Tokoname for brick.

17 Sep 2012

Demolition in progress.

1 Dec 2014

Study on paints.

14 Dec 2014

Study on brick laying.

22 Jan 2014

Installing insulation.

27 Feb 2015

Arrive of bricks.

27 Feb 2015

Study on brick .

2 Mar 2015

Mixing the composite.

2 Mar 2015

Spreading clay.

2 May 2015

Tiles embedded.

2 May 2015

Self building...

8 May 2015

More building...

8 May 2015

Near completion.

Construction Process

Blooming House with wild flowers, Seoul, South Korea

studio-GAON

Idea Sketch

Elevation East

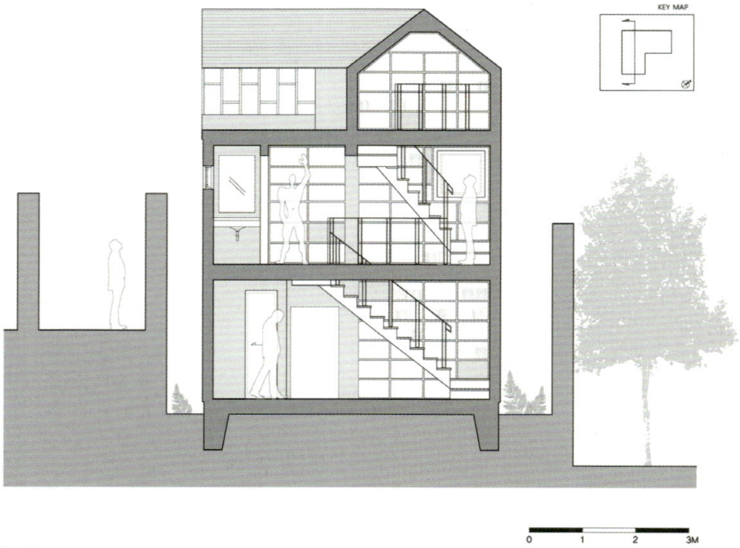

Section A

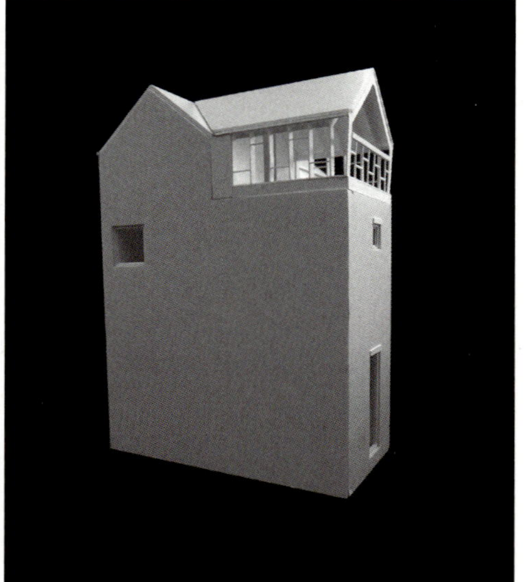

Study Modeling

© Youngchae Park

© Youngchae Park

© Youngchae Park

© Youngchae Park

© Youngchae Park

© Youngchae Park

© Youngchae Park

Nguyen Tien Thanh

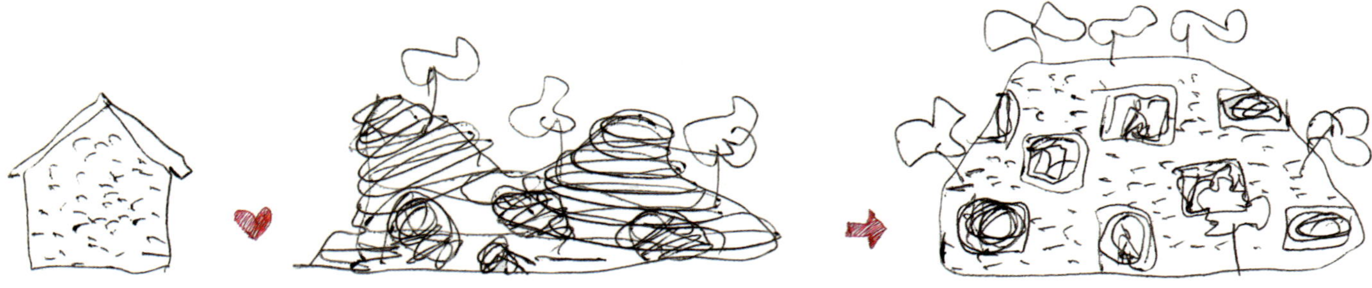

Design Process

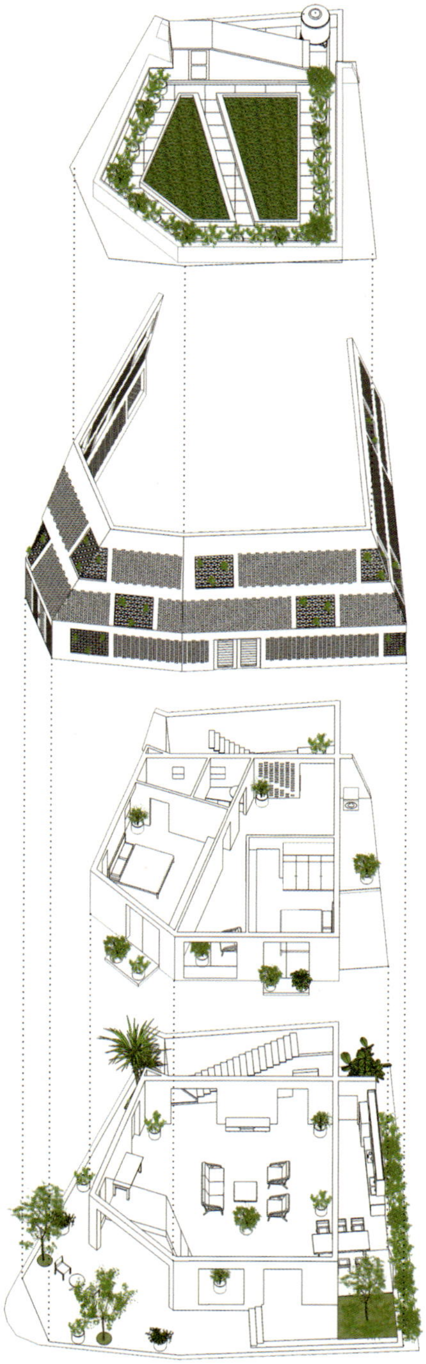

Diagram

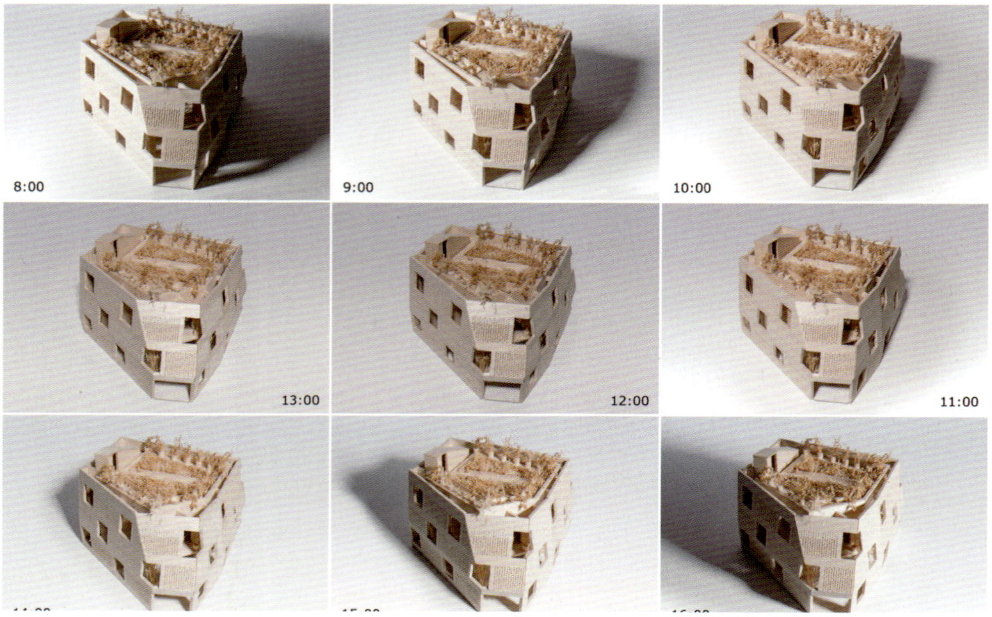
8:00
9:00
10:00
13:00
12:00
11:00

Study Modeling

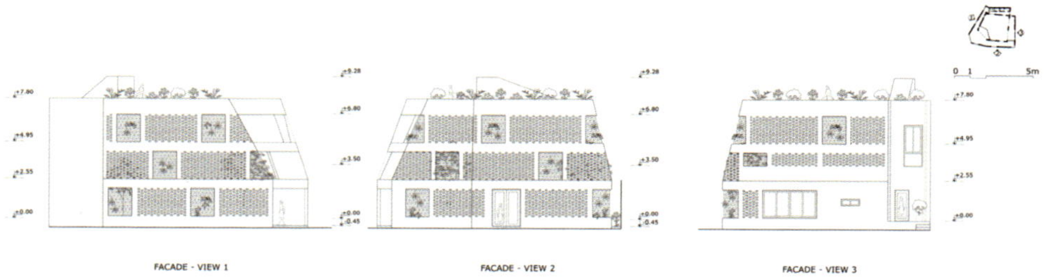
0 1 5m
FACADE - VIEW 1
FACADE - VIEW 2
FACADE - VIEW 3

Facade

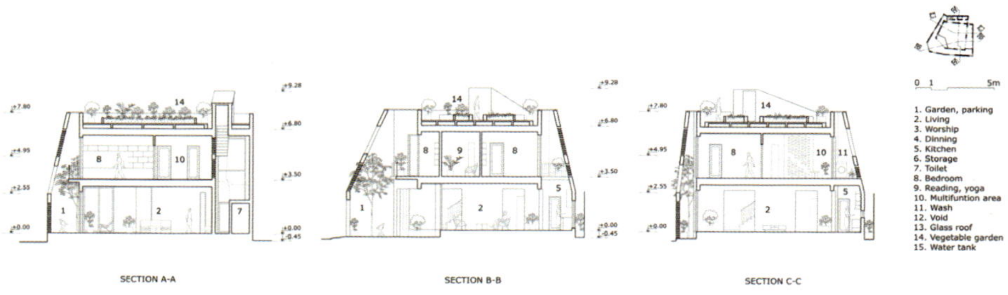
0 1 5m
1. Garden, parking
2. Living
3. Worship
4. Dinning
5. Kitchen
6. Storage
7. Toilet
8. Bedroom
9. Reading, yoga
10. Multifuntion area
11. Wash
12. Void
13. Glass roof
14. Vegetable garden
15. Water tank
SECTION A-A
SECTION B-B
SECTION C-C

Section

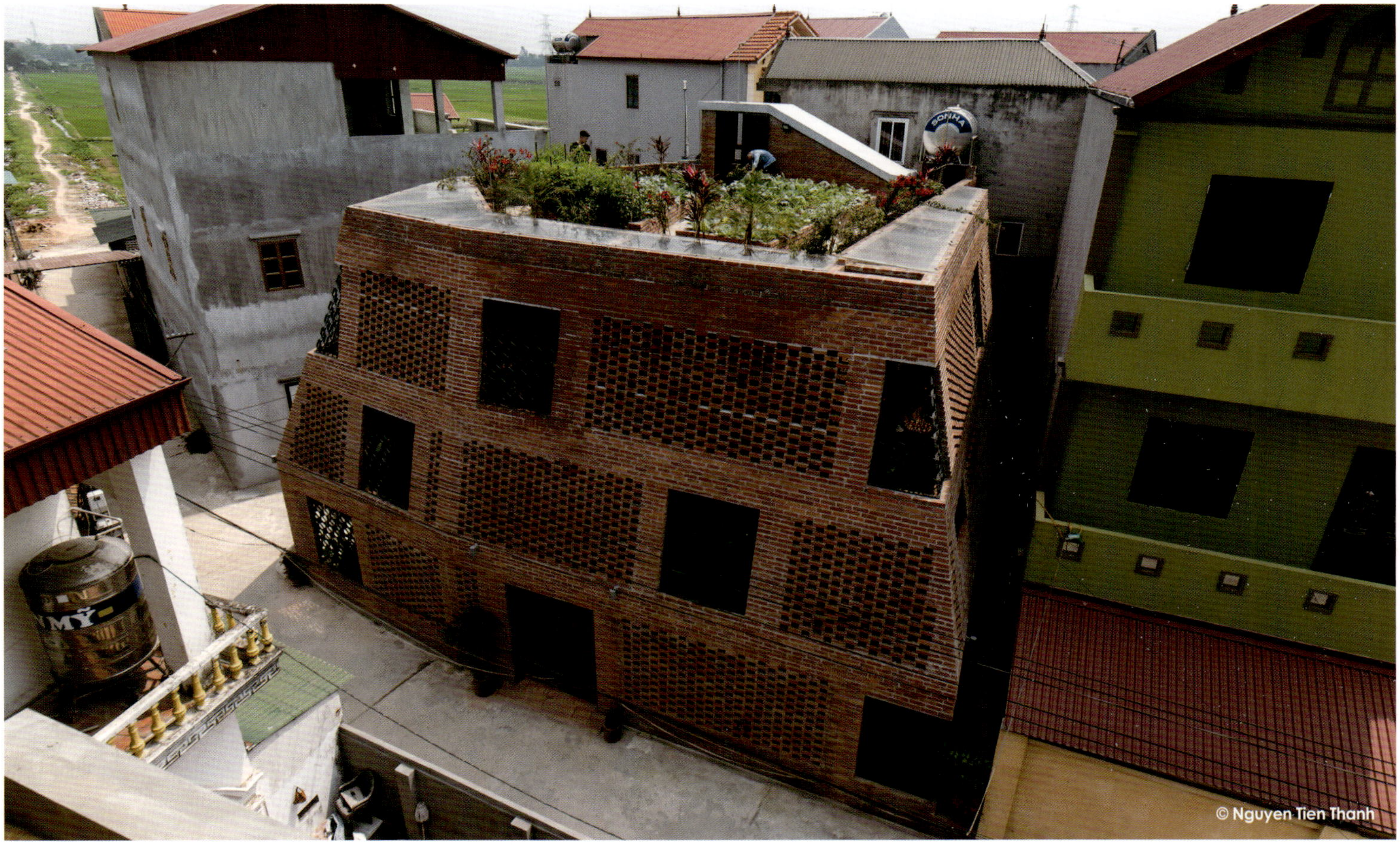

© Nguyen Tien Thanh

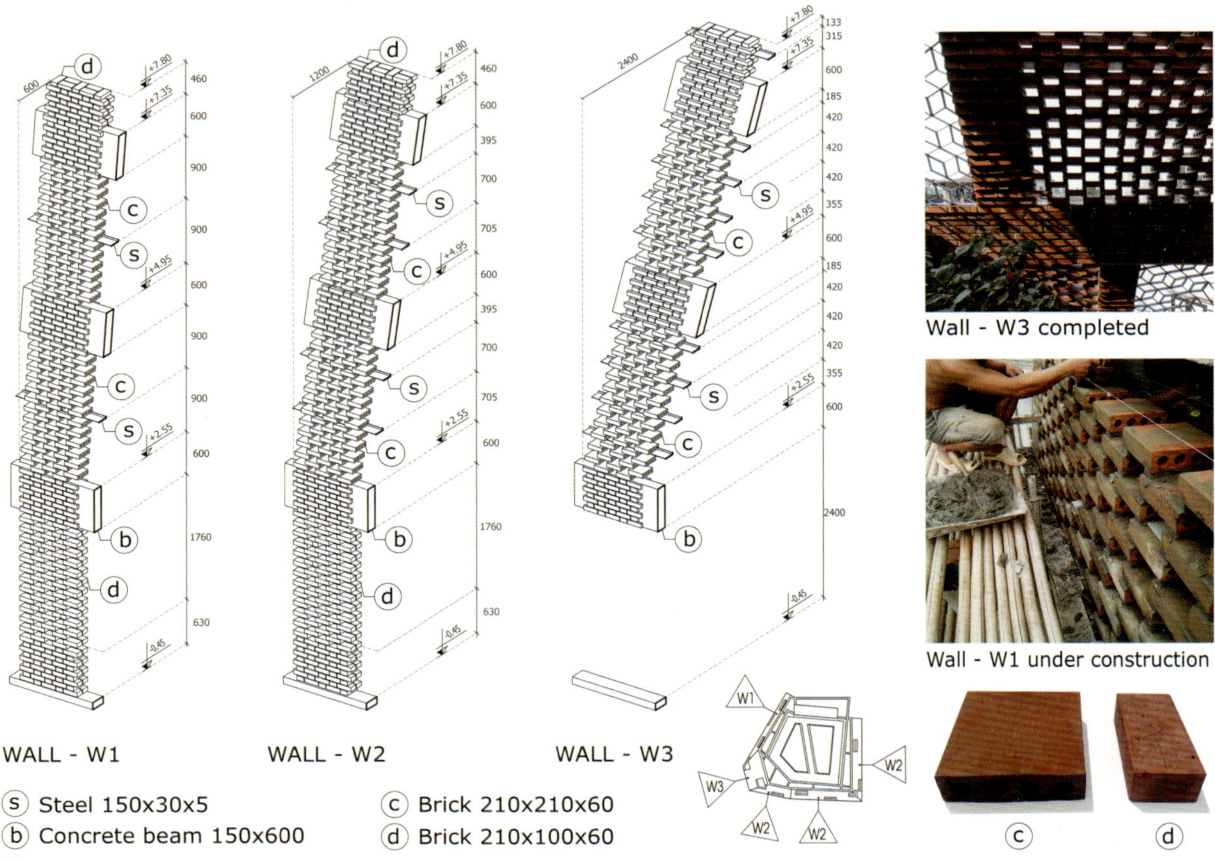

Construction Detail

© Nguyen Tien Thanh

© Nguyen Tien Thanh

© Nguyen Tien Thanh

© Nguyen Tien Thanh

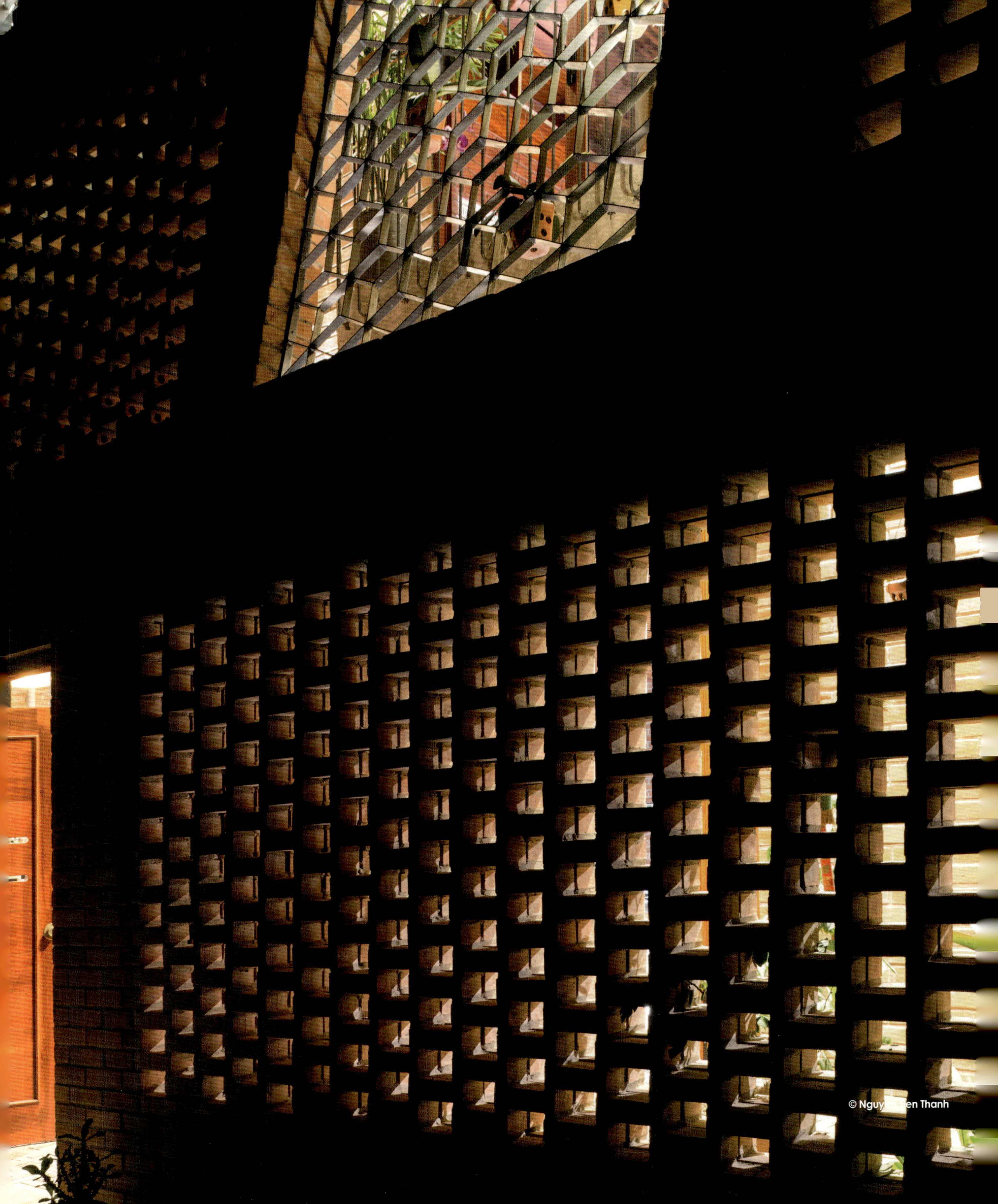

© Nguyen Tien Thanh

Dr. Sudhakar Jadhav`s Residence, Maharashtra, India

Sunil Patil and Associates

Ground Plan

First Floor

iA House, Madrid, Spain

LANDÍNEZ+REY arquitectos

Elevation

Section

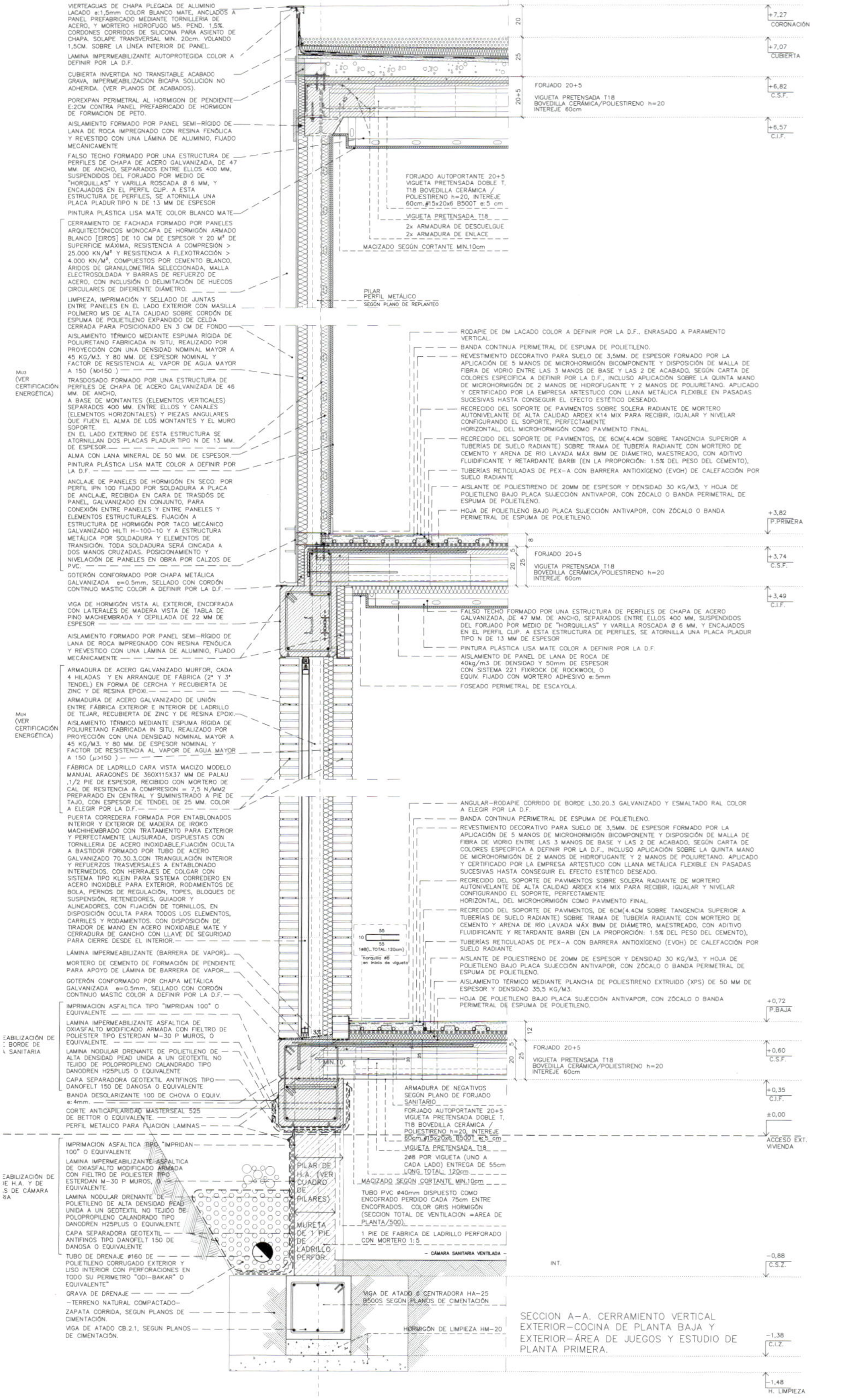

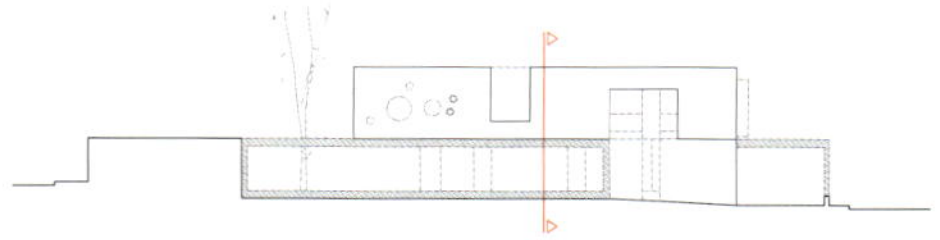

VIERTEAGUAS DE CHAPA PLEGADA DE ALUMINIO LACADO e:1,5mm COLOR BLANCO MATE, ANCLADOS A PANEL PREFABRICADO MEDIANTE TORNILLERIA DE ACERO, Y MORTERO HIDROFUGO M5. PEND. 1,5% CORDONES CORRIDOS DE SILICONA PARA ASIENTO DE CHAPA. SOLAPE TRANSVERSAL MIN. 20cm. VOLANDO 1,5CM. SOBRE LA LINEA INTERIOR DE PANEL.

LAMINA IMPERMEABILIZANTE AUTOPROTEGIDA COLOR A DEFINIR POR LA D.F.

CUBIERTA INVERTIDA NO TRANSITABLE ACABADO GRAVA, IMPERMEABILIZACION BICAPA SOLUCION NO ADHERIDA. (VER PLANOS DE ACABADOS).

POREXPAN PERIMETRAL AL HORMIGON DE PENDIENTE E:2CM CONTRA PANEL PREFABRICADO DE HORMIGON DE FORMACION DE PETO.

AISLAMIENTO FORMADO POR PANEL SEMI-RIGIDO DE LANA DE ROCA IMPREGNADO CON RESINA FENÓLICA Y REVESTIDO CON UNA LÁMINA DE ALUMINIO, FIJADO MECÁNICAMENTE

FALSO TECHO FORMADO POR UNA ESTRUCTURA DE PERFILES DE CHAPA DE ACERO GALVANIZADA, DE 47 MM. DE ANCHO, SEPARADOS ENTRE ELLOS 400 MM, SUSPENDIDOS DEL FORJADO POR MEDIO DE "HORQUILLAS" Y VARILLA ROSCADA Ø 6 MM, Y ENCAJADOS EN EL PERFIL CLIP. A ESTA ESTRUCTURA DE PERFILES, SE ATORNILLA UNA PLACA PLADUR TIPO N DE 13 MM DE ESPESOR

PINTURA PLÁSTICA LISA MATE COLOR BLANCO MATE

CERRAMIENTO DE FACHADA FORMADO POR PANELES ARQUITECTÓNICOS MONOCAPA DE HORMIGÓN ARMADO BLANCO [EIROS] DE 10 CM DE ESPESOR Y 20 M² DE SUPERFICIE MÁXIMA, RESISTENCIA A COMPRESIÓN > 25.000 KN/M² Y RESISTENCIA A FLEXOTRACCIÓN > 4.000 KN/M², COMPUESTOS POR CEMENTO BLANCO, ÁRIDOS DE GRANULOMETRÍA SELECCIONADA, MALLA ELECTROSOLDADA Y BARRAS DE REFUERZO DE ACERO, CON INCLUSIÓN O DELIMITACIÓN DE HUECOS CIRCULARES DE DIFERENTE DIÁMETRO.

LIMPIEZA, IMPRIMACIÓN Y SELLADO DE JUNTAS ENTRE PANELES EN EL LADO EXTERIOR CON MASILLA POLÍMERO MS DE ALTA CALIDAD SOBRE CORDÓN DE ESPUMA DE POLIETILENO EXPANDIDO DE CELDA CERRADA PARA POSICIONADO EN 3 CM DE FONDO

Mxx (VER CERTIFICACIÓN ENERGÉTICA)

AISLAMIENTO TÉRMICO MEDIANTE ESPUMA RÍGIDA DE POLIURETANO FABRICADA IN SITU, REALIZADO POR PROYECCIÓN CON UNA DENSIDAD NOMINAL MAYOR A 45 KG/M3. Y 80 MM. DE ESPESOR NOMINAL Y FACTOR DE RESISTENCIA AL VAPOR DE AGUA MAYOR A 150 (µ≥150)

TRASDOSADO FORMADO POR UNA ESTRUCTURA DE PERFILES DE CHAPA DE ACERO GALVANIZADA DE 46 MM. DE ANCHO, A BASE DE MONTANTES (ELEMENTOS VERTICALES) SEPARADOS 400 MM. ENTRE ELLOS Y CANALES (ELEMENTOS HORIZONTALES) Y PIEZAS ANGULARES QUE FIJEN EL ALMA DE LOS MONTANTES Y EL MURO SOPORTE. EN EL LADO EXTERNO DE ESTA ESTRUCTURA SE ATORNILLAN DOS PLACAS PLADUR TIPO N DE 13 MM. DE ESPESOR

ALMA CON LANA MINERAL DE 50 MM. DE ESPESOR

PINTURA PLÁSTICA LISA MATE COLOR A DEFINIR POR LA D.F.

ANCLAJE DE PANELES DE HORMIGÓN EN SECO: POR PERFIL IPN 100 FIJADO POR SOLDADURA A PLACA DE ANCLAJE, RECIBIDA EN CARA DE TRASDÓS DE PANEL, GALVANIZADO EN CONJUNTO, PARA CONEXIÓN ENTRE PANELES Y ENTRE PANELES Y ELEMENTOS ESTRUCTURALES. FIJACIÓN A ESTRUCTURA DE HORMIGÓN POR TACO MECÁNICO GALVANIZADO HILTI H-100-10 Y A ESTRUCTURA METÁLICA POR SOLDADURA Y ELEMENTOS DE TRANSICIÓN. TODA SOLDADURA SERÁ CINCADA A DOS MANOS CRUZADAS. POSICIONAMIENTO Y NIVELACIÓN DE PANELES EN OBRA POR CALZOS DE PVC.

GOTERÓN CONFORMADO POR CHAPA METÁLICA GALVANIZADA e=0.5mm, SELLADO CON CORDÓN CONTINUO MASTIC COLOR A DEFINIR POR LA D.F.

VIGA DE HORMIGÓN VISTA AL EXTERIOR, ENCOFRADA CON LATERALES DE MADERA VISTA DE TABLA DE PINO MACHIEMBRADA Y CEPILLADA DE 22 MM DE ESPESOR

AISLAMIENTO FORMADO POR PANEL SEMI-RÍGIDO DE LANA DE ROCA IMPREGNADO CON RESINA FENÓLICA Y REVESTIDO CON UNA LÁMINA DE ALUMINIO, FIJADO MECÁNICAMENTE

ARMADURA DE ACERO GALVANIZADO MURFOR, CADA 4 HILADAS Y EN ARRANQUE DE FÁBRICA (2ª Y 3ª TENDEL) EN FORMA DE CERCHA Y RECUBIERTA DE ZINC Y DE RESINA EPOXI

ARMADURA DE ACERO GALVANIZADO DE UNIÓN ENTRE FÁBRICA EXTERIOR E INTERIOR DE LADRILLO DE TEJAR, RECUBIERTA DE ZINC Y DE RESINA EPOXI

Mxx (VER CERTIFICACIÓN ENERGÉTICA)

AISLAMIENTO TÉRMICO MEDIANTE ESPUMA RÍGIDA DE POLIURETANO FABRICADA IN SITU, REALIZADO POR PROYECCIÓN CON UNA DENSIDAD NOMINAL MAYOR A 45 KG/M3. Y 80 MM. DE ESPESOR NOMINAL Y FACTOR DE RESISTENCIA AL VAPOR DE AGUA MAYOR A 150 (µ≥150)

FÁBRICA DE LADRILLO CARA VISTA MACIZO MODELO MANUAL ARAGONÉS DE 360X115X37 MM DE PALAU 1/2 PIE DE ESPESOR, RECIBIDO CON MORTERO DE CAL DE RESISTENCIA A COMPRESIÓN = 7,5 N/MM2 PREPARADO EN CENTRAL Y SUMINISTRADO A PIE DE TAJO, CON ESPESOR DE TENDEL DE 25 MM. COLOR A ELEGIR POR LA D.F.

PUERTA CORREDERA FORMADA POR ENTABLONADOS INTERIOR Y EXTERIOR DE MADERA DE IROKO MACHIHEMBRADO CON TRATAMIENTO PARA EXTERIOR Y PERFECTAMENTE LAUSURADA, DISPUESTAS CON TORNILLERIA DE ACERO INOXIDABLE,FIJACIÓN OCULTA A BASTIDOR FORMADO POR TUBO DE ACERO GALVANIZADO 70.30.3,CON TRIANGULACIÓN INTERIOR Y REFUERZOS TRASVERSALES A ENTABLONADO INTERMEDIOS, CON HERRAJES DE COLGAR CON SISTEMA TIPO KLEIN PARA SISTEMA CORREDERO EN ACERO INOXIDBLE PARA EXTERIOR, RODAMIENTOS DE BOLA, PERNOS DE REGULACIÓN, TOPES, BLOQUES DE SUSPENSIÓN, RETENEDORES, GUIADOR Y ALINEADORES, CON FIJACIÓN DE TORNILLOS, EN DISPOSICIÓN OCULTA PARA TODOS LOS ELEMENTOS, CARRILES Y RODAMIENTOS. CON DISPOSICIÓN DE TIRADOR DE MANO EN ACERO INOXIDABLE MATE Y CERRADURA DE GANCHO CON LLAVE DE SEGURIDAD PARA CIERRE DESDE EL INTERIOR.

LÁMINA IMPERMEABILIZANTE (BARRERA DE VAPOR)

MORTERO DE CEMENTO DE FORMACIÓN DE PENDIENTE PARA APOYO DE LÁMINA DE BARRERA DE VAPOR

GOTERÓN CONFORMADO POR CHAPA METÁLICA GALVANIZADA e=0.5mm, SELLADO CON CORDÓN CONTINUO MASTIC COLOR A DEFINIR POR LA D.F.

IMPRIMACION ASFALTICA TIPO "IMPRIDAN 100" O EQUIVALENTE

LAMINA IMPERMEABILIZANTE ASFALTICA DE OXIASFALTO MODIFICADO ARMADA CON FIELTRO DE POLIESTER TIPO ESTERDAN M-30 P MUROS, O EQUIVALENTE.

LAMINA NODULAR DRENANTE DE POLIETILENO DE ALTA DENSIDAD PEAD UNIDA A UN GEOTEXTIL NO TEJIDO DE POLOPROPILENO CALANDRADO TIPO DANODREN H25PLUS O EQUIVALENTE

CAPA SEPARADORA GEOTEXTIL ANTIFINOS TIPO DANOFELT 150 DE DANOSA O EQUIVALENTE

BANDA DESOLARIZANTE 100 DE CHOVA O EQUIV. e: 4mm.

CORTE ANTICAPILARIDAD MASTERSEAL 525 DE BETTOR O EQUIVALENTE

PERFIL METALICO PARA FIJACION LAMINAS

…EABILIZACIÓN DE …BORDE DE …SANITARIA

IMPRIMACION ASFALTICA TIPO "IMPRIDAN 100" O EQUIVALENTE

LAMINA IMPERMEABILIZANTE ASFALTICA DE OXIASFALTO MODIFICADO ARMADA CON FIELTRO DE POLIESTER TIPO ESTERDAN M-30 P MUROS, O EQUIVALENTE.

…EABILIZACIÓN DE …E H.A. Y DE …S DE CÁMARA …A

LAMINA NODULAR DRENANTE DE POLIETILENO DE ALTA DENSIDAD PEAD UNIDA A UN GEOTEXTIL NO TEJIDO DE POLOPROPILENO CALANDRADO TIPO DANODREN H25PLUS O EQUIVALENTE

CAPA SEPARADORA GEOTEXTIL ANTIFINOS TIPO DANOFELT 150 DE DANOSA O EQUIVALENTE

TUBO DE DRENAJE #160 DE POLIETILENO CORRUGADO EXTERIOR Y LISO INTERIOR CON PERFORACIONES EN TODO SU PERIMETRO "ODI-BAKAR" O EQUIVALENTE"

GRAVA DE DRENAJE

-TERRENO NATURAL COMPACTADO-

ZAPATA CORRIDA, SEGUN PLANOS DE CIMENTACIÓN.

VIGA DE ATADO CB.2.1, SEGUN PLANOS DE CIMENTACIÓN.

FORJADO AUTOPORTANTE 20+5 VIGUETA PRETENSADA DOBLE T, T18 BOVEDILLA CERÁMICA / POLIESTIRENO h=20, INTEREJE 60cm ø15x20x6 B500T e:5 cm

VIGUETA PRETENSADA T18

2x ARMADURA DE DESCUELGUE 2x ARMADURA DE ENLACE

MACIZADO SEGÚN CORTANTE MIN.10cm

PILAR PERFIL METÁLICO SEGÚN PLANO DE REPLANTEO

FORJADO 20+5 VIGUETA PRETENSADA T18 BOVEDILLA CERÁMICA/POLIESTIRENO h=20 INTEREJE 60cm

+7,27 CORONACIÓN

+7,07 CUBIERTA

+6,82 C.S.F.

+6,57 C.I.F.

RODAPIE DE DM LACADO COLOR A DEFINIR POR LA D.F., ENRASADO A PARAMENTO VERTICAL.

BANDA CONTINUA PERIMETRAL DE ESPUMA DE POLIETILENO.

REVESTIMIENTO DECORATIVO PARA SUELO DE 3,5MM. DE ESPESOR FORMADO POR LA APLICACIÓN DE 5 MANOS DE MICROHORMIGÓN BICOMPONENTE Y DISPOSICIÓN DE MALLA DE FIBRA DE VIDRIO ENTRE LAS 3 MANOS DE BASE Y LAS 2 DE ACABADO, SEGÚN CARTA DE COLORES ESPECIFICA A DEFINIR POR LA D.F., INCLUSO APLICACIÓN SOBRE LA QUINTA MANO DE MICROHORMIGÓN DE 2 MANOS DE HIDROFUGANTE Y 2 MANOS DE POLIURETANO. APLICADO Y CERTIFICADO POR LA EMPRESA ARTESTUCO CON LLANA METÁLICA FLEXIBLE EN PASADAS SUCESIVAS HASTA CONSEGUIR EL EFECTO ESTÉTICO DESEADO.

RECRECIDO DEL SOPORTE DE PAVIMENTOS SOBRE SOLERA RADIANTE DE MORTERO AUTONIVELANTE DE ALTA CALIDAD ARDEX K14 MIX PARA RECIBIR, IGUALAR Y NIVELAR CONFIGURANDO EL SOPORTE, PERFECTAMENTE HORIZONTAL, DEL MICROHORMIGÓN COMO PAVIMENTO FINAL.

RECRECIDO DEL SOPORTE DE PAVIMENTOS, DE 6CM(4.4CM SOBRE TANGENCIA SUPERIOR A TUBERÍAS DE SUELO RADIANTE) SOBRE TRAMA DE TUBERÍA RADIANTE CON MORTERO DE CEMENTO Y ARENA DE RÍO LAVADA MÁX 8MM DE DIÁMETRO, MAESTREADO, CON ADITIVO FLUIDIFICANTE Y RETARDANTE BARBI (EN LA PROPORCIÓN: 1.5% DEL PESO DEL CEMENTO).

TUBERÍAS RETICULADAS DE PEX-A CON BARRERA ANTIOXÍGENO (EVOH) DE CALEFACCIÓN POR SUELO RADIANTE

AISLANTE DE POLIESTIRENO DE 20MM DE ESPESOR Y DENSIDAD 30 KG/M3, Y HOJA DE POLIETILENO BAJO PLACA SUJECCIÓN ANTIVAPOR, CON ZÓCALO O BANDA PERIMETRAL DE ESPUMA DE POLIETILENO.

HOJA DE POLIETILENO BAJO PLACA SUJECCIÓN ANTIVAPOR, CON ZÓCALO O BANDA PERIMETRAL DE ESPUMA DE POLIETILENO.

+3,82 P.PRIMERA

FORJADO 20+5 VIGUETA PRETENSADA T18 BOVEDILLA CERÁMICA/POLIESTIRENO h=20 INTEREJE 60cm

+3,74 C.S.F.

+3,49 C.I.F.

FALSO TECHO FORMADO POR UNA ESTRUCTURA DE PERFILES DE CHAPA DE ACERO GALVANIZADA, DE 47 MM. DE ANCHO, SEPARADOS ENTRE ELLOS 400 MM, SUSPENDIDOS DEL FORJADO POR MEDIO DE "HORQUILLAS" Y VARILLA ROSCADA Ø 6 MM, Y ENCAJADOS EN EL PERFIL CLIP. A ESTA ESTRUCTURA DE PERFILES, SE ATORNILLA UNA PLACA PLADUR TIPO N DE 13 MM DE ESPESOR

PINTURA PLÁSTICA LISA MATE COLOR A DEFINIR POR LA D.F.

AISLAMIENTO DE PANEL DE LANA DE ROCA 40kg/m3 DE DENSIDAD Y 50mm DE ESPESOR CON SISTEMA 221 FIXROCK DE ROCKWOOL O EQUIV. FIJADO CON MORTERO ADHESIVO e:5mm

FOSEADO PERIMETRAL DE ESCAYOLA.

ANGULAR-RODAPIE CORRIDO DE BORDE L30.20.3 GALVANIZADO Y ESMALTADO RAL COLOR A ELEGIR POR LA D.F.

BANDA CONTINUA PERIMETRAL DE ESPUMA DE POLIETILENO.

REVESTIMIENTO DECORATIVO PARA SUELO DE 3,5MM. DE ESPESOR FORMADO POR LA APLICACIÓN DE 5 MANOS DE MICROHORMIGÓN BICOMPONENTE Y DISPOSICIÓN DE MALLA DE FIBRA DE VIDRIO ENTRE LAS 3 MANOS DE BASE Y LAS 2 DE ACABADO, SEGÚN CARTA DE COLORES ESPECIFICA A DEFINIR POR LA D.F., INCLUSO APLICACIÓN SOBRE LA QUINTA MANO DE MICROHORMIGÓN DE 2 MANOS DE HIDROFUGANTE Y 2 MANOS DE POLIURETANO. APLICADO Y CERTIFICADO POR LA EMPRESA ARTESTUCO CON LLANA METÁLICA FLEXIBLE EN PASADAS SUCESIVAS HASTA CONSEGUIR EL EFECTO ESTÉTICO DESEADO.

RECRECIDO DEL SOPORTE DE PAVIMENTOS SOBRE SOLERA RADIANTE DE MORTERO AUTONIVELANTE DE ALTA CALIDAD ARDEX K14 MIX PARA RECIBIR, IGUALAR Y NIVELAR CONFIGURANDO EL SOPORTE, PERFECTAMENTE HORIZONTAL, DEL MICROHORMIGÓN COMO PAVIMENTO FINAL.

RECRECIDO DEL SOPORTE DE PAVIMENTOS, DE 6CM(4.4CM SOBRE TANGENCIA SUPERIOR A TUBERÍAS DE SUELO RADIANTE) SOBRE TRAMA DE TUBERÍA RADIANTE CON MORTERO DE CEMENTO Y ARENA DE RÍO LAVADA MÁX 8MM DE DIÁMETRO, MAESTREADO, CON ADITIVO FLUIDIFICANTE Y RETARDANTE BARBI (EN LA PROPORCIÓN: 1.5% DEL PESO DEL CEMENTO).

TUBERÍAS RETICULADAS DE PEX-A CON BARRERA ANTIOXÍGENO (EVOH) DE CALEFACCIÓN POR SUELO RADIANTE

AISLANTE DE POLIESTIRENO DE 20MM DE ESPESOR Y DENSIDAD 30 KG/M3, Y HOJA DE POLIETILENO BAJO PLACA SUJECCIÓN ANTIVAPOR, CON ZÓCALO O BANDA PERIMETRAL DE ESPUMA DE POLIETILENO.

AISLAMIENTO TÉRMICO MEDIANTE PLANCHA DE POLIESTIRENO EXTRUIDO (XPS) DE 50 MM DE ESPESOR Y DENSIDAD 35,5 KG/M3.

HOJA DE POLIETILENO BAJO PLACA SUJECCIÓN ANTIVAPOR, CON ZÓCALO O BANDA PERIMETRAL DE ESPUMA DE POLIETILENO.

+0,72 P.BAJA

FORJADO 20+5 VIGUETA PRETENSADA T18 BOVEDILLA CERÁMICA/POLIESTIRENO h=20 INTEREJE 60cm

+0,60 C.S.F.

+0,35 C.I.F.

±0,00

ACCESO EXT. VIVIENDA

ARMADURA DE NEGATIVOS SEGÚN PLANO DE FORJADO SANITARIO

FORJADO AUTOPORTANTE 20+5 VIGUETA PRETENSADA DOBLE T, T18 BOVEDILLA CERÁMICA / POLIESTIRENO h=20, INTEREJE 60cm ø15x20x6 B500T e:5 cm

VIGUETA PRETENSADA T18

2ø8 POR VIGUETA (UNO A CADA LADO) ENTREGA DE 55cm LONG. TOTAL: 120cm

MACIZADO SEGÚN CORTANTE MIN.10cm

TUBO PVC ø40mm DISPUESTO COMO ENCOFRADO PERDIDO CADA 75cm ENTRE ENCOFRADOS. COLOR GRIS HORMIGÓN (SECCION TOTAL DE VENTILACION =AREA DE PLANTA/500)

1 PIE DE FABRICA DE LADRILLO PERFORADO CON MORTERO 1:5

PILAR DE H.A. (VER CUADRO DE PILARES)

MURETA DE 1 PIE DE LADRILLO PERFOR.

- CÁMARA SANITARIA VENTILADA -

INT.

-0,88 C.S.Z.

VIGA DE ATADO & CENTRADORA HA-25 B500S SEGÚN PLANOS DE CIMENTACIÓN

HORMIGÓN DE LIMPIEZA HM-20

-1,38 C.I.Z.

-1,48 H. LIMPIEZA

SECCION A-A. CERRAMIENTO VERTICAL EXTERIOR-COCINA DE PLANTA BAJA Y EXTERIOR-ÁREA DE JUEGOS Y ESTUDIO DE PLANTA PRIMERA.

SECCIÓN AA'. CERRAMIENTO EXTERIOR

Section Detail

© raúl del VALLE

© raúl del VALLE

raúl del VALLE

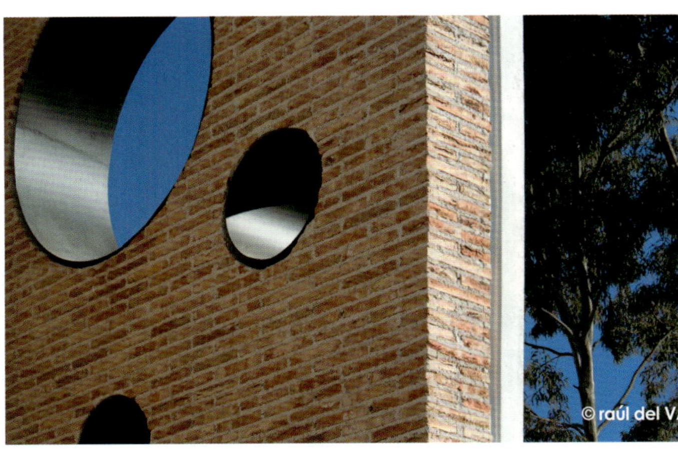

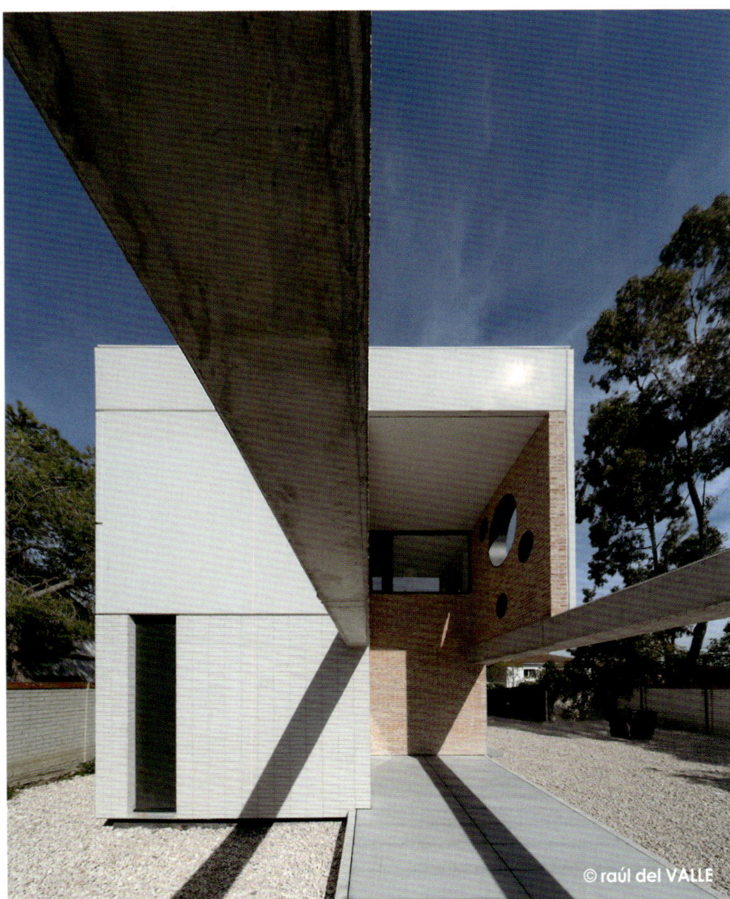

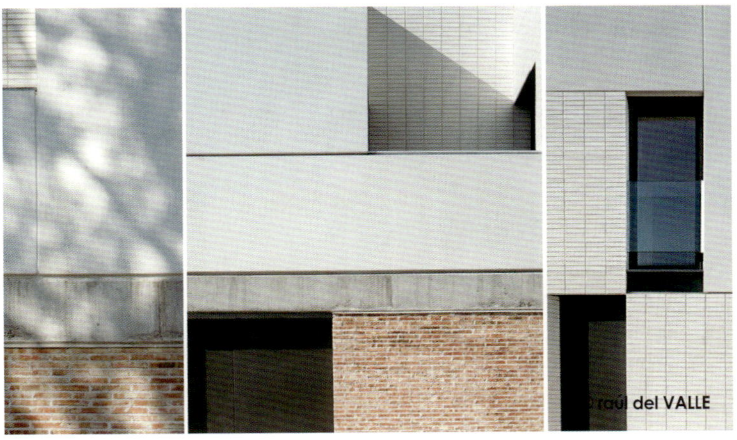

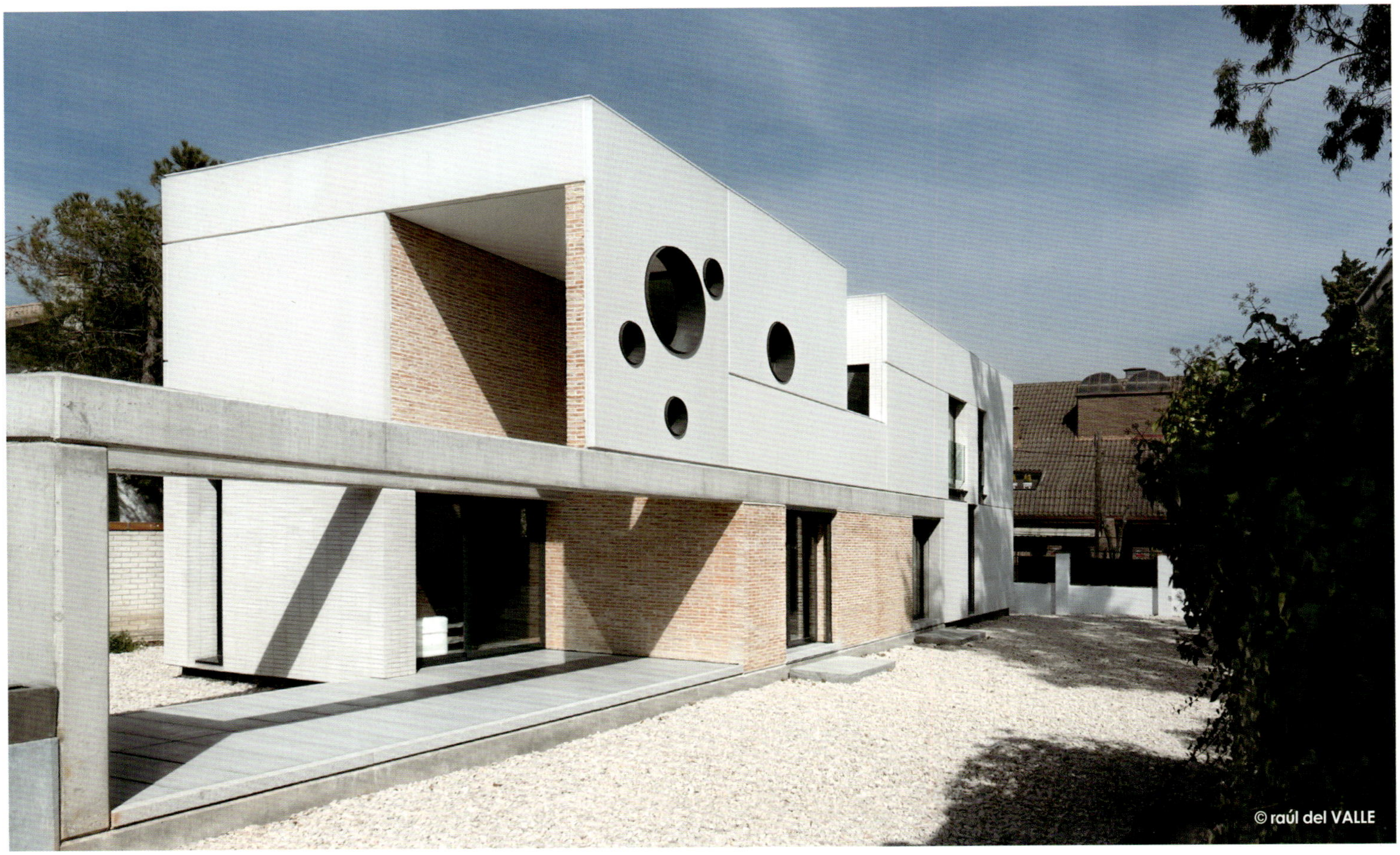
© raúl del VALLE

©raúl del VALLE

LT House, Dong Nai Province, Vietnam
Tropical Space Co.,Ltd

© Trieu Chien

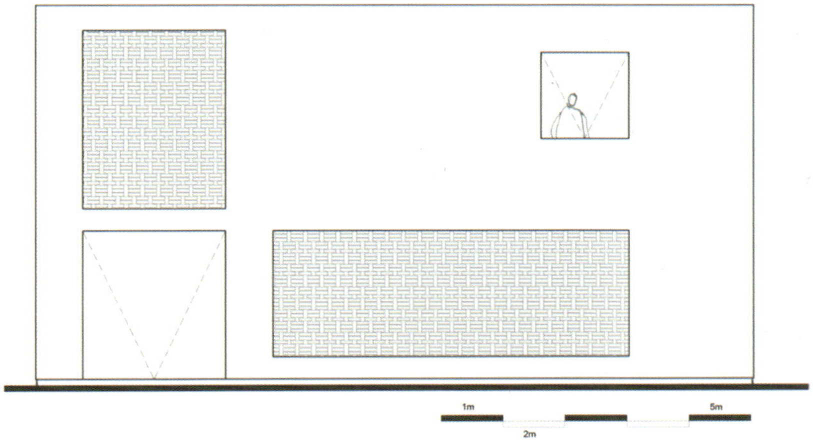

1m 5m
2m

Elevation A

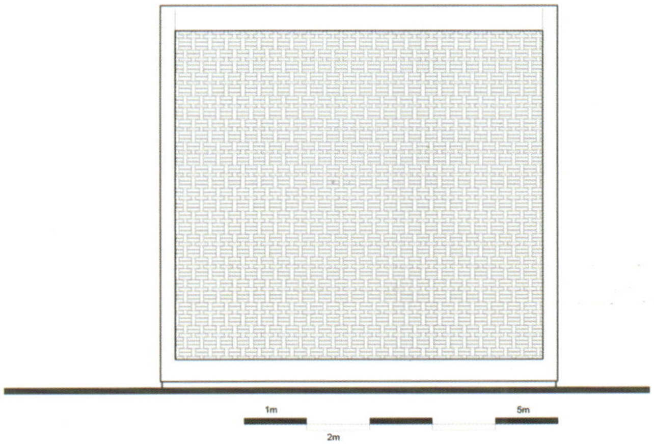

1m 5m
2m

Elevation B

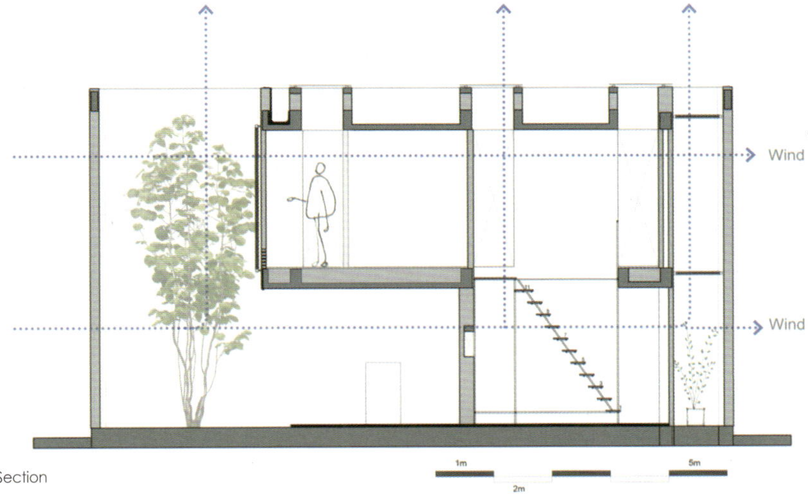

Wind

Wind

Views from the Window

Section

1m 5m
2m

1m 5m
2m

© Trieu Chien

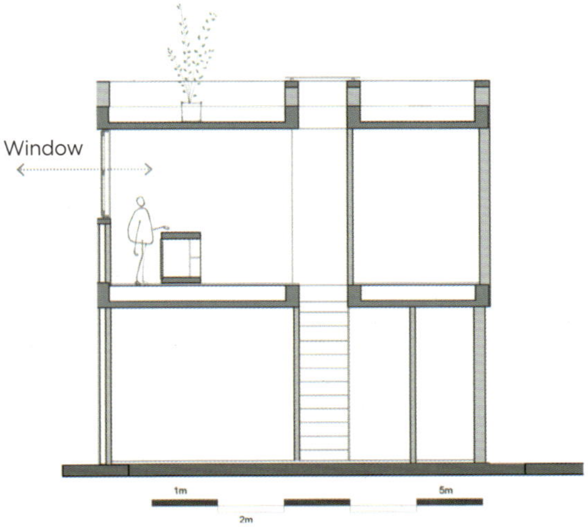

© Trieu Chien

Renovation House RS, Seoul, South Korea

SUPA Schweitzer Song

© Kyungsub Shin

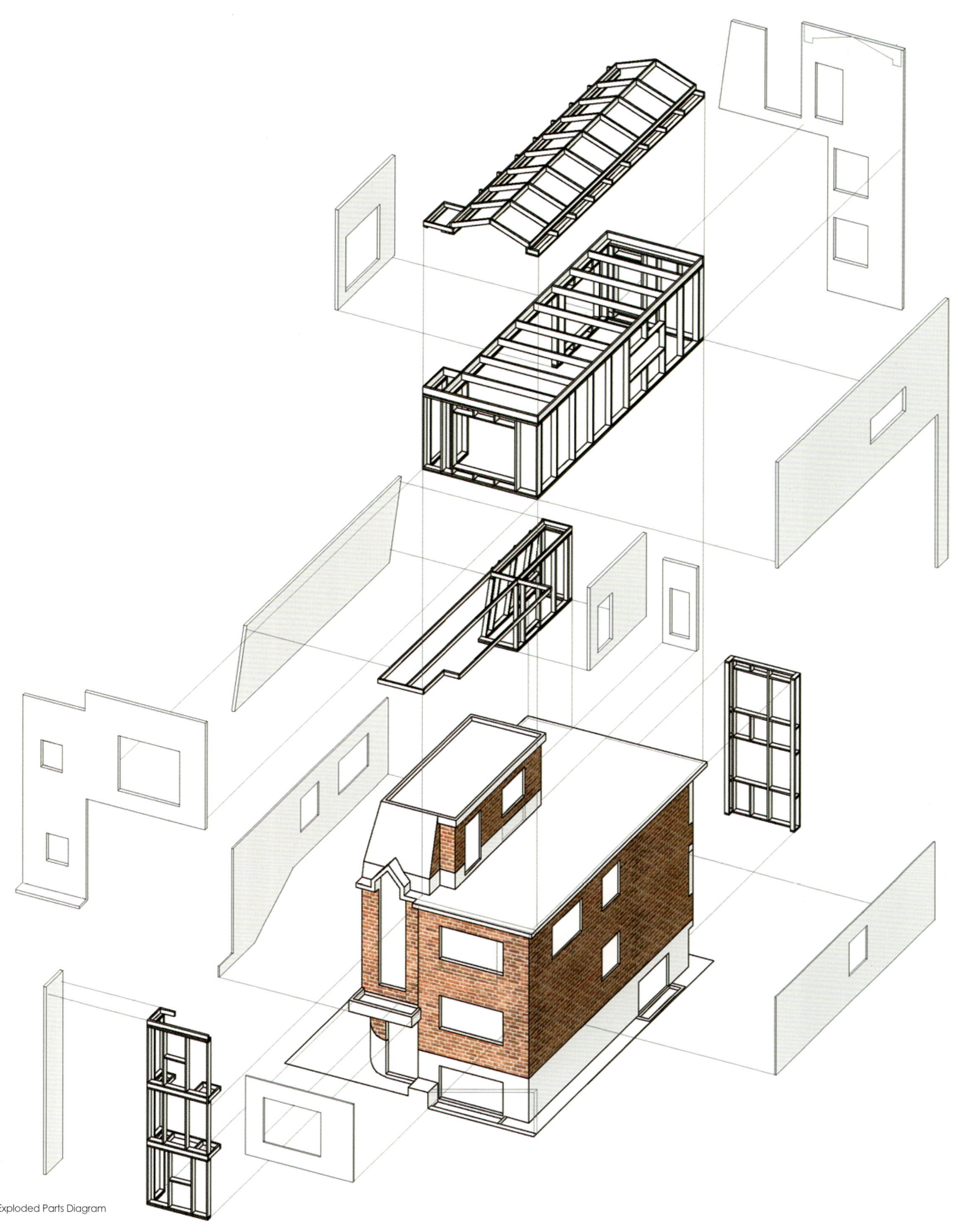

Exploded Parts Diagram

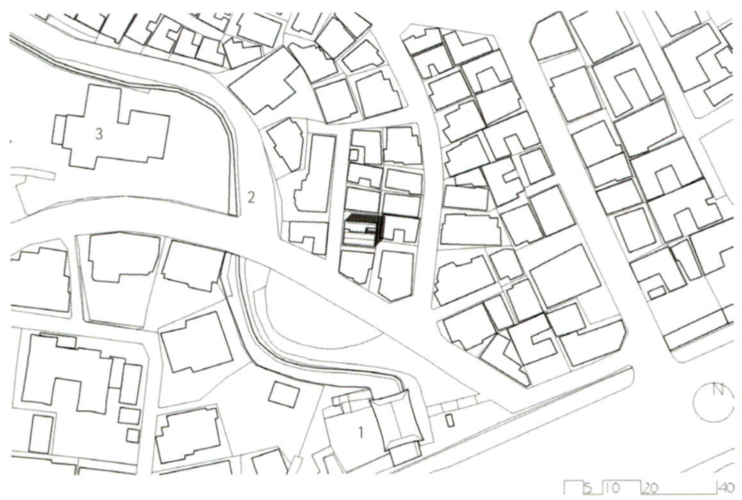

Renovation

Extension

Existing

SITE PLAN

1. Hyehwa gate
2. Historic city wall
3. Seoul major's residence

LAYER

Layering: To express the different stages of the transformation of the existing house into something new while respecting the old we layered the interventions visibly on the façade. While the façade of the remaining unit was just painted white, the renovated part was plastered over. The wooden extension part again was plastered slightly projecting out of the existing parts.

FLOOR PLAN

Second Floor

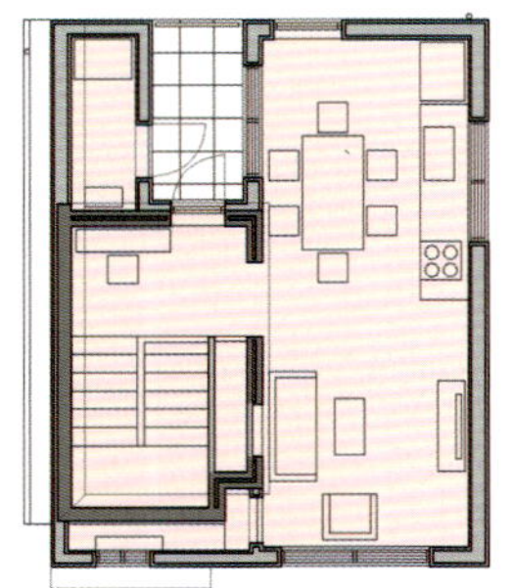

Third Floor

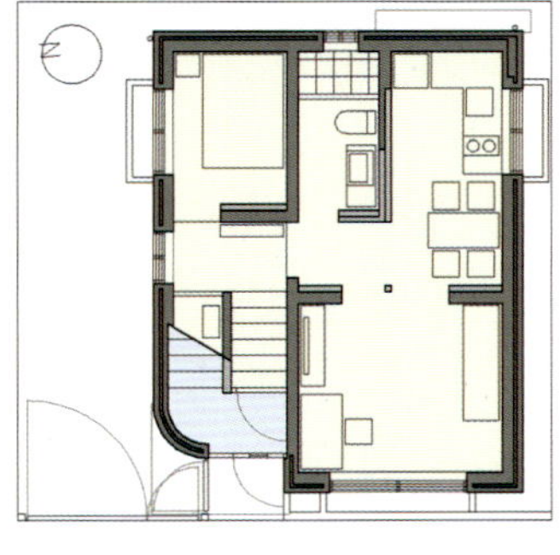

Basement Floor

First Floor

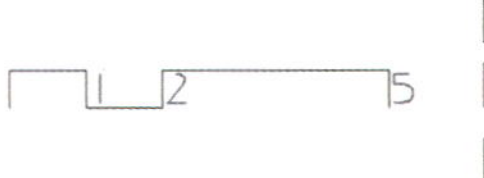

Unit 1 / Basement

Unit 2 / 1st Floor

Existing Brick Structure

Unit 3 / Extension

Stairwell

Extension Wood Structure

UNIT DIAGRAM

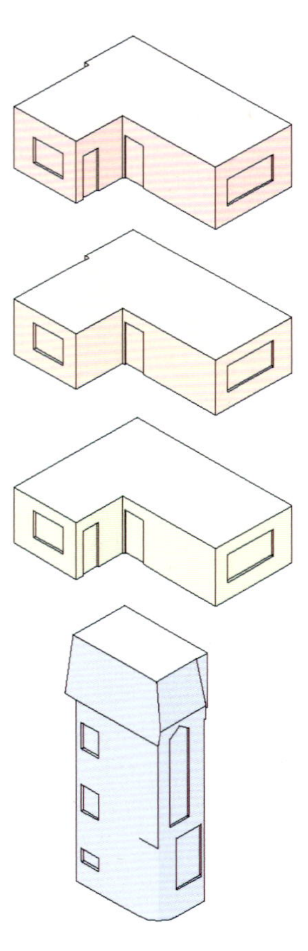

Before

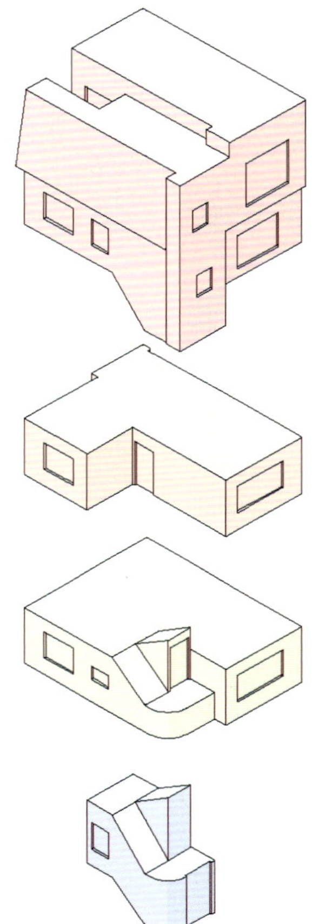

After

© Kyungsub Shin

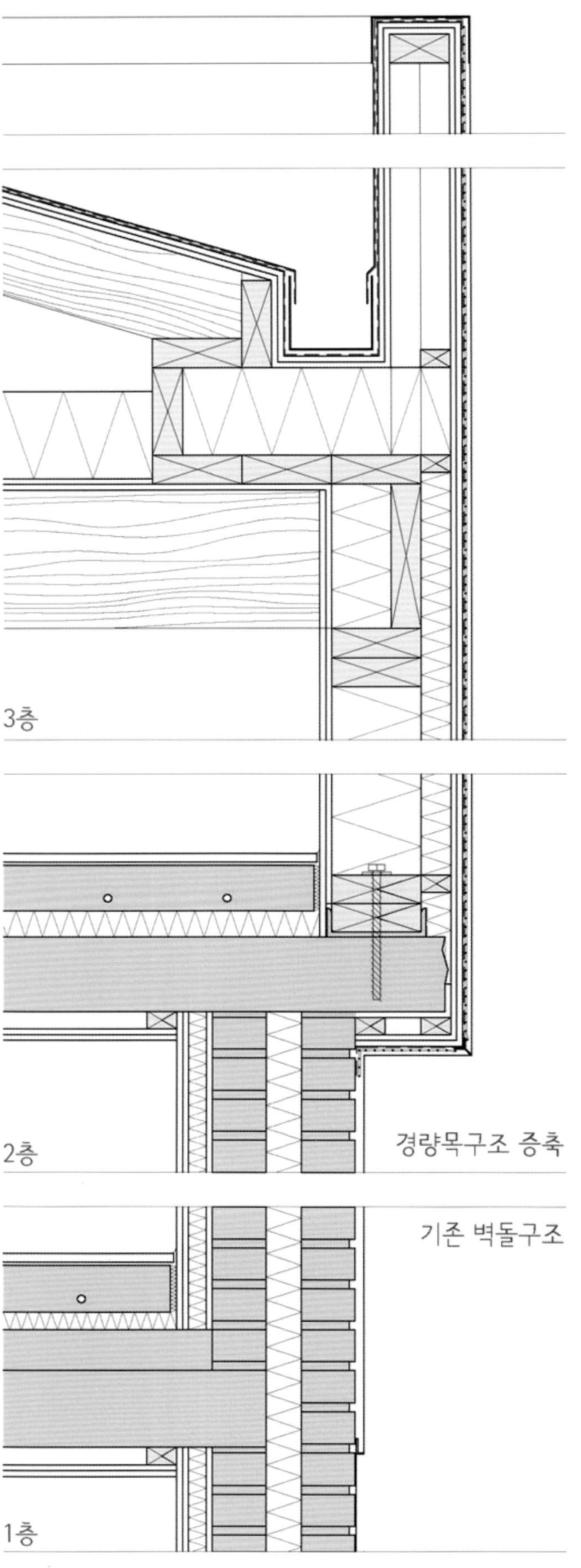

Section Detail

© Kyungsub Shin

Seaside Single House, Toscana, Italy

modostudio

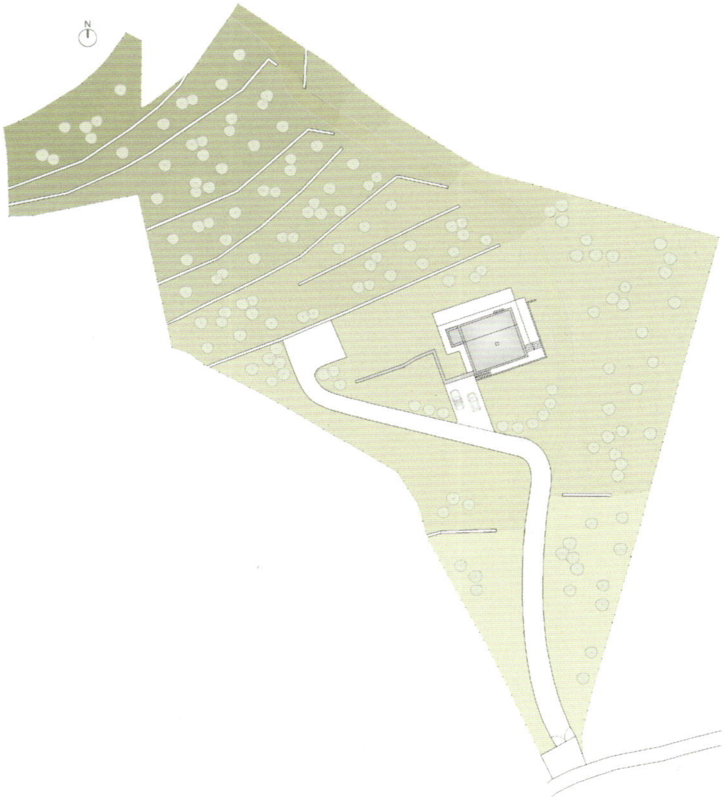

FLOOR PLAN

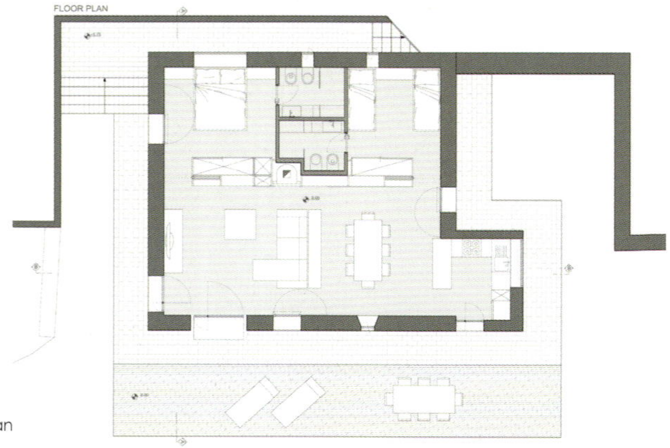

Floor Plan

ROOF PLAN

General Plan

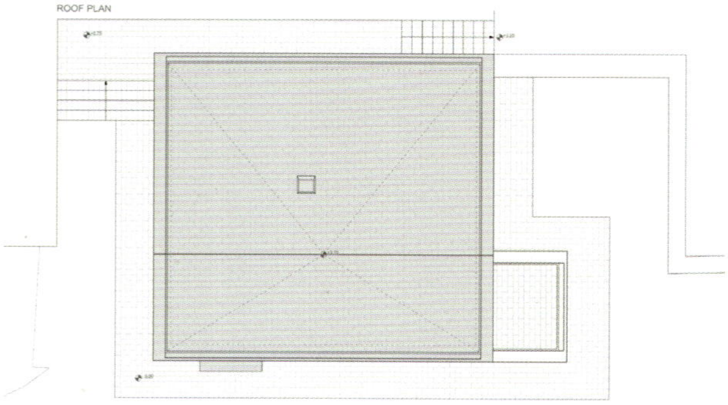

Roof Plan

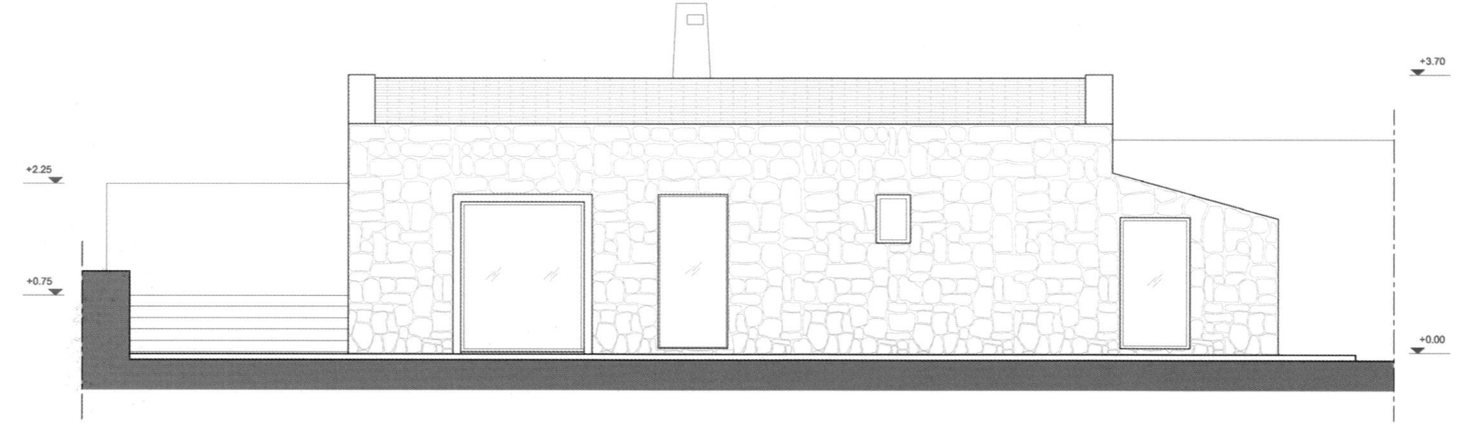

North Elevation

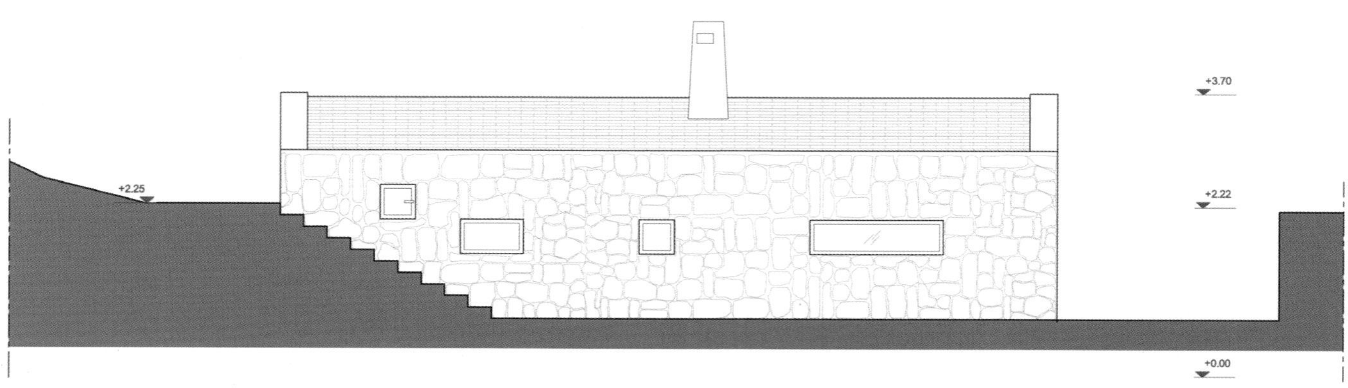

South Elevation

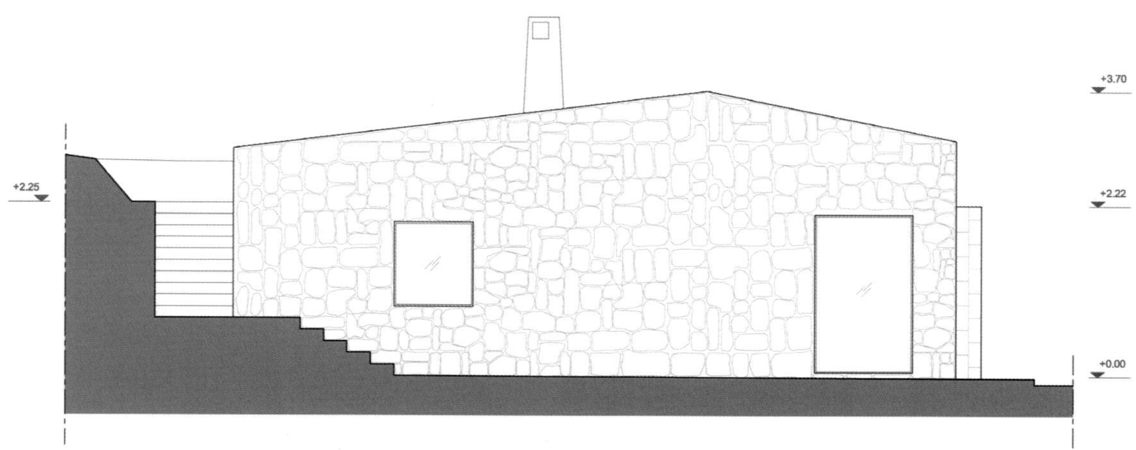

East Elevation

Termitary House, Da Nang City, Vietnam
Tropical Space Co.,Ltd
© Oki Hiroyuki

Elevation A

Elevation B

The sky was visible from the position of working area through the Toplight

Section

Section Diagram

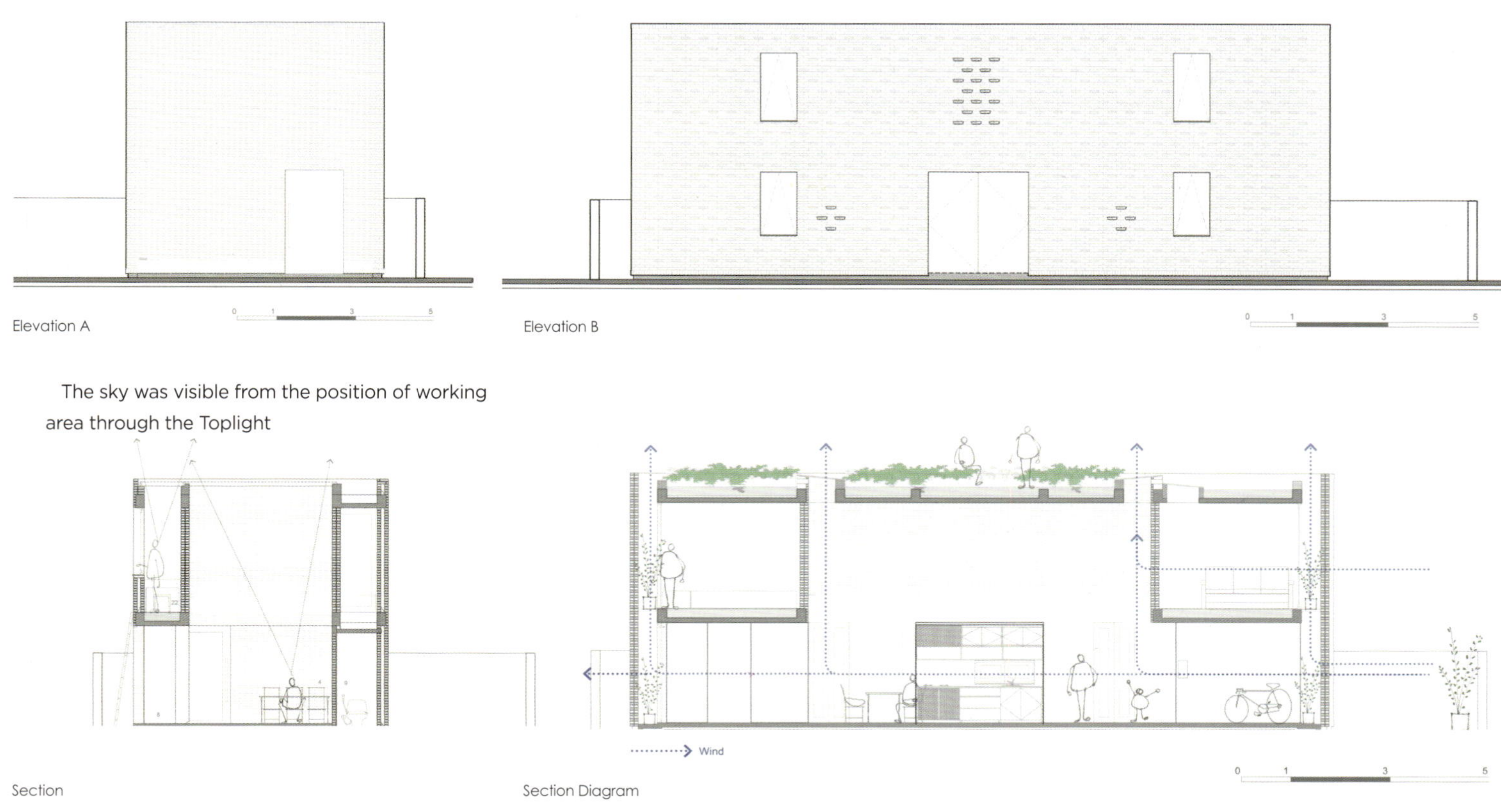

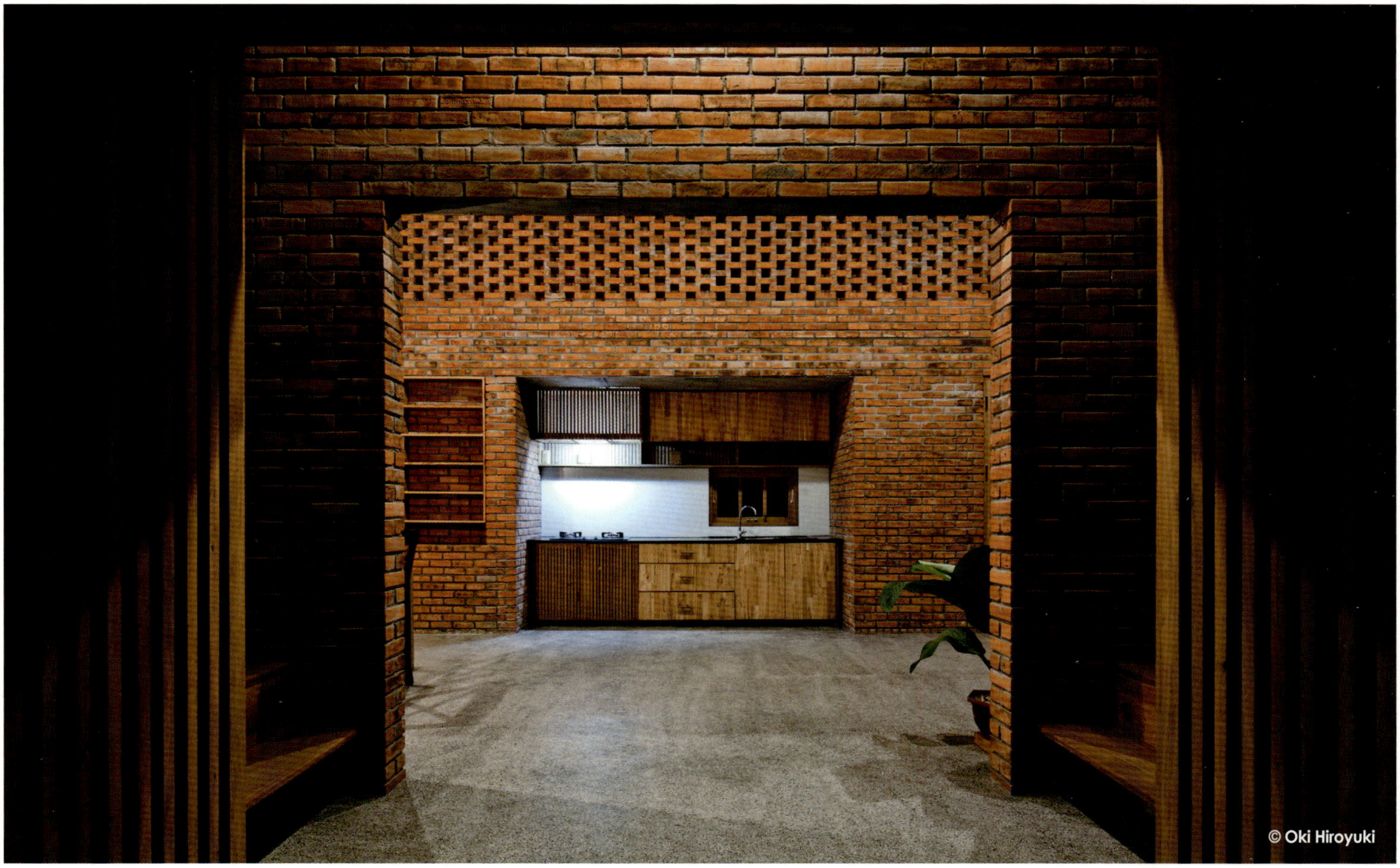

© Oki Hiroyuki

© Oki Hiroyuki

© Oki Hiroyuki

© Oki Hiroyuki

© Oki Hiroyuki

© Oki Hiroyuki

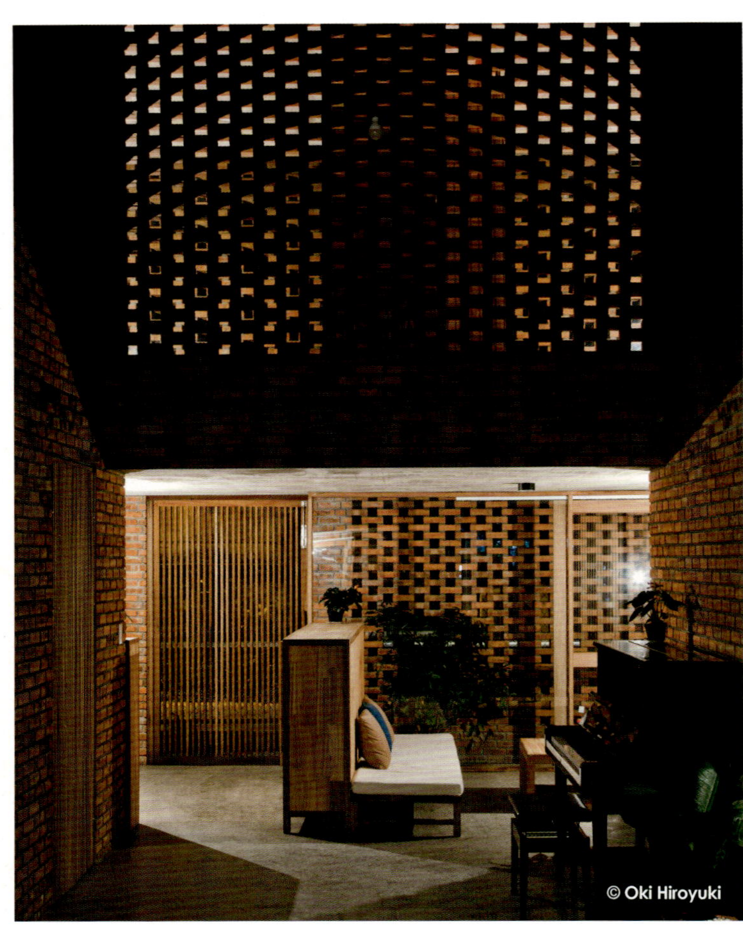

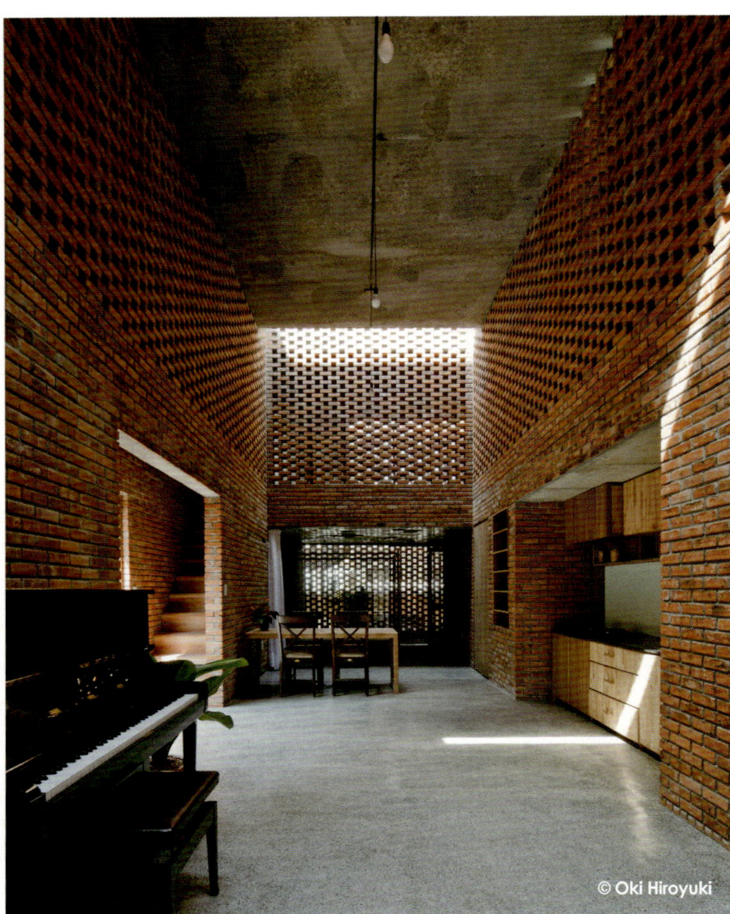

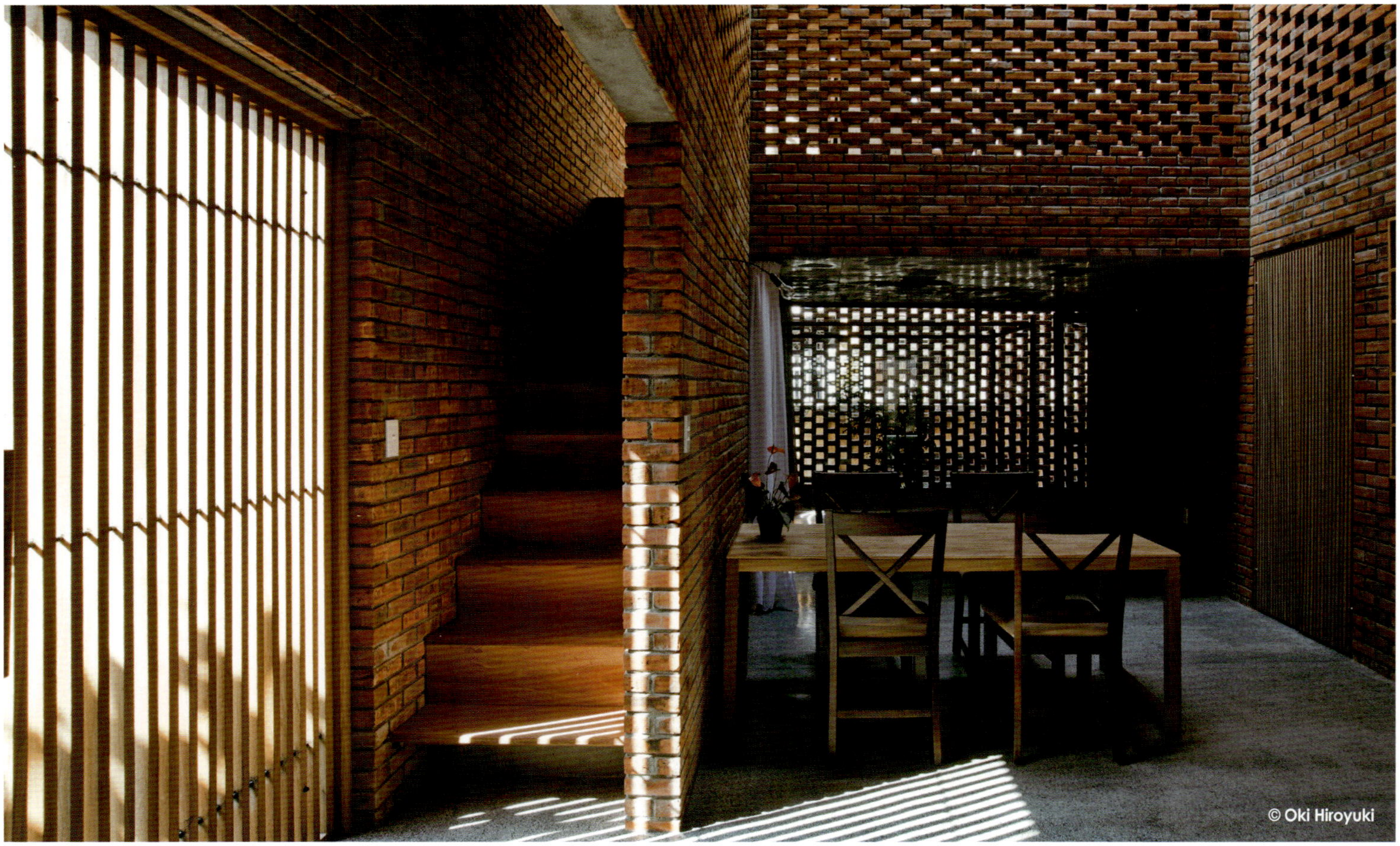
© Oki Hiroyuki

Villa - Villa, Inner Mongolia, China

nARCHITECTS

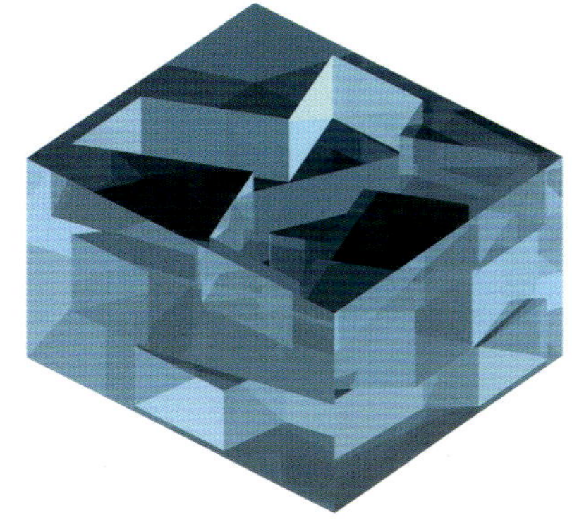

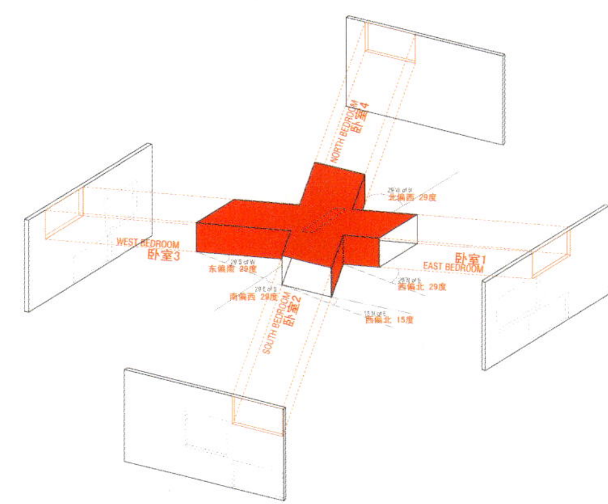

Conceived as a house within a house, our villa combines two distinct spatial and thermal conditions.

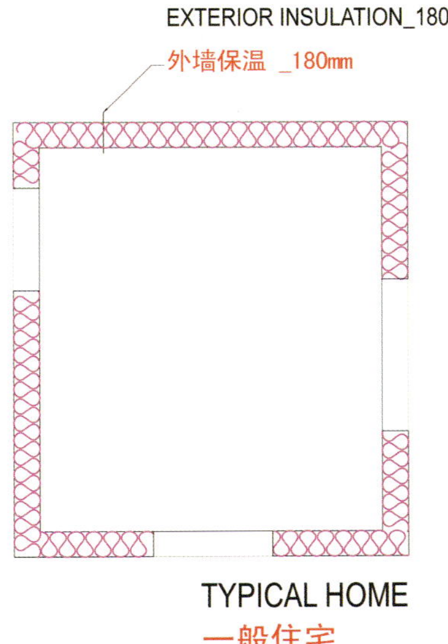

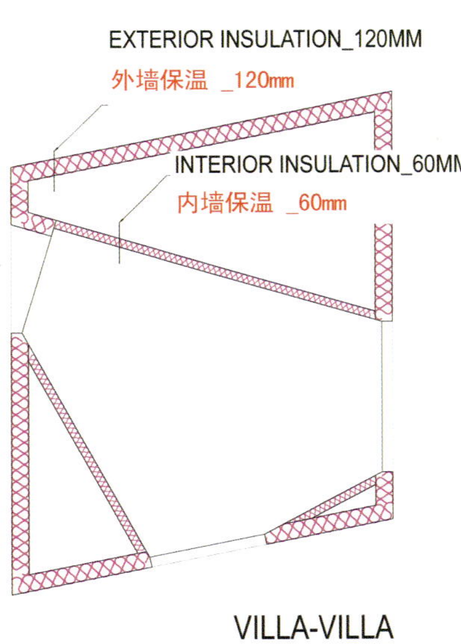

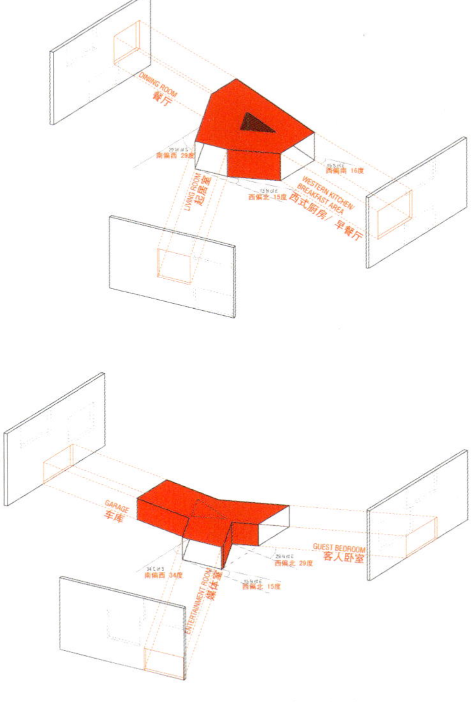

A layered strategy responds to the extreme swings of the Ordos desert climate.

Three single-storey volumes stack on top of each other....

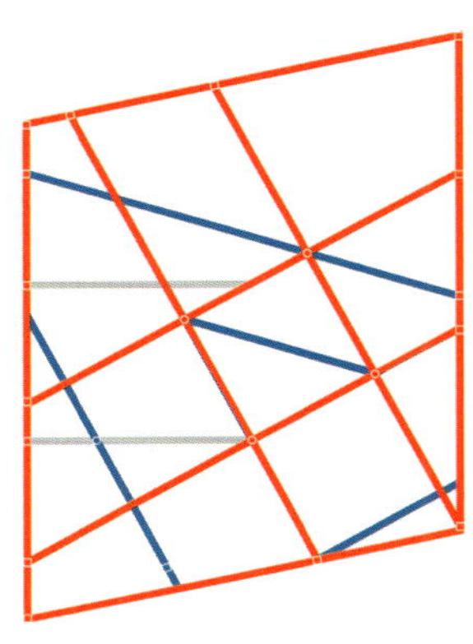

Structural Grid

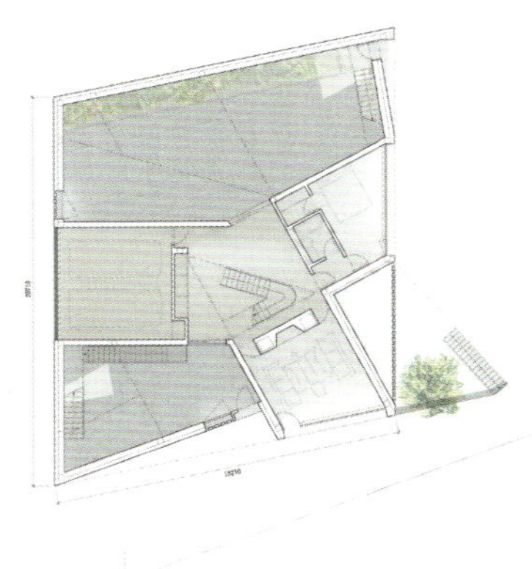

Ground Floor +/ 0m

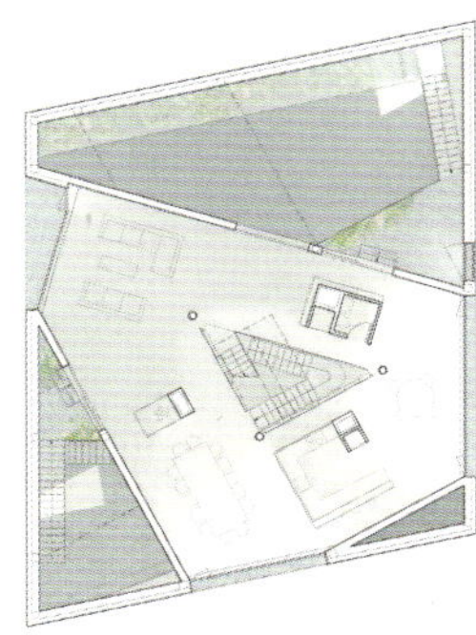

Second Floor + 3.8m

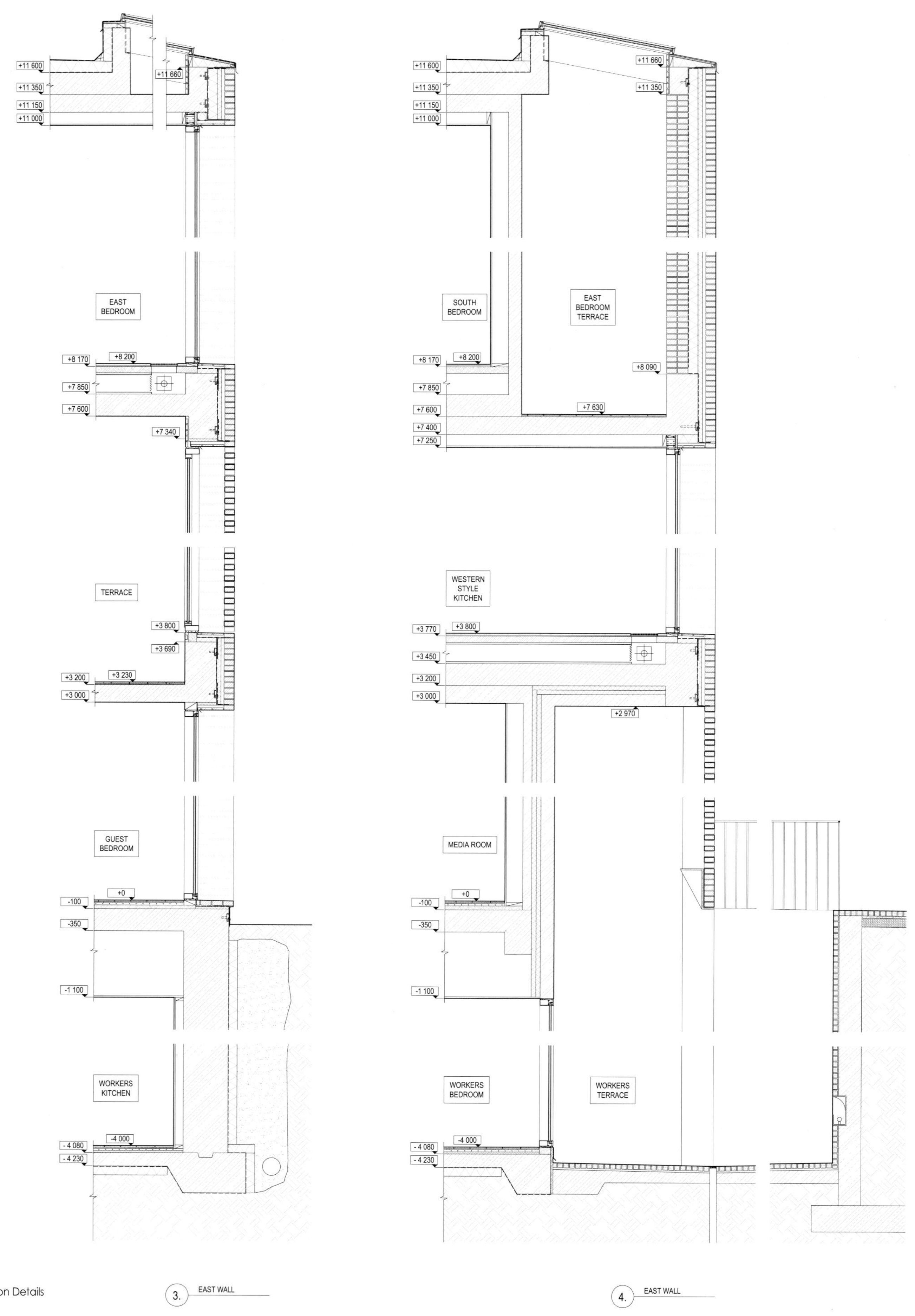

Wall Section Details

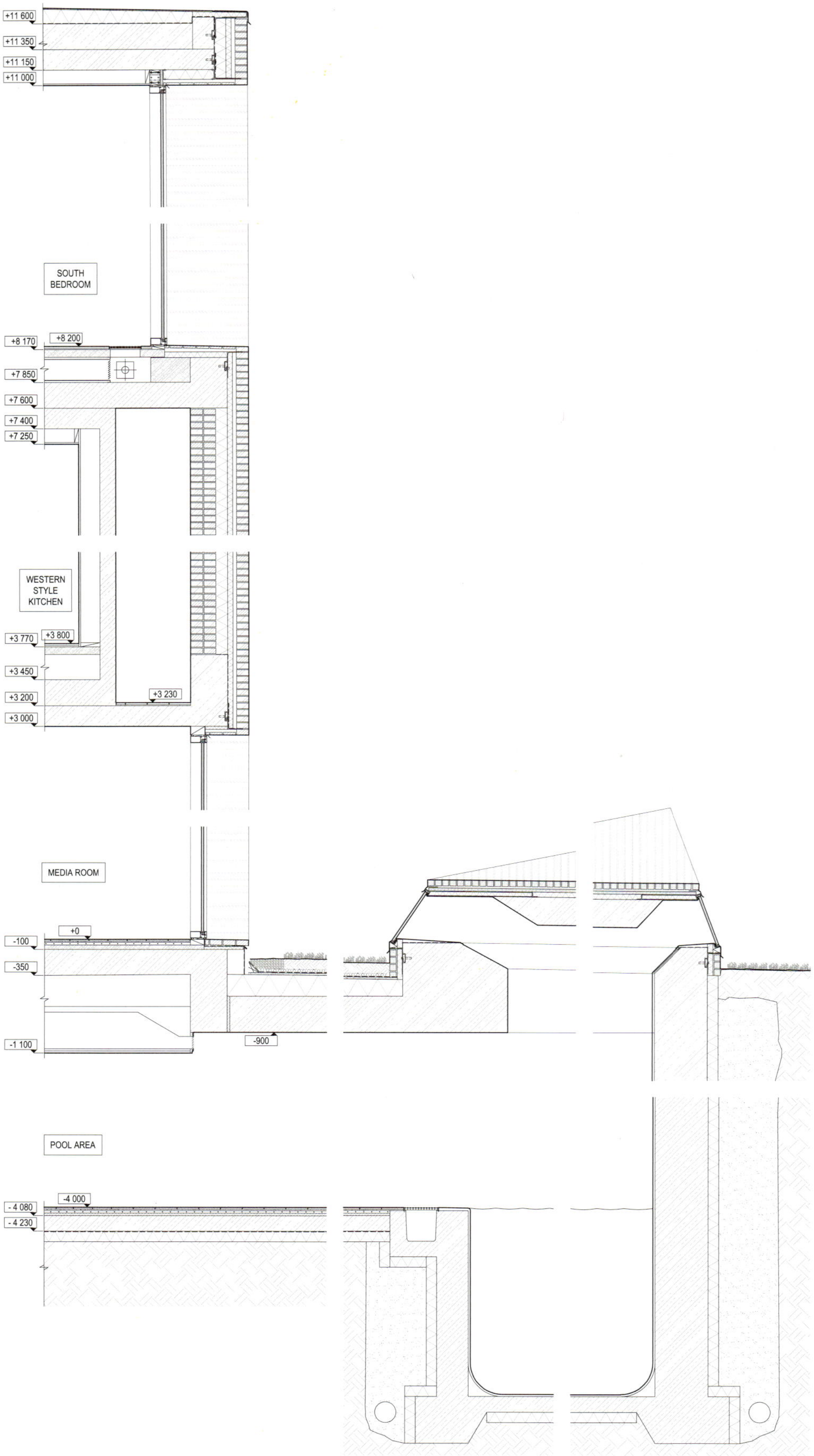

+11 600
+11 350
+11 150
+11 000

SOUTH BEDROOM

+8 170
+8 200
+7 850
+7 600
+7 400
+7 250

WESTERN STYLE KITCHEN

+3 770
+3 800
+3 450
+3 200
+3 230
+3 000

MEDIA ROOM

+0
-100
-350
-1 100
-900

POOL AREA

-4 000
-4 080
-4 230

Wall Section Details (1.) SOUTH WALL

Y House, Beijing, China
Hironori Matsubara

© Misae Hiromatsu

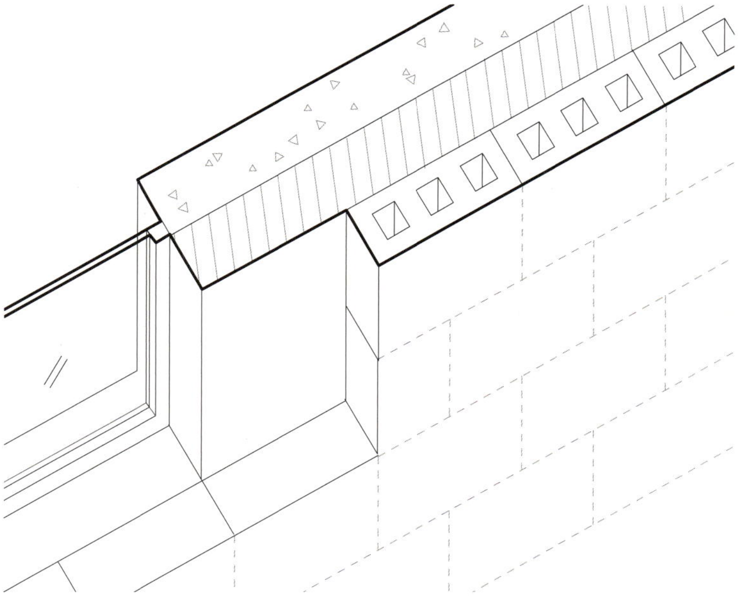

Construction Process

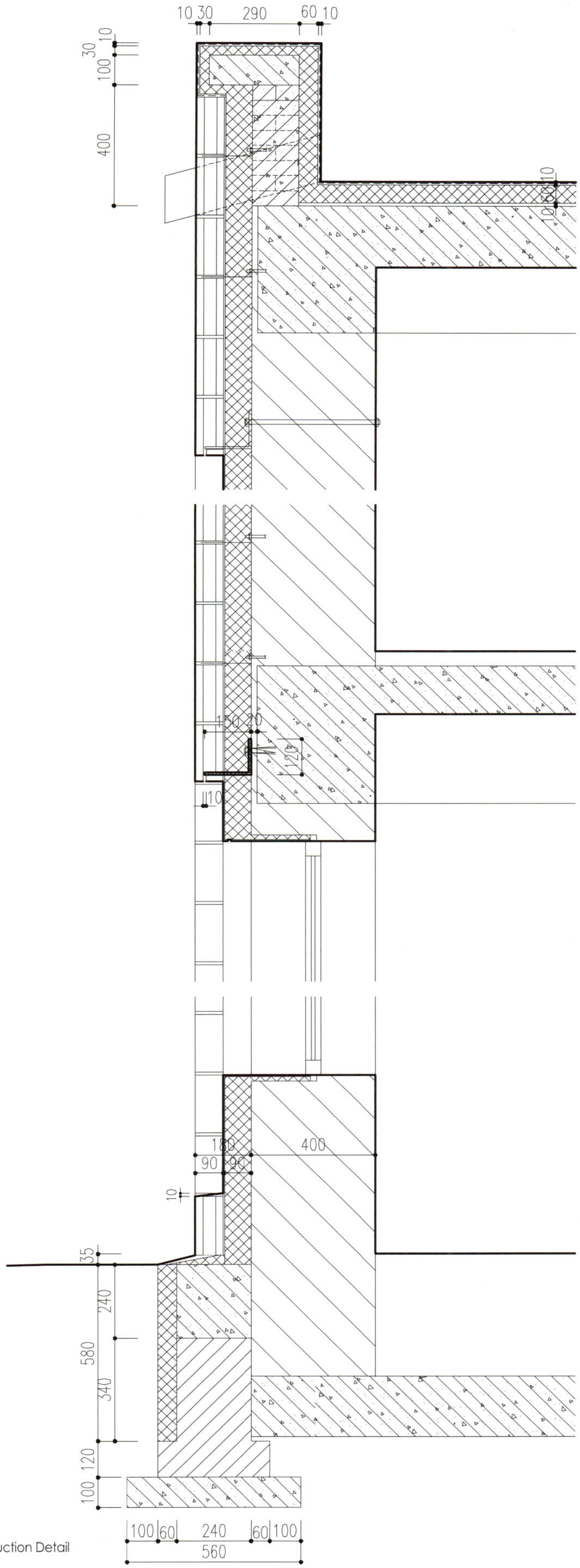

Construction Detail

Construction Process

© Misae Hiromatsu

© Misae Hiromatsu

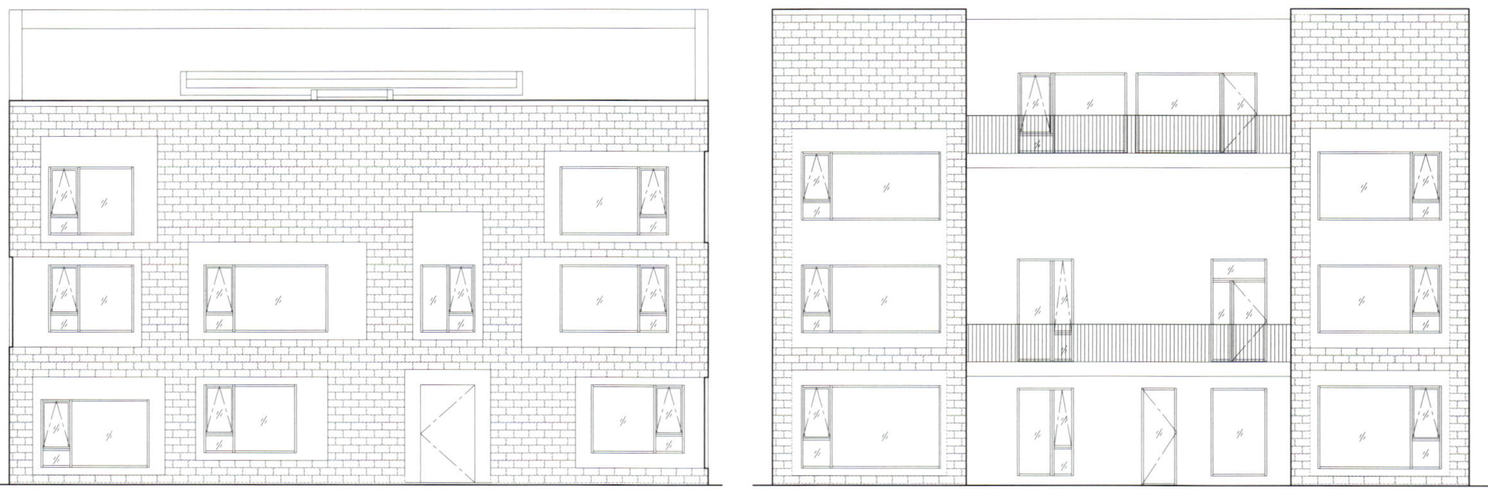

South Elevation

North Elevation

© Misae Hiromatsu

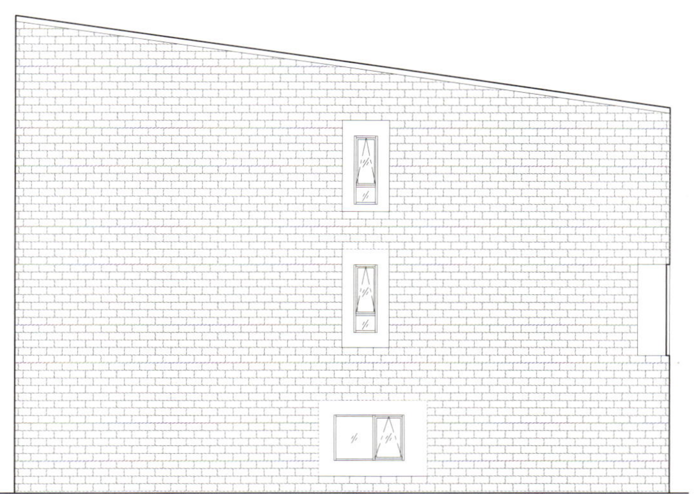

West Elevation

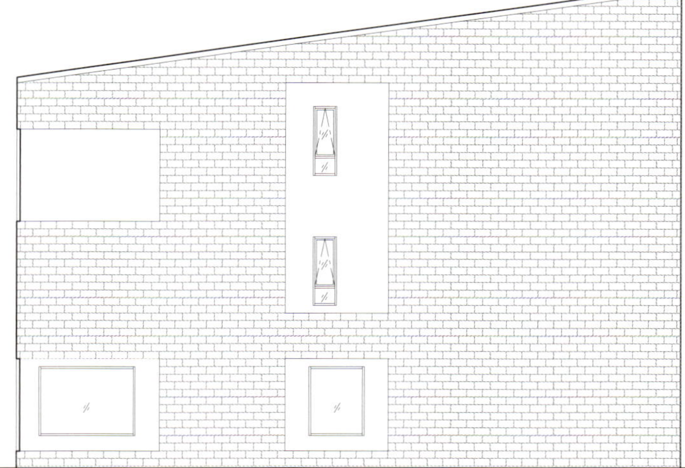

East Elevation

© Misae Hiromatsu

Asma Bahçeler Residences, İzmir, Turkey
M artı D Mimarlık
© ZM Yasa Photography

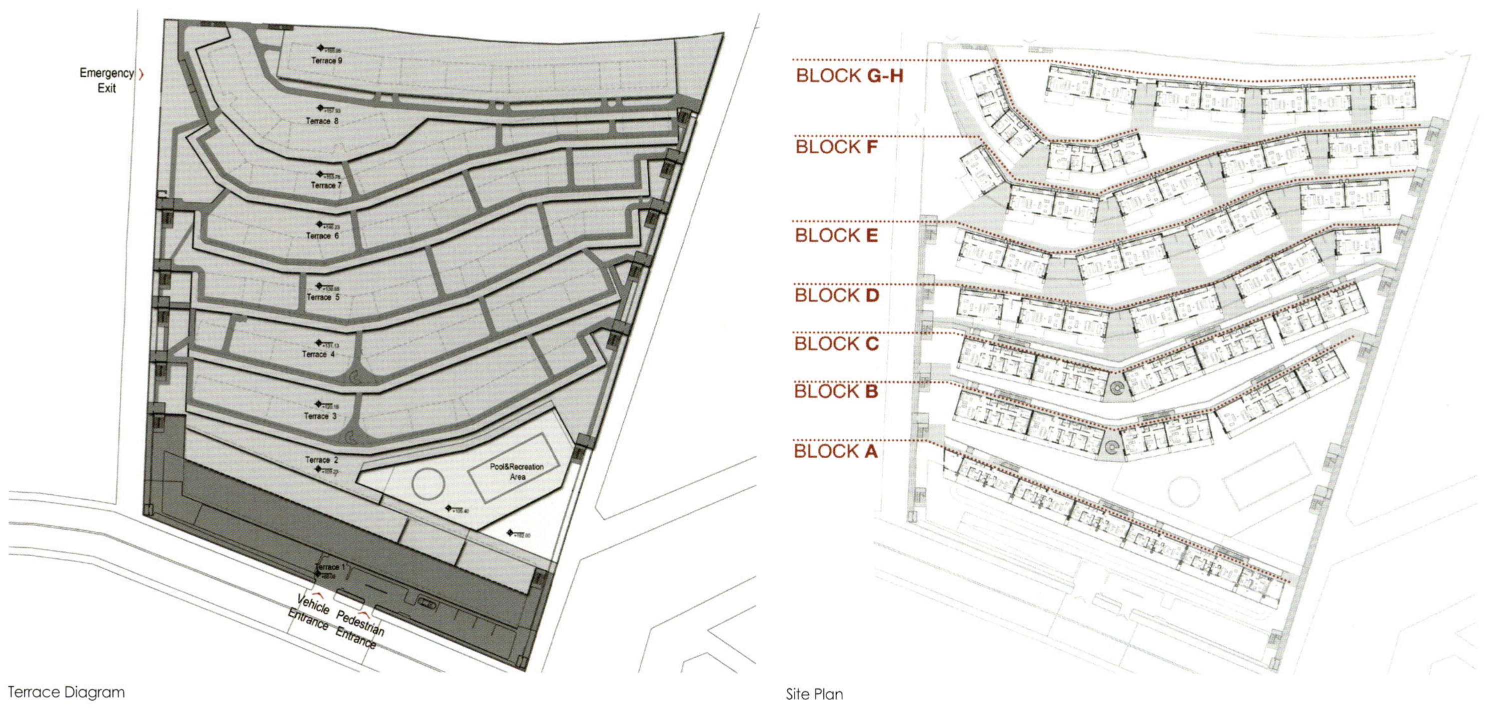

Emergency Exit
Terrace 9
Terrace 8
Terrace 7
Terrace 6
Terrace 5
Terrace 4
Terrace 3
Terrace 2
Pool&Recreation Area
Terrace 1
Vehicle Entrance
Pedestrian Entrance

Terrace Diagram

BLOCK G-H
BLOCK F
BLOCK E
BLOCK D
BLOCK C
BLOCK B
BLOCK A

Site Plan

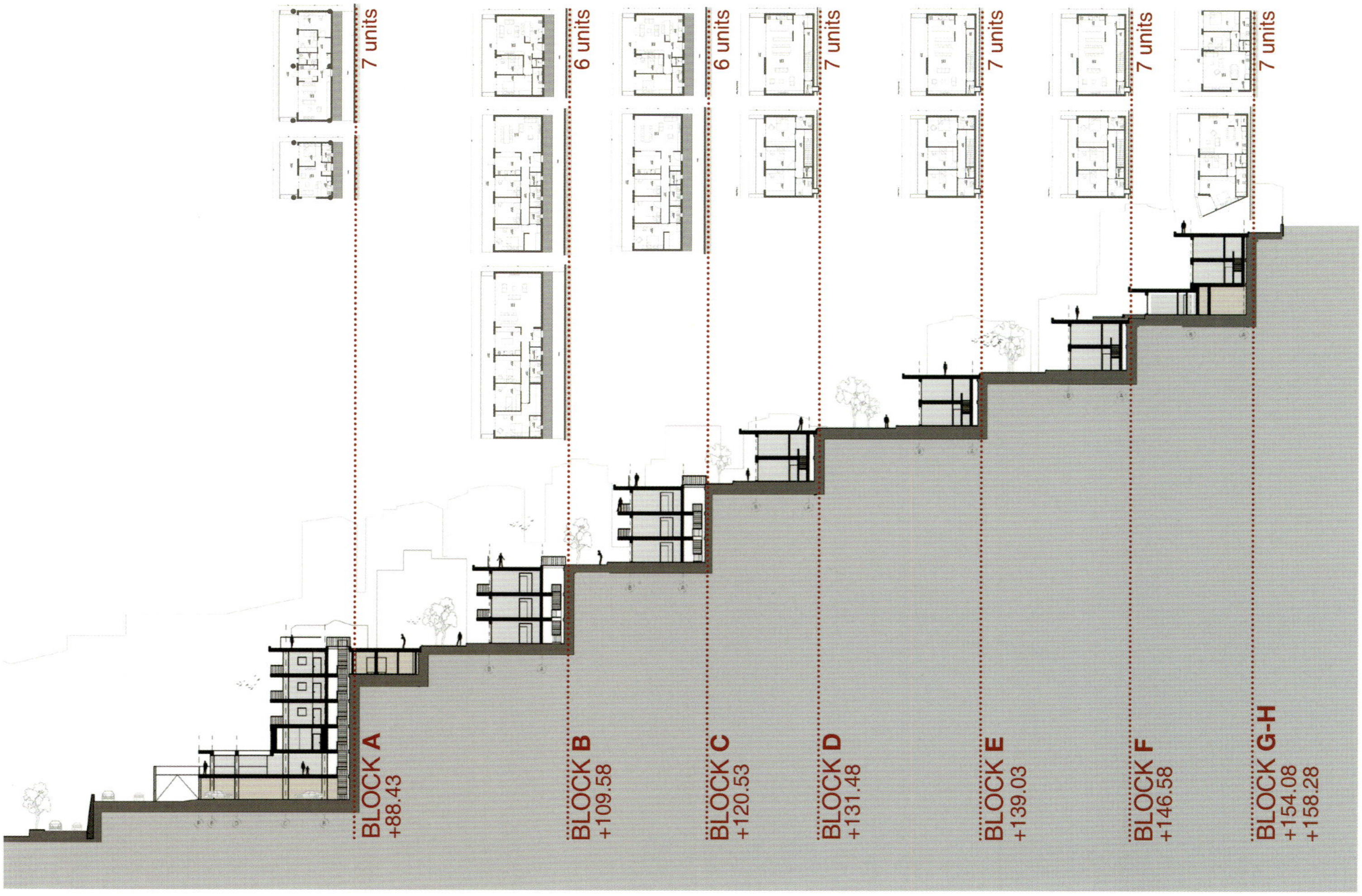

7 units
6 units
6 units
7 units
7 units
7 units
7 units
BLOCK A +88.43
BLOCK B +109.58
BLOCK C +120.53
BLOCK D +131.48
BLOCK E +139.03
BLOCK F +146.58
BLOCK G-H +154.08 +158.28

Building Diagram

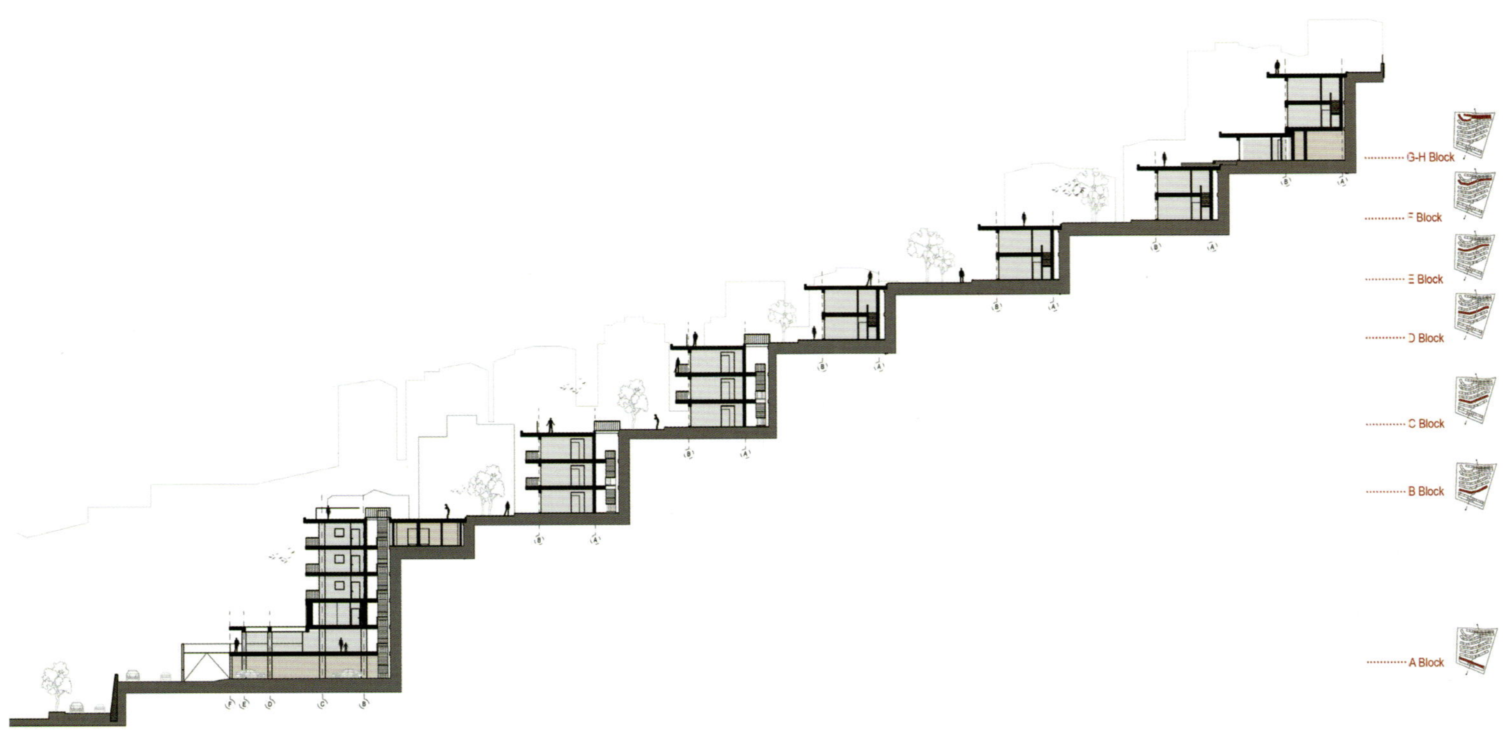

Section

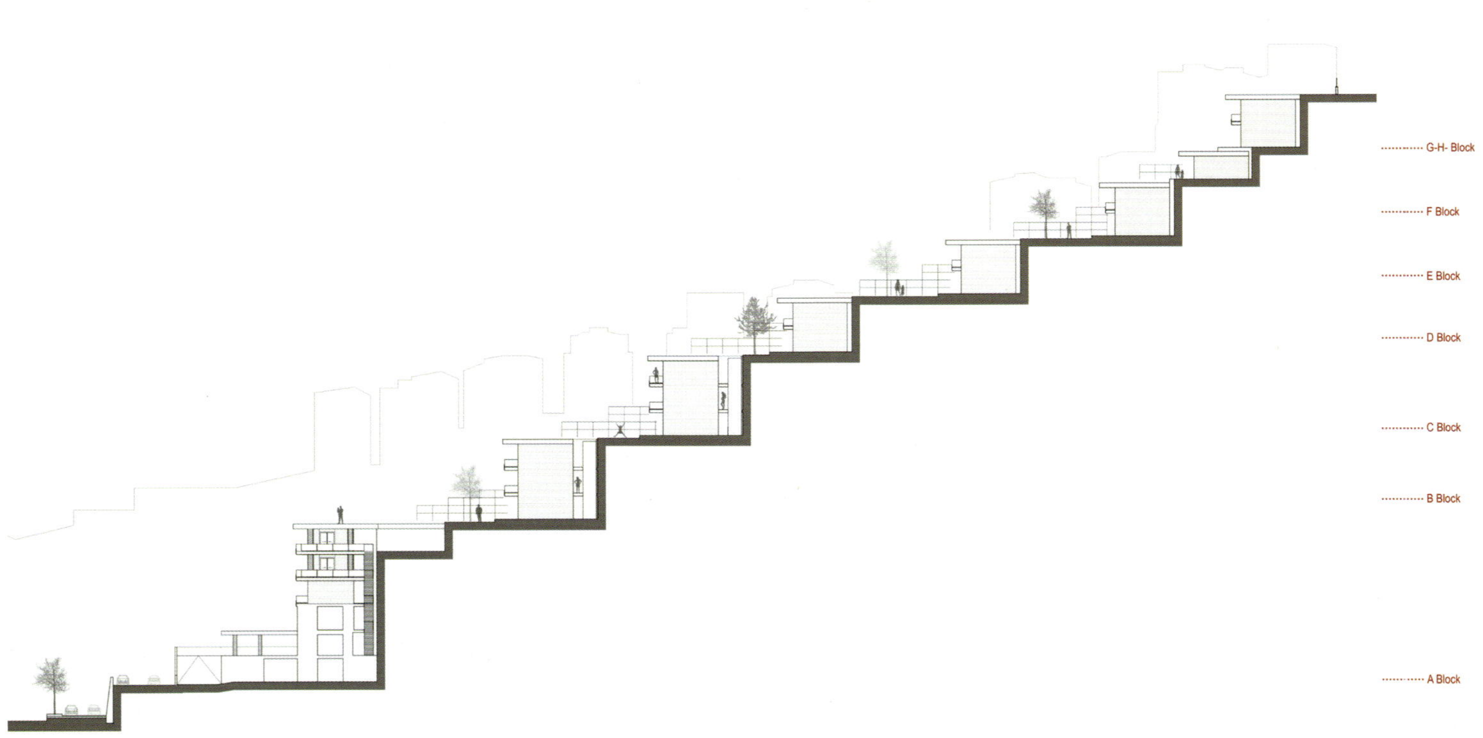

Elevation

© ZM Yasa Photography

© ZM Yasa Photography

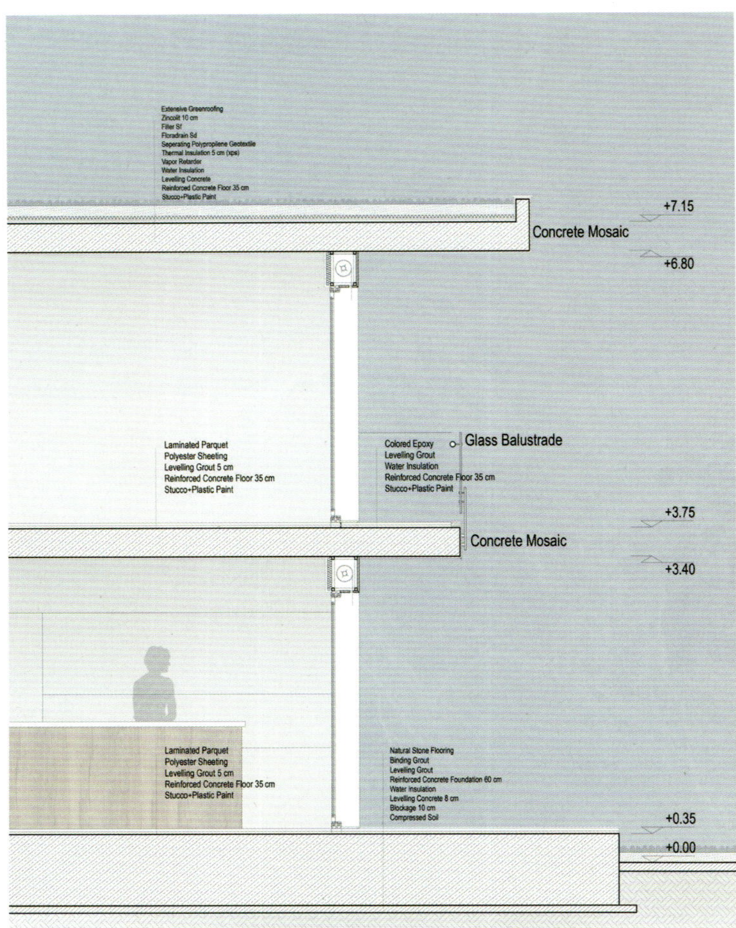

System Section

Elevation

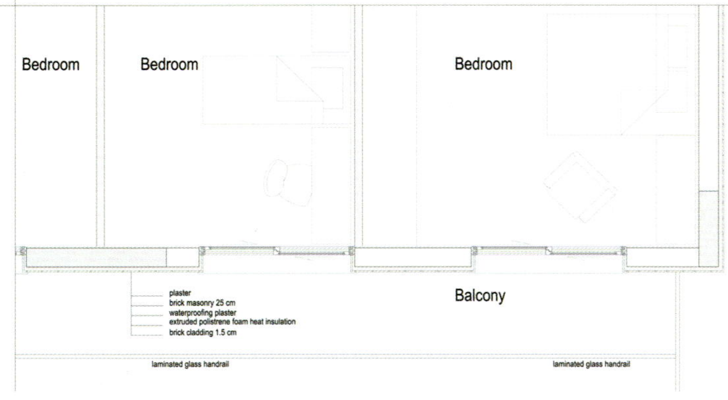

Floor Plan Detail

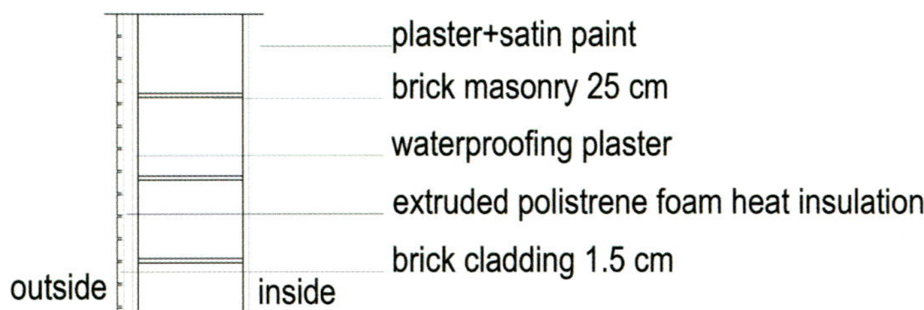

System Detail

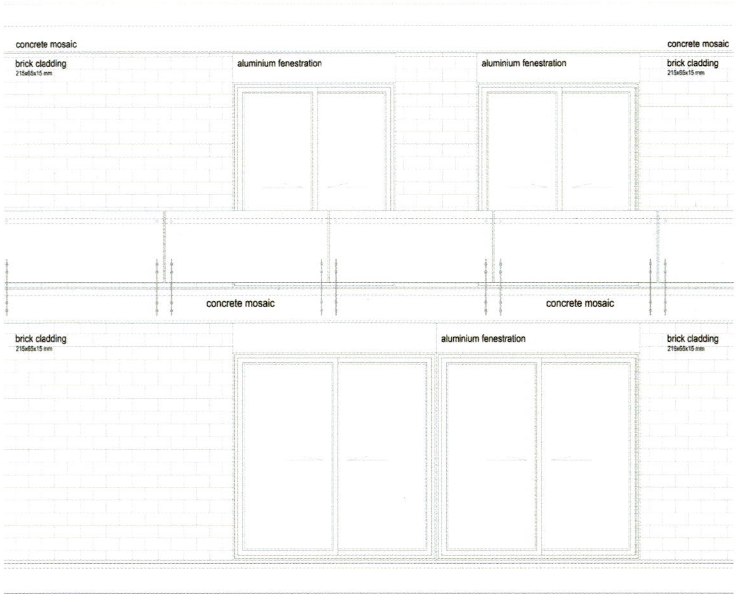

Section Detail

© ZM Yasa Photography

© ZM Yasa Photography

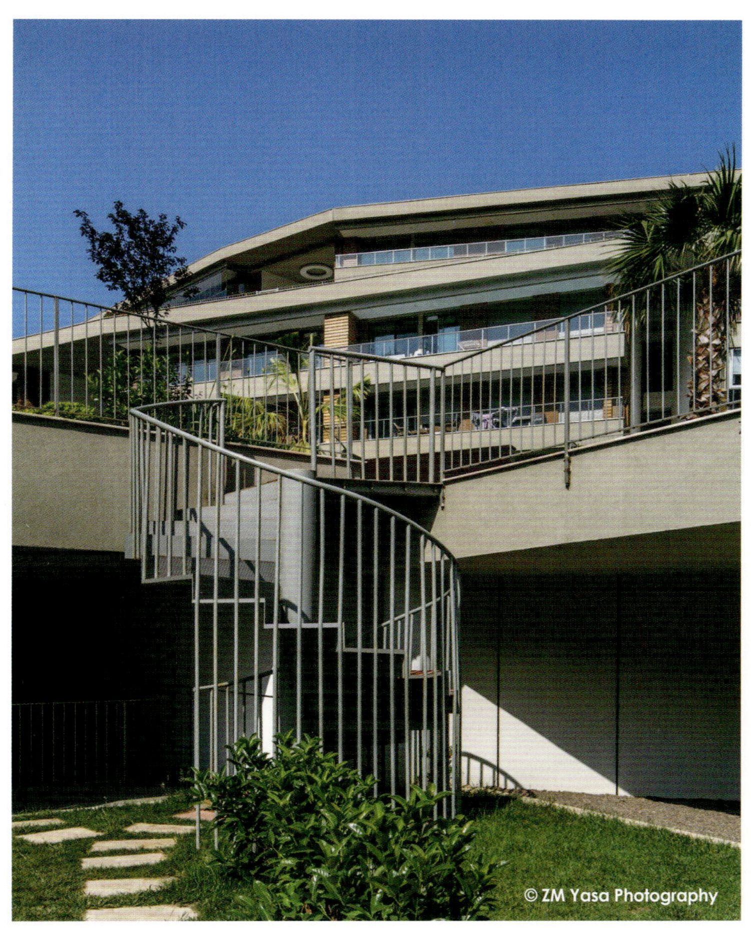

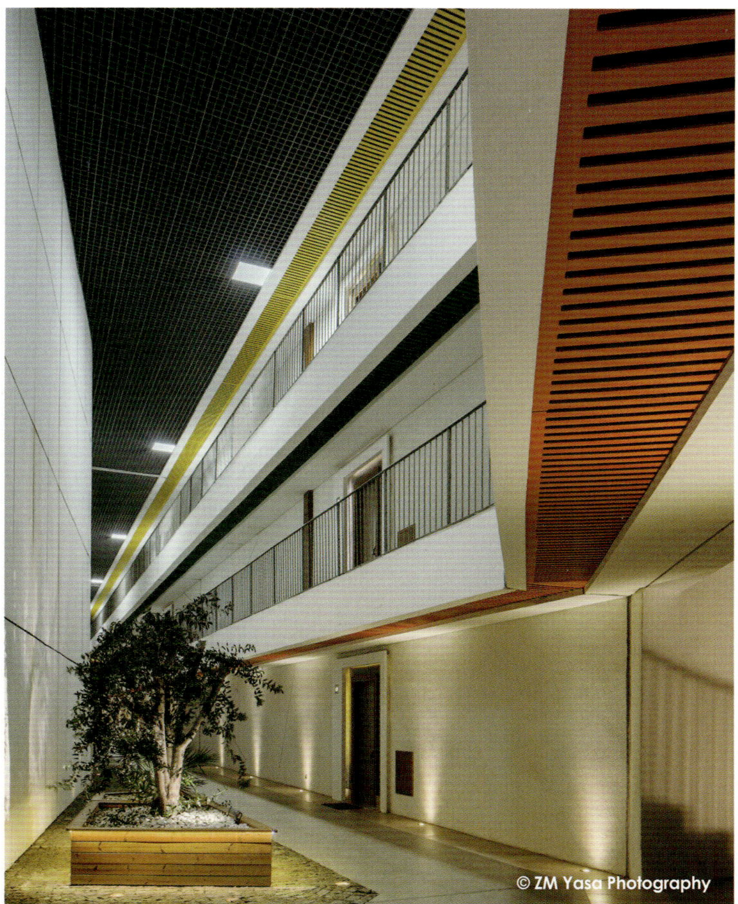

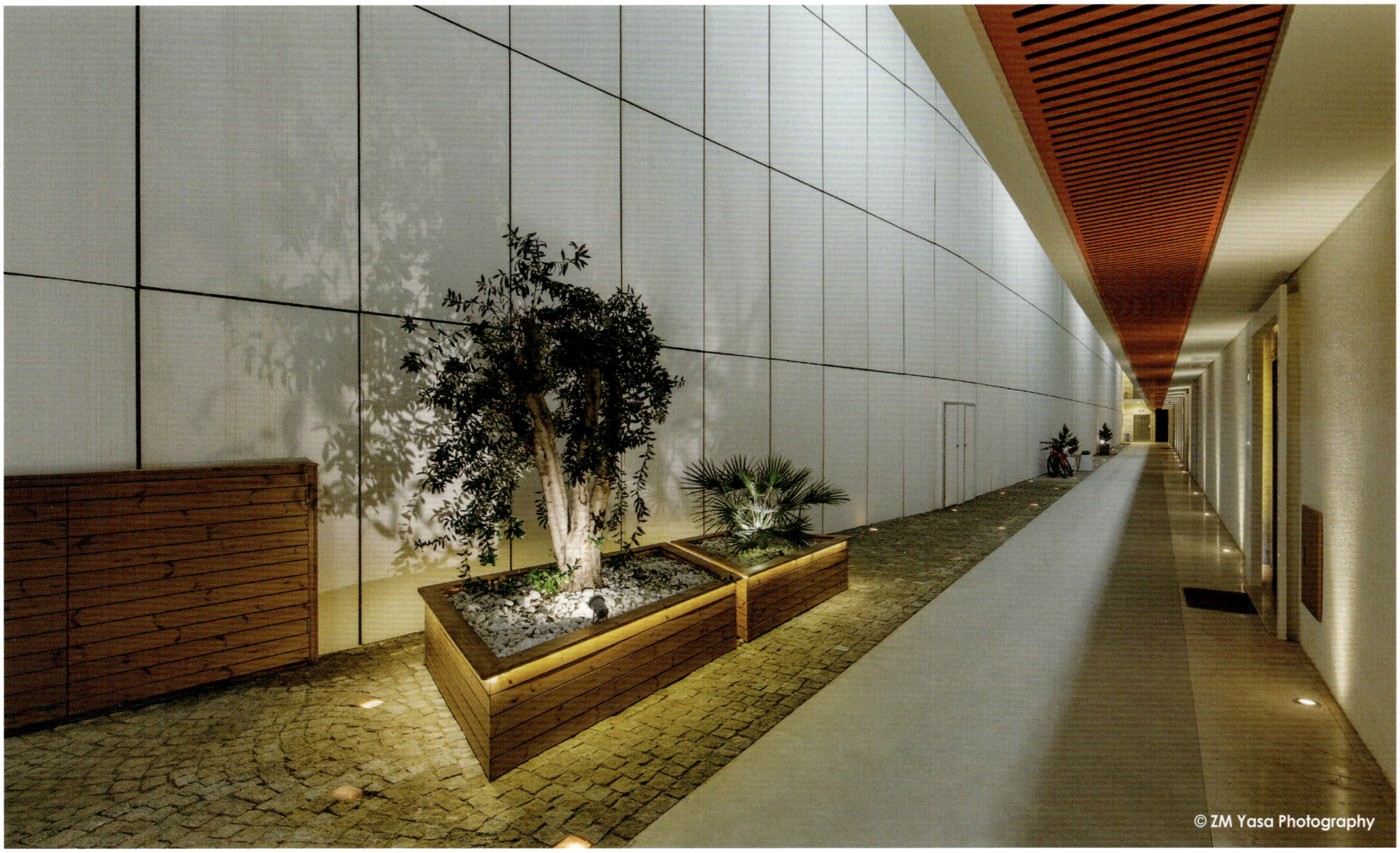
© ZM Yasa Photography

© ZM Yasa Photography

Black & White Twins, Blaricum, The Netherlands

Casanova+Hernandez Architects

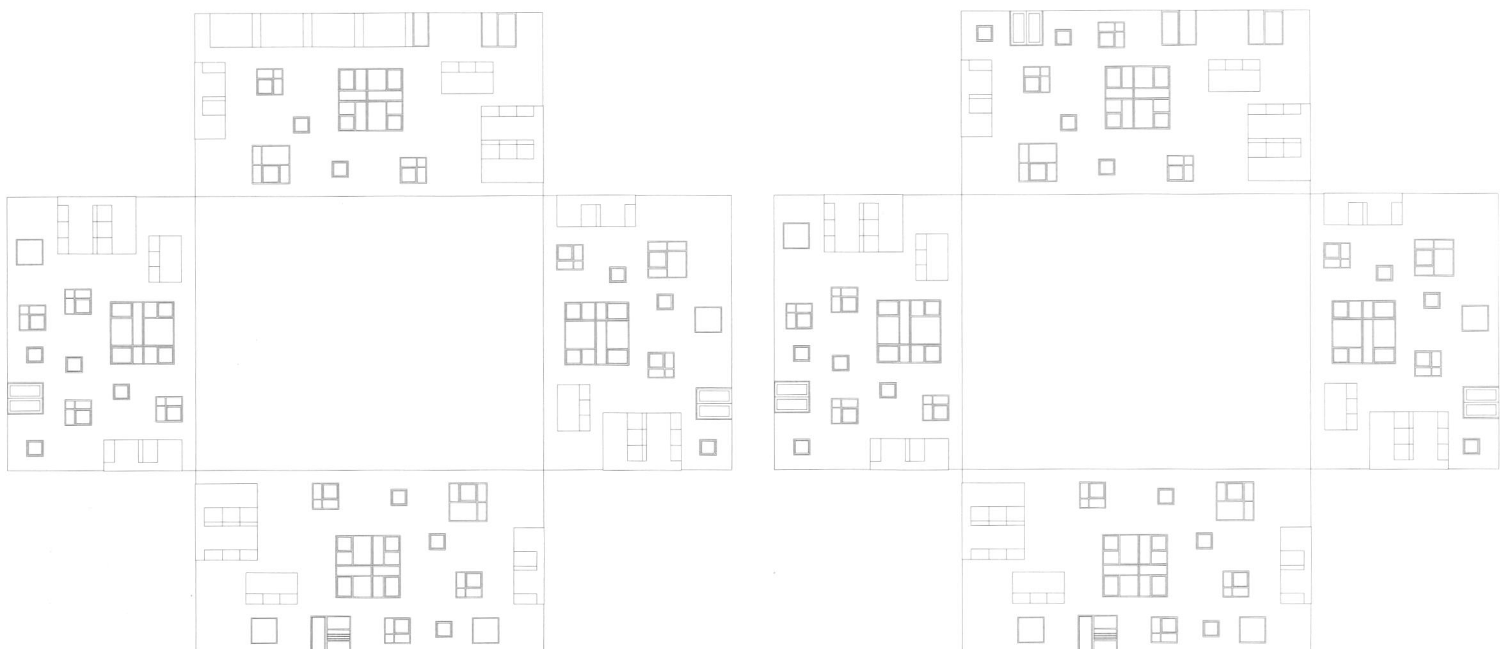

Block A Developed Elevations

Block B Developed Elevations

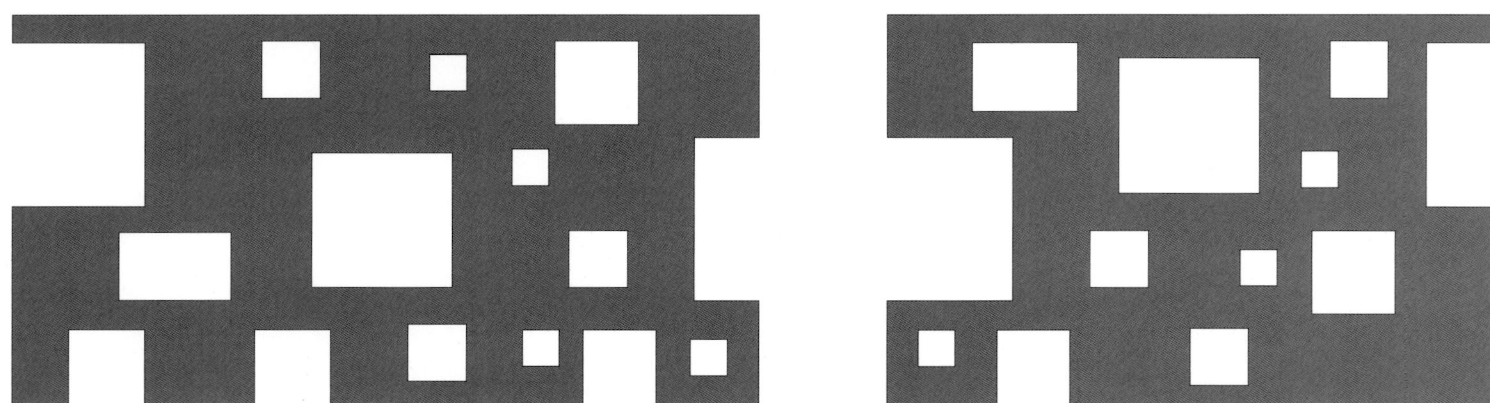

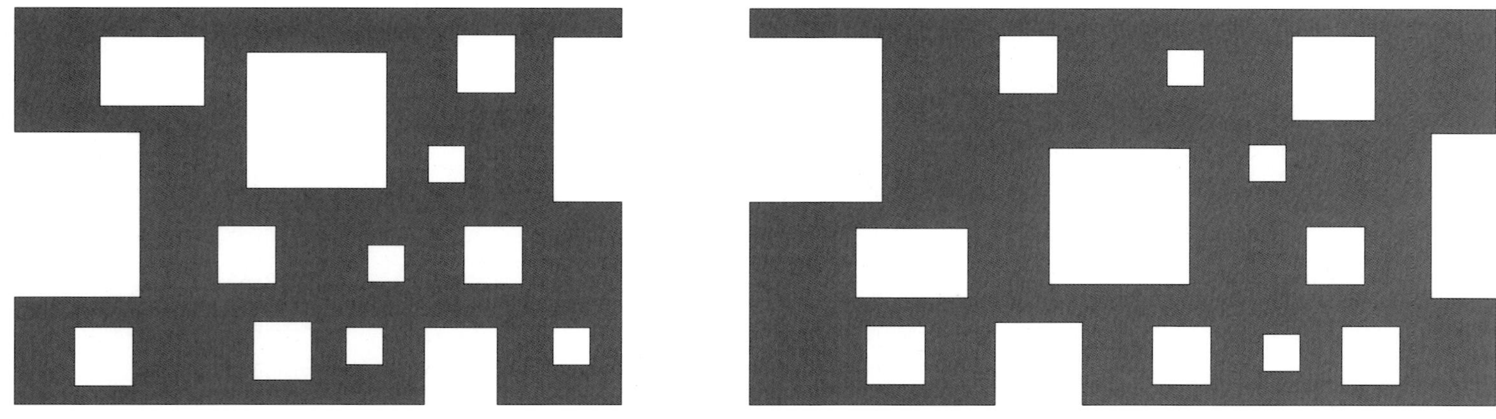

Elevations

© Casanova + Hernandez architects Christian Richters

© Casanova + Hernandez architects Christian Richters

Concrete Modeling

Steel Modeling

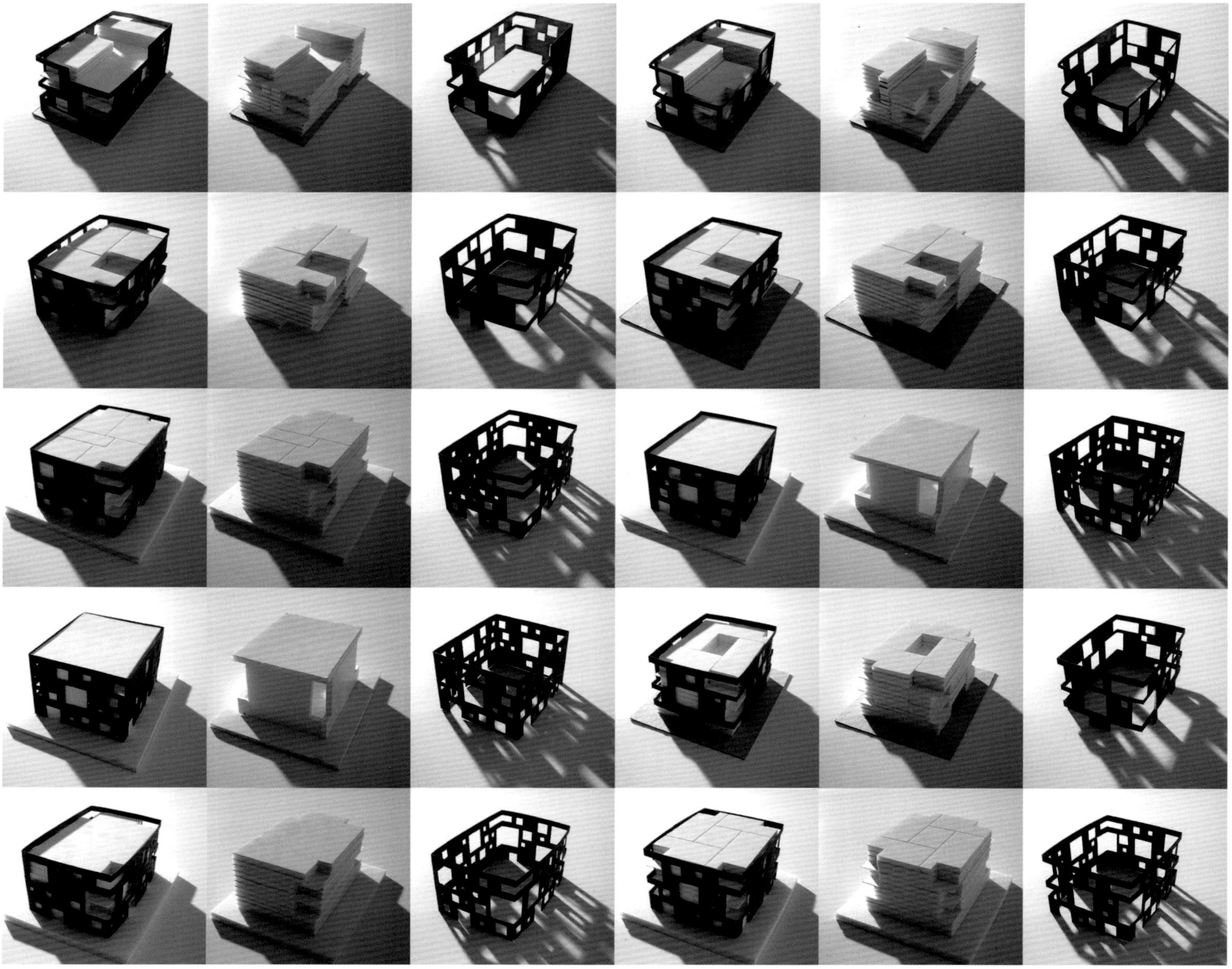

Photo Process

B05 Tin Tin Tower, Rotterdam, the Netherlands

NL Architects

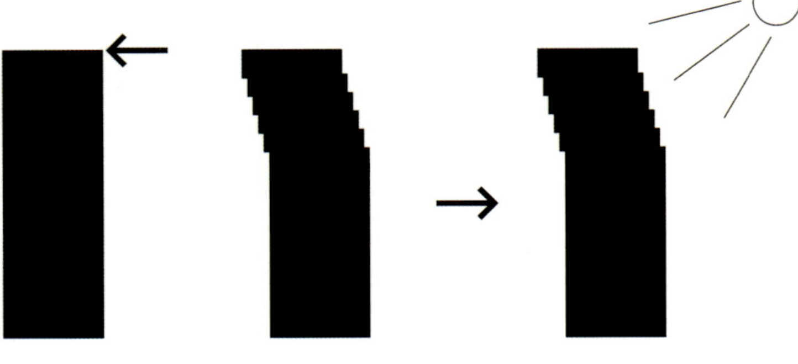

Diagram - Sun

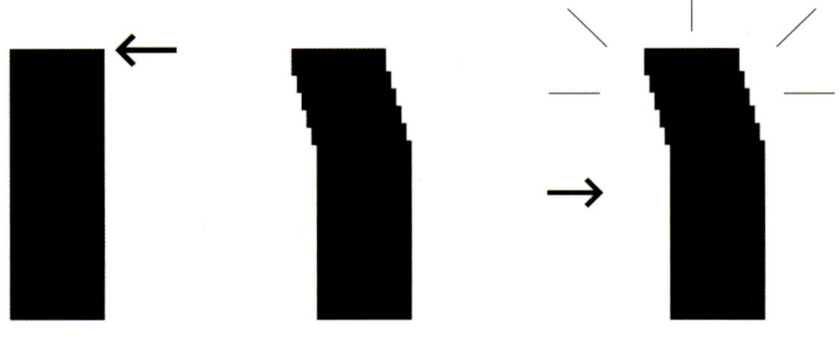

Diagram - Shape

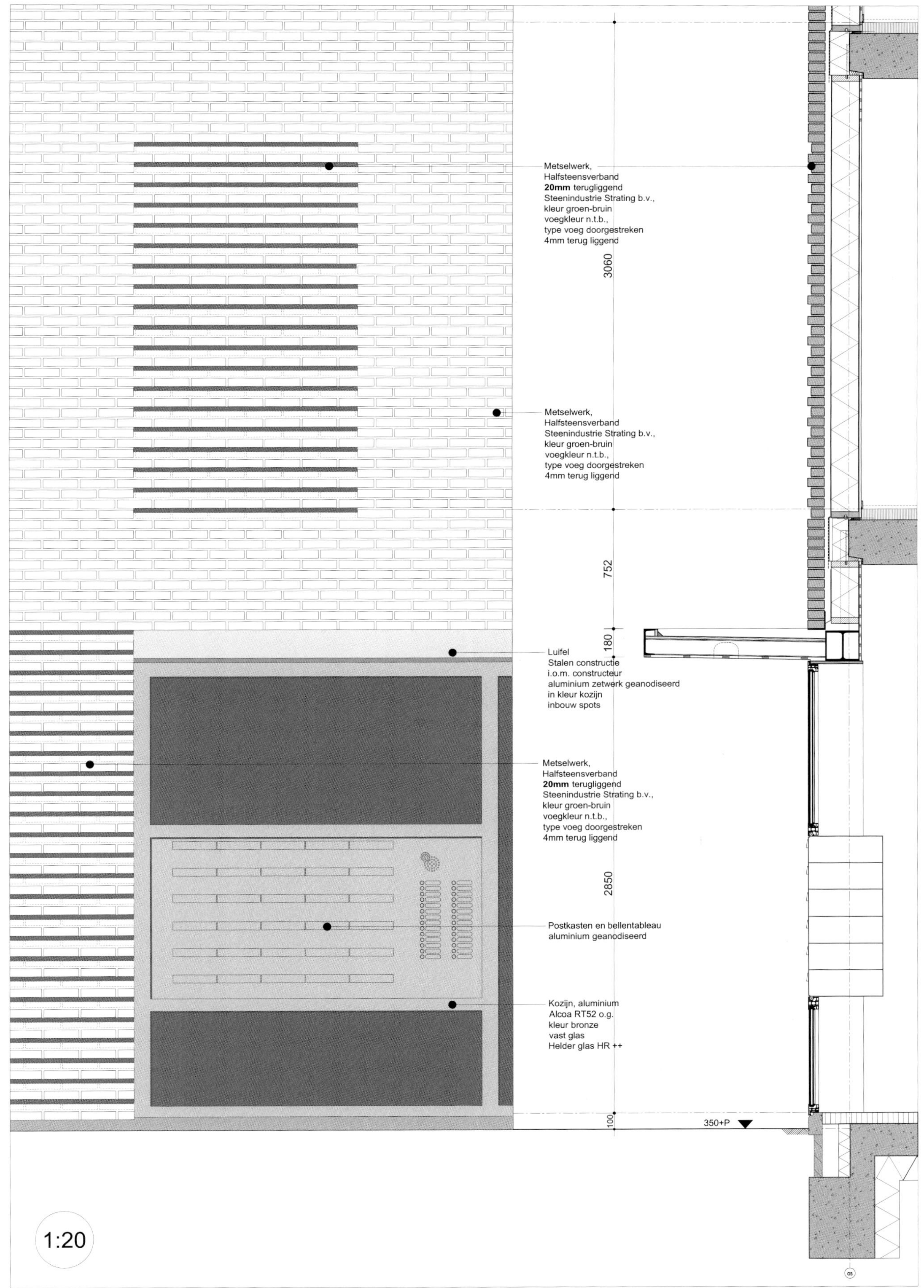
Metselwerk,
Halfsteensverband
20mm terugliggend
Steenindustrie Strating b.v.,
kleur groen-bruin
voegkleur n.t.b.,
type voeg doorgestreken
4mm terug liggend
3060
Metselwerk,
Halfsteensverband
Steenindustrie Strating b.v.,
kleur groen-bruin
voegkleur n.t.b.,
type voeg doorgestreken
4mm terug liggend
752
180
Luifel
Stalen constructie
i.o.m. constructeur
aluminium zetwerk geanodiseerd
in kleur kozijn
inbouw spots
Metselwerk,
Halfsteensverband
20mm terugliggend
Steenindustrie Strating b.v.,
kleur groen-bruin
voegkleur n.t.b.,
type voeg doorgestreken
4mm terug liggend
2850
Postkasten en bellentableau
aluminium geanodiseerd
Kozijn, aluminium
Alcoa RT52 o.g.
kleur bronze
vast glas
Helder glas HR ++
100
350+P
1:20

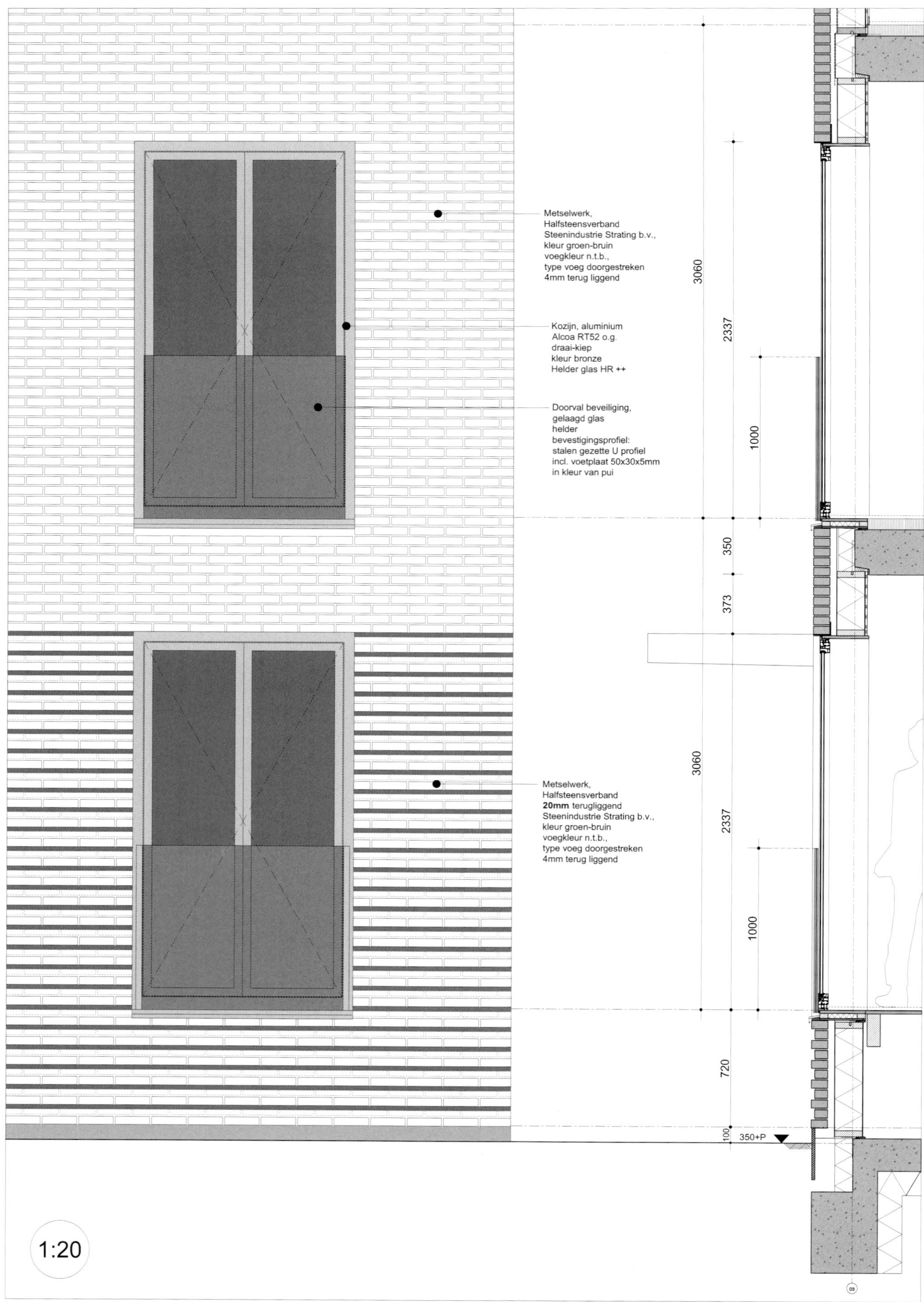
Metselwerk,
Halfsteensverband
Steenindustrie Strating b.v.,
kleur groen-bruin
voegkleur n.t.b.,
type voeg doorgestreken
4mm terug liggend
Kozijn, aluminium
Alcoa RT52 o.g.
draai-kiep
kleur bronze
Helder glas HR ++
Doorval beveiliging,
gelaagd glas
helder
bevestigingsprofiel:
stalen gezette U profiel
incl. voetplaat 50x30x5mm
in kleur van pui
Metselwerk,
Halfsteensverband
20mm terugliggend
Steenindustrie Strating b.v.,
kleur groen-bruin
voegkleur n.t.b.,
type voeg doorgestreken
4mm terug liggend
3060
2337
1000
350
373
3060
2337
1000
720
100
350+P
1:20

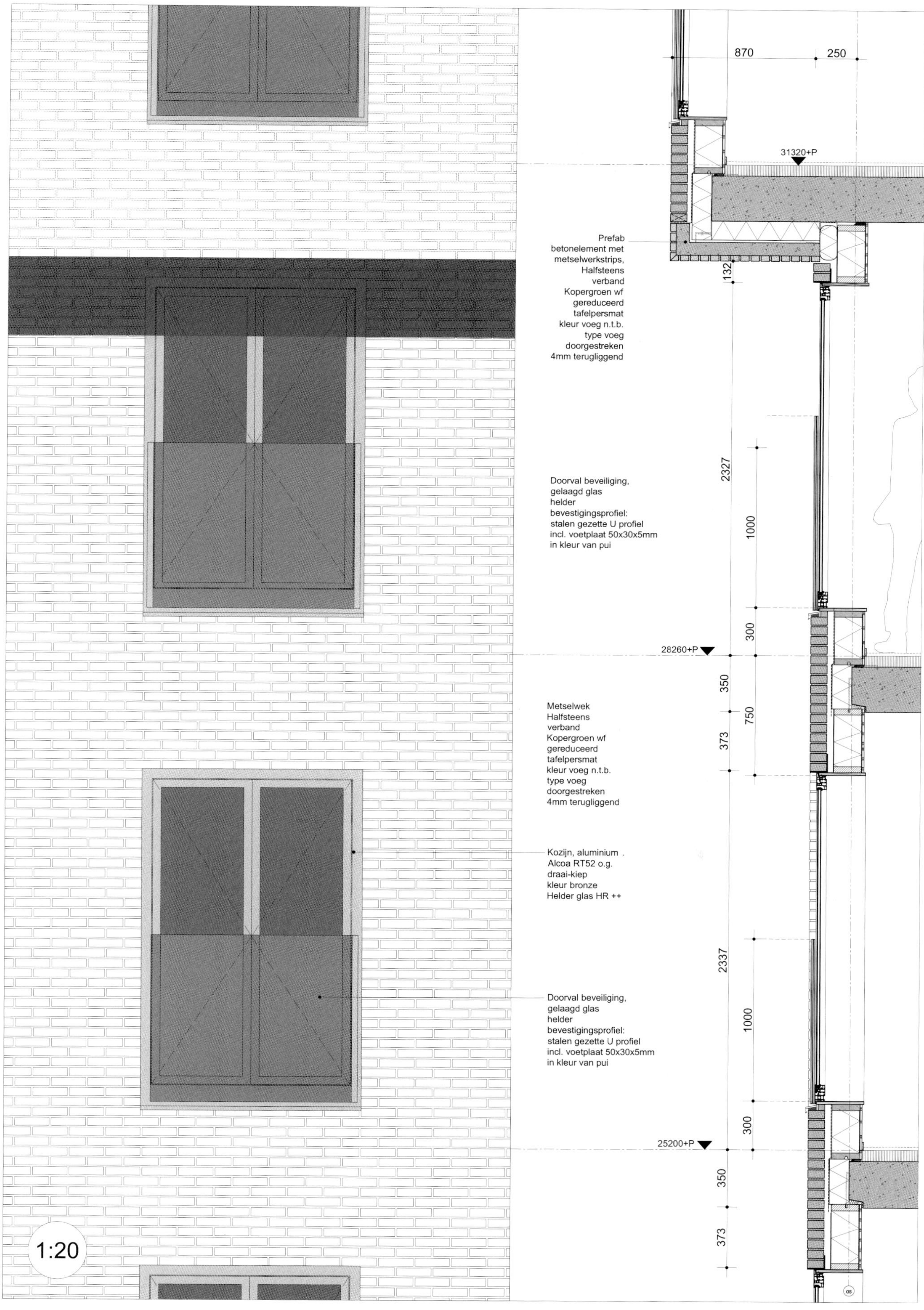

870
250
31320+P
Prefab
betonelement met
metselwerkstrips,
Halfsteens
verband
Kopergroen wf
gereduceerd
tafelpersmat
kleur voeg n.t.b.
type voeg
doorgestreken
4mm terugliggend
132
2327
1000
Doorval beveiliging,
gelaagd glas
helder
bevestigingsprofiel:
stalen gezette U profiel
incl. voetplaat 50x30x5mm
in kleur van pui
300
28260+P
350
750
373
Metselwek
Halfsteens
verband
Kopergroen wf
gereduceerd
tafelpersmat
kleur voeg n.t.b.
type voeg
doorgestreken
4mm terugliggend
Kozijn, aluminium .
Alcoa RT52 o.g.
draai-kiep
kleur bronze
Helder glas HR ++
2337
1000
Doorval beveiliging,
gelaagd glas
helder
bevestigingsprofiel:
stalen gezette U profiel
incl. voetplaat 50x30x5mm
in kleur van pui
300
25200+P
350
373
1:20

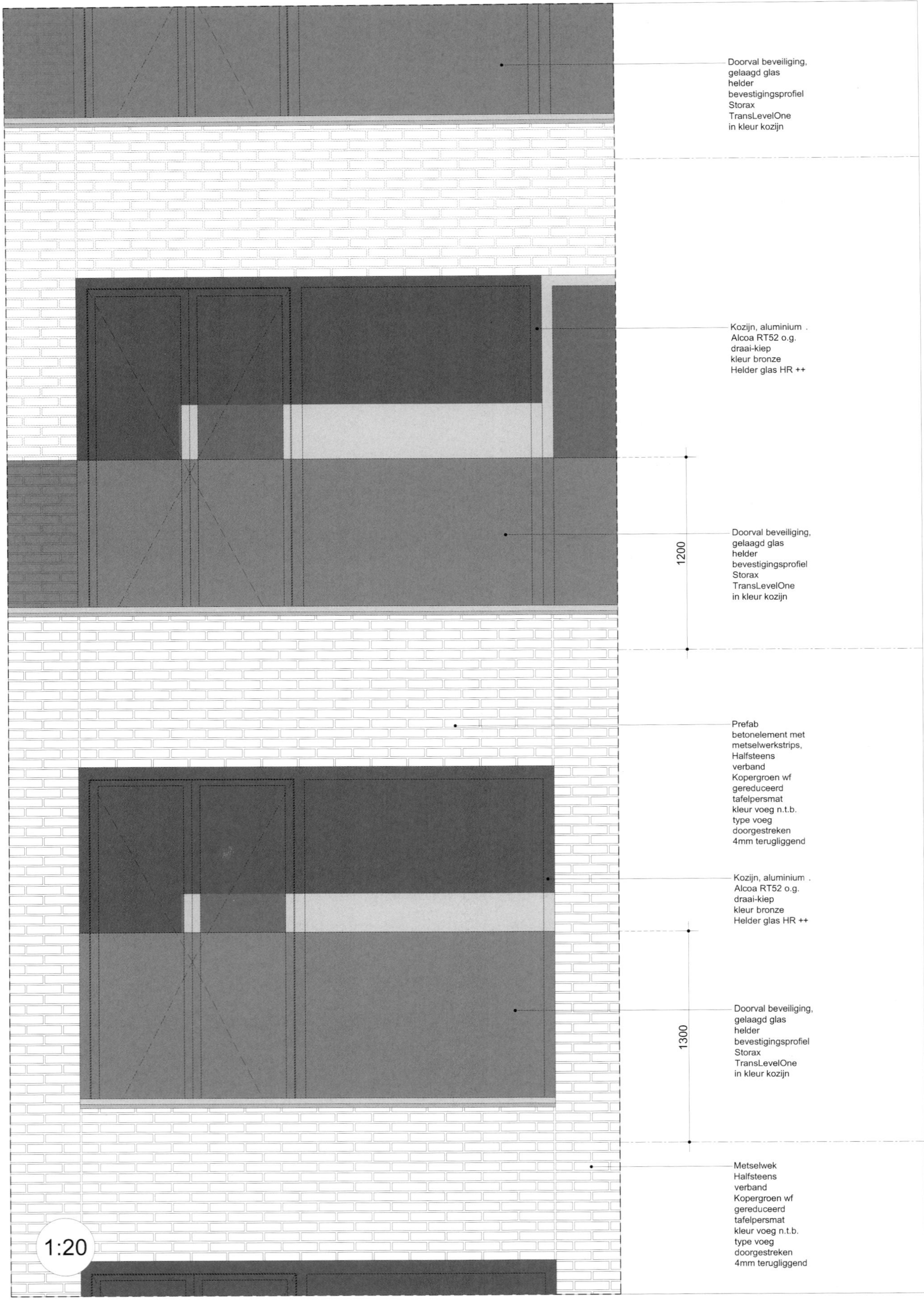

Doorval beveiliging,
gelaagd glas
helder
bevestigingsprofiel
Storax
TransLevelOne
in kleur kozijn
Kozijn, aluminium .
Alcoa RT52 o.g.
draai-kiep
kleur bronze
Helder glas HR ++
1200
Doorval beveiliging,
gelaagd glas
helder
bevestigingsprofiel
Storax
TransLevelOne
in kleur kozijn
Prefab
betonelement met
metselwerkstrips,
Halfsteens
verband
Kopergroen wf
gereduceerd
tafelpersmat
kleur voeg n.t.b.
type voeg
doorgestreken
4mm terugliggend
Kozijn, aluminium .
Alcoa RT52 o.g.
draai-kiep
kleur bronze
Helder glas HR ++
1300
Doorval beveiliging,
gelaagd glas
helder
bevestigingsprofiel
Storax
TransLevelOne
in kleur kozijn
Metselwek
Halfsteens
verband
Kopergroen wf
gereduceerd
tafelpersmat
kleur voeg n.t.b.
type voeg
doorgestreken
4mm terugliggend
1:20

Fairyland Guorui, Beijing, China

UNStudio

© Edmon Leong

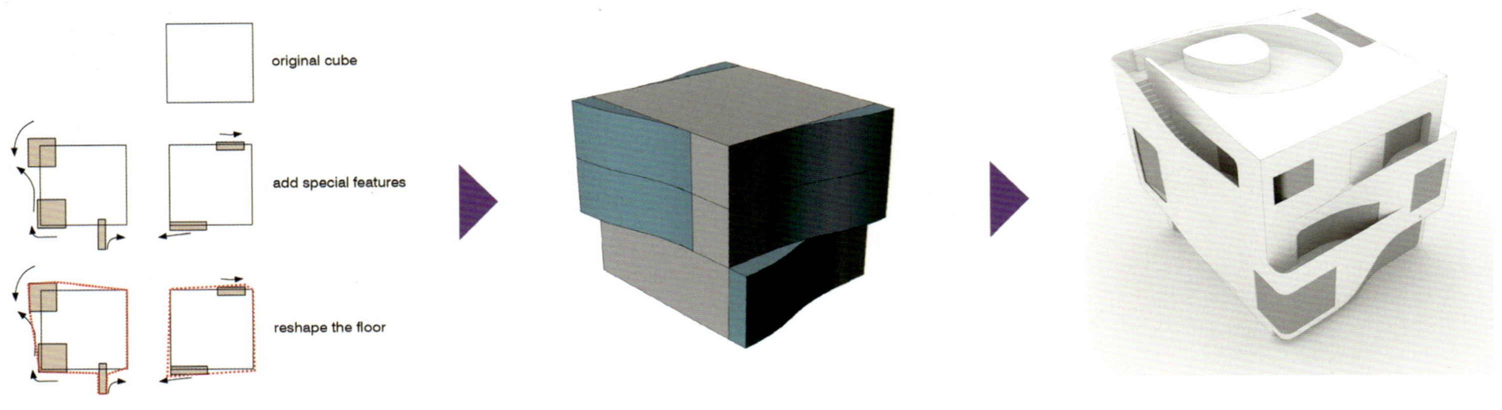

设定元素融入别墅的原则
Establishing Feature Parameter

元素主宰立面变化
Facade Undulate from adding Features

最后形成形体
Resulting in Final Prototypes

Design Concept

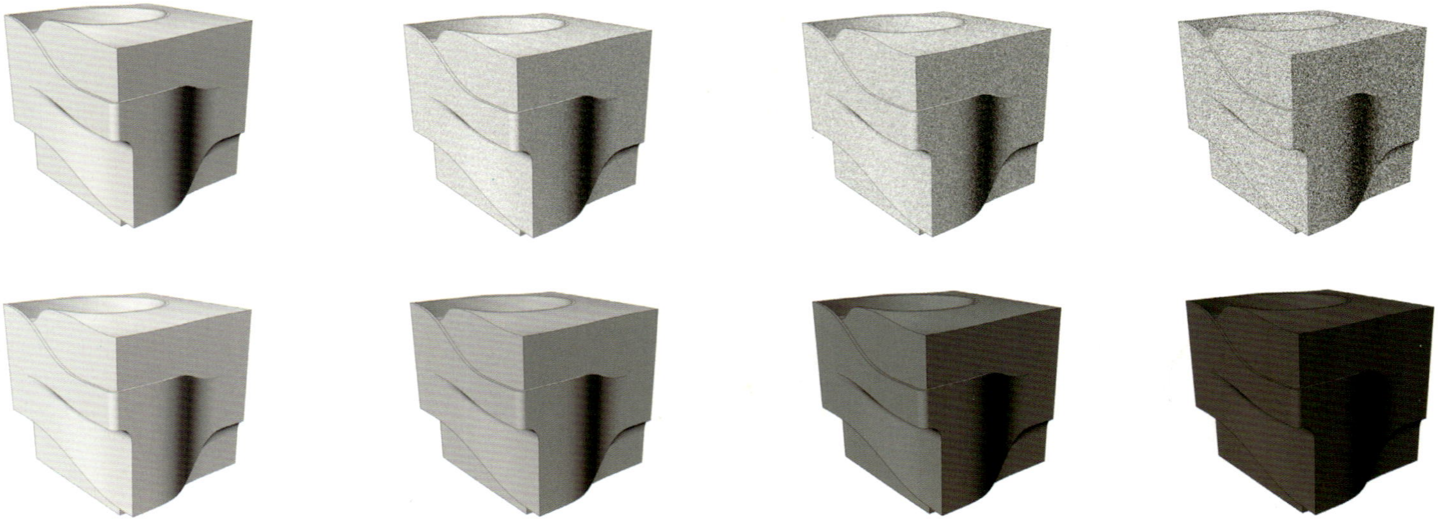

Material Varity

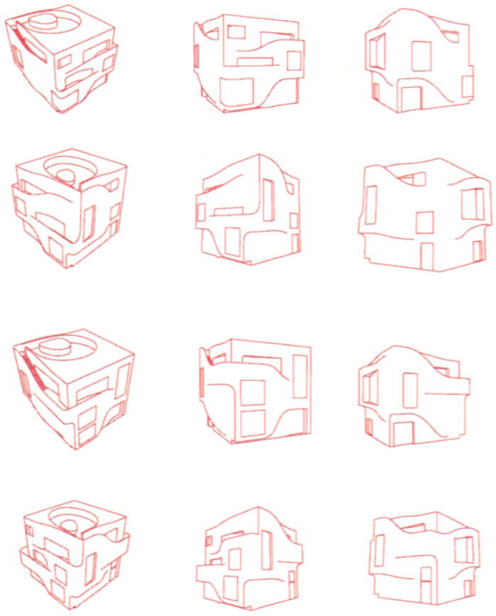

Make2d Cover

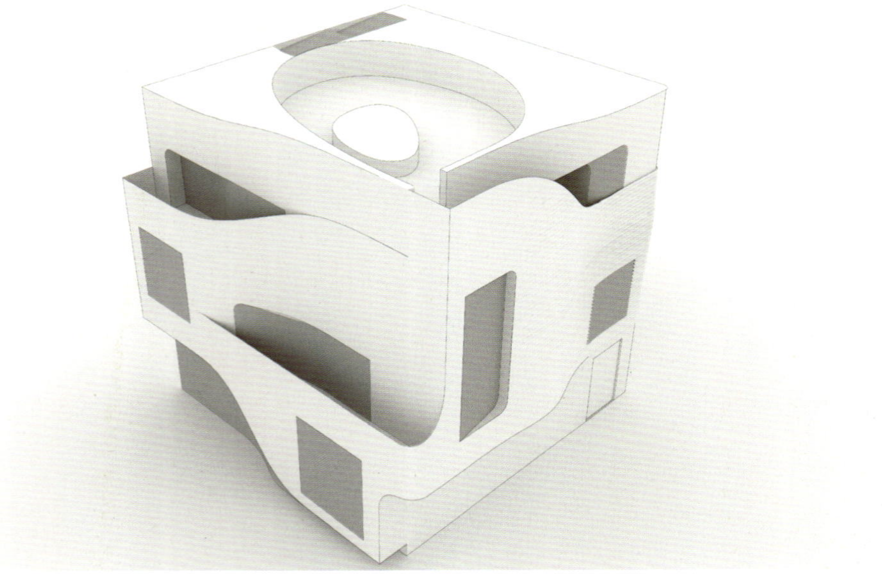

Modeling

© Edmon Leong

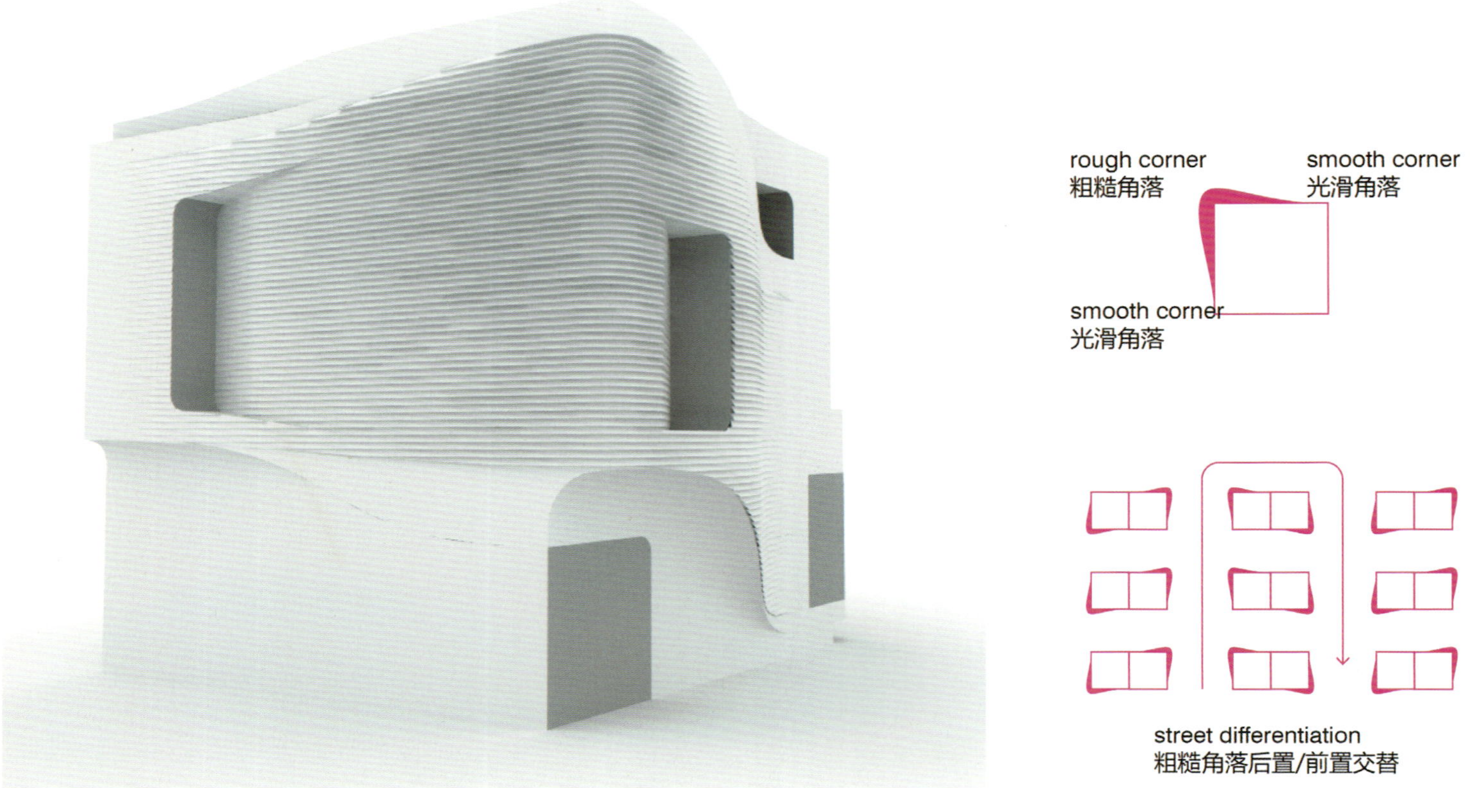

Stone Texture Emphasis

鸟瞰 Birdeye View

Concept Plan - Birds-eye view

平滑面 Rough Surface

从平滑（左）到粗糙（右）的一个过渡
Transformation from Smooth (left) to Rought (right)

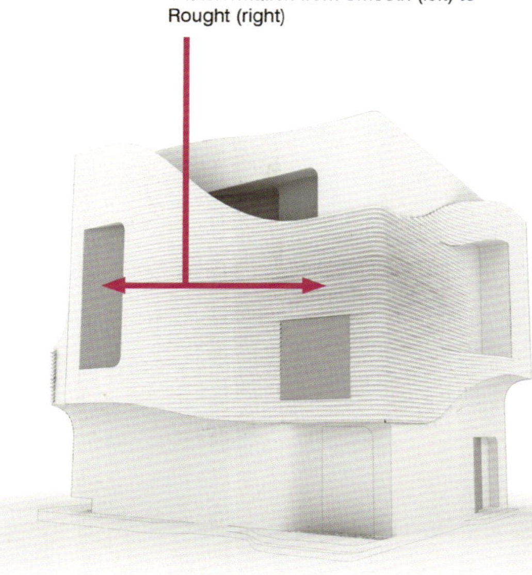

粗糙面 Rough Surface

Stone Textures

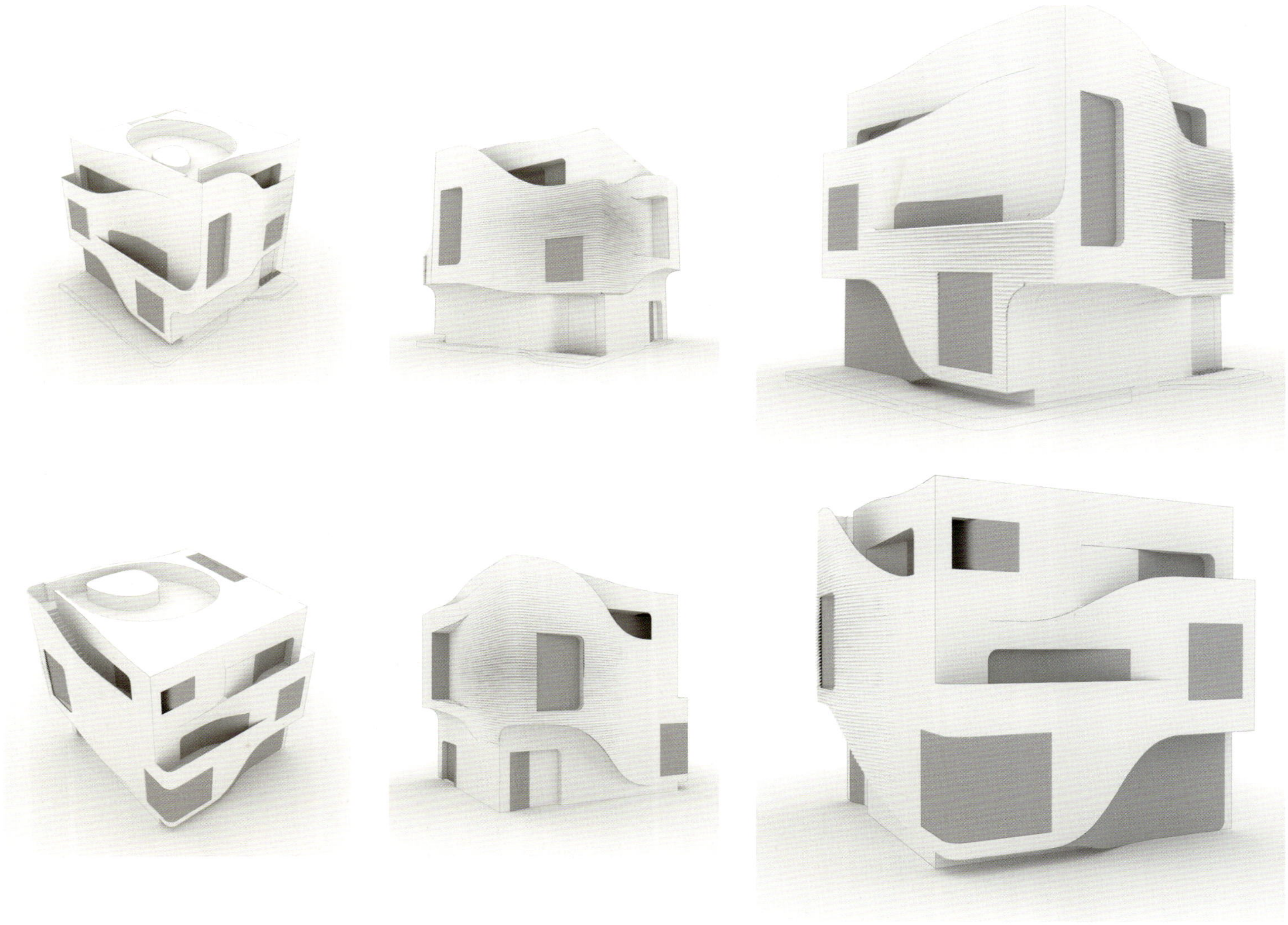

Prototype

© Edmon Leong

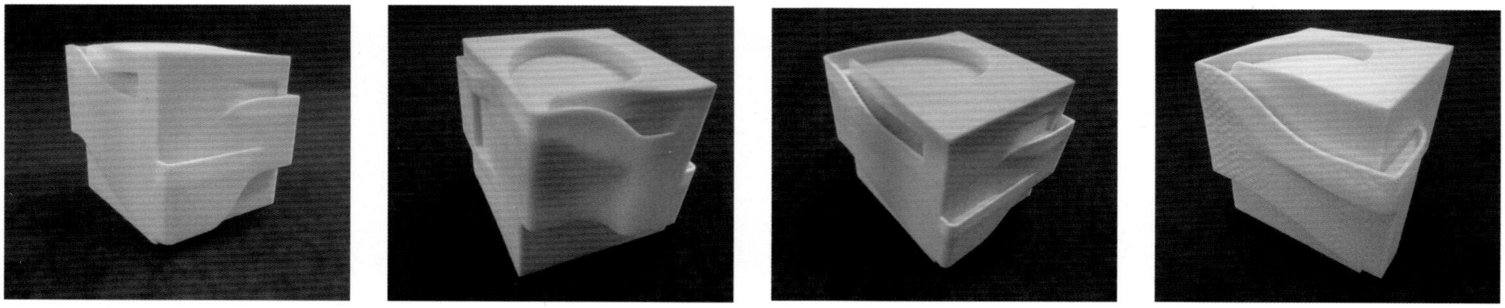

Study Modeling

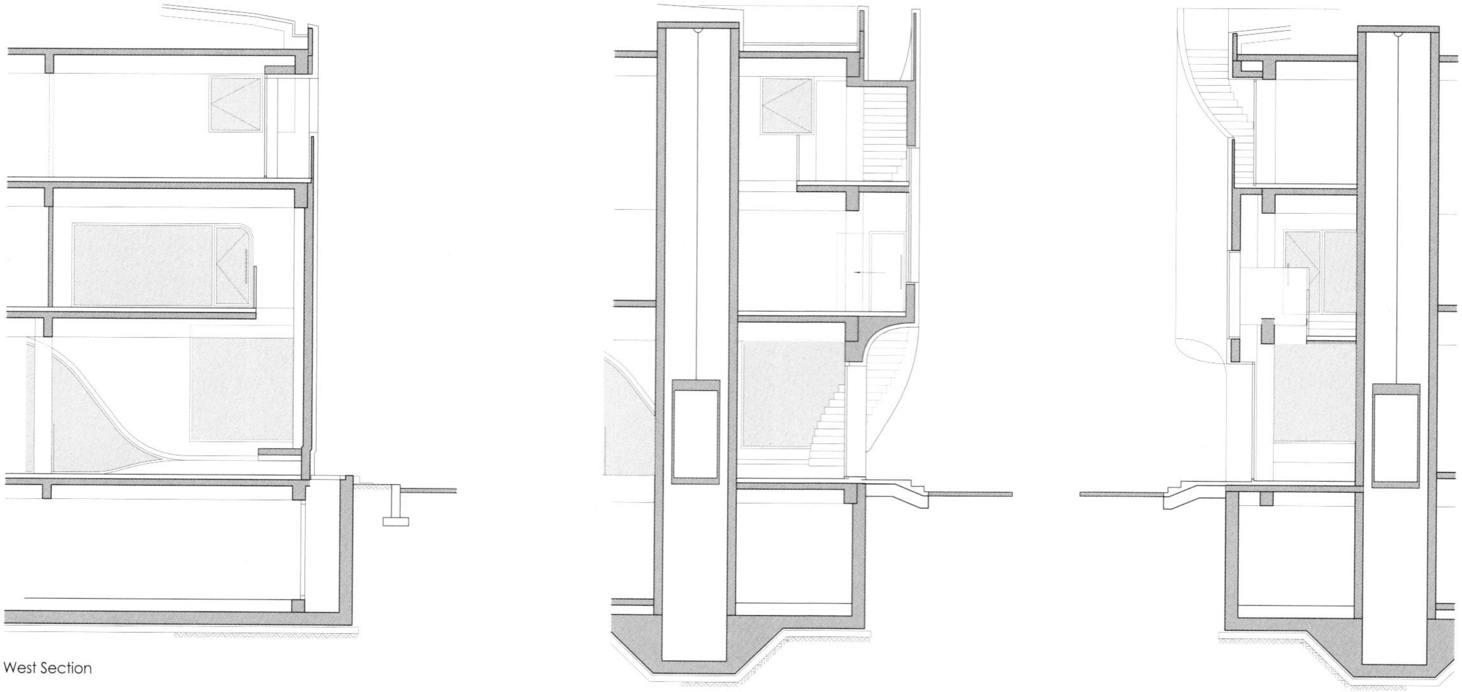

West Section

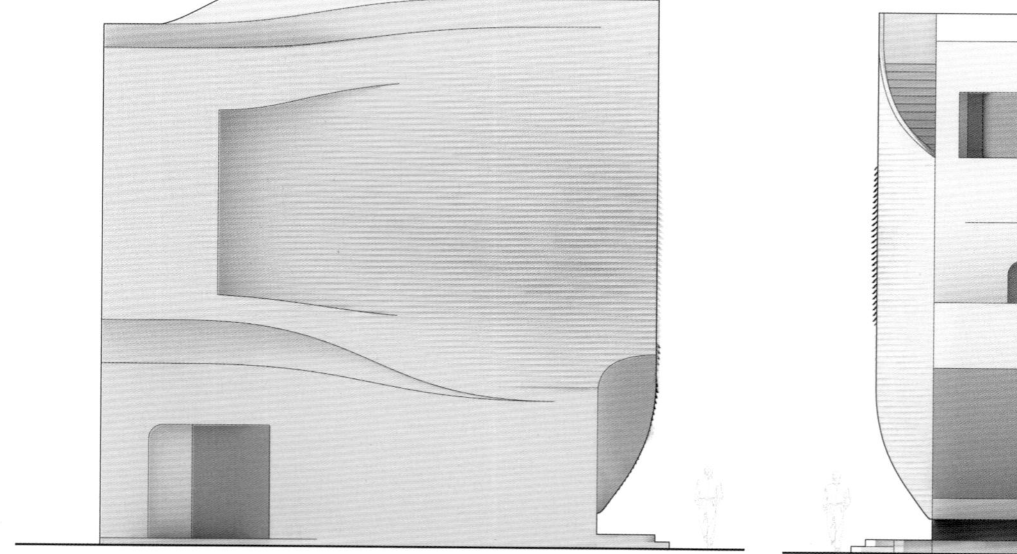

North Elevation

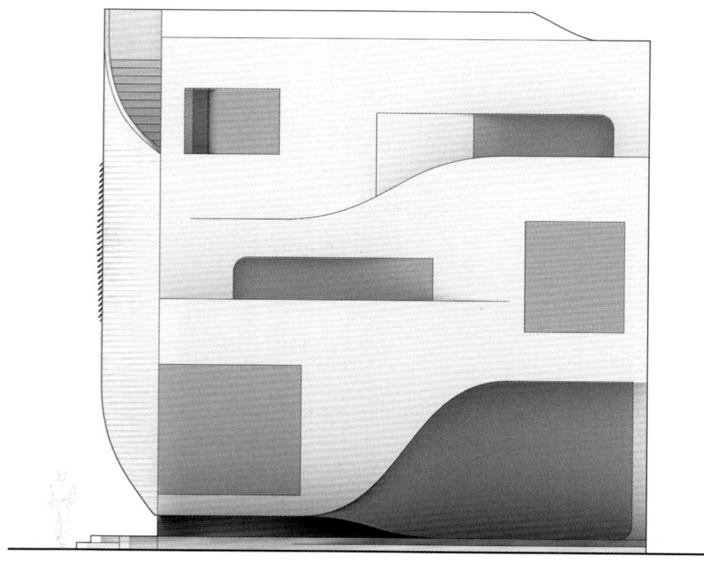

South Elevation

© Edmon Leong

© Edmon Leong

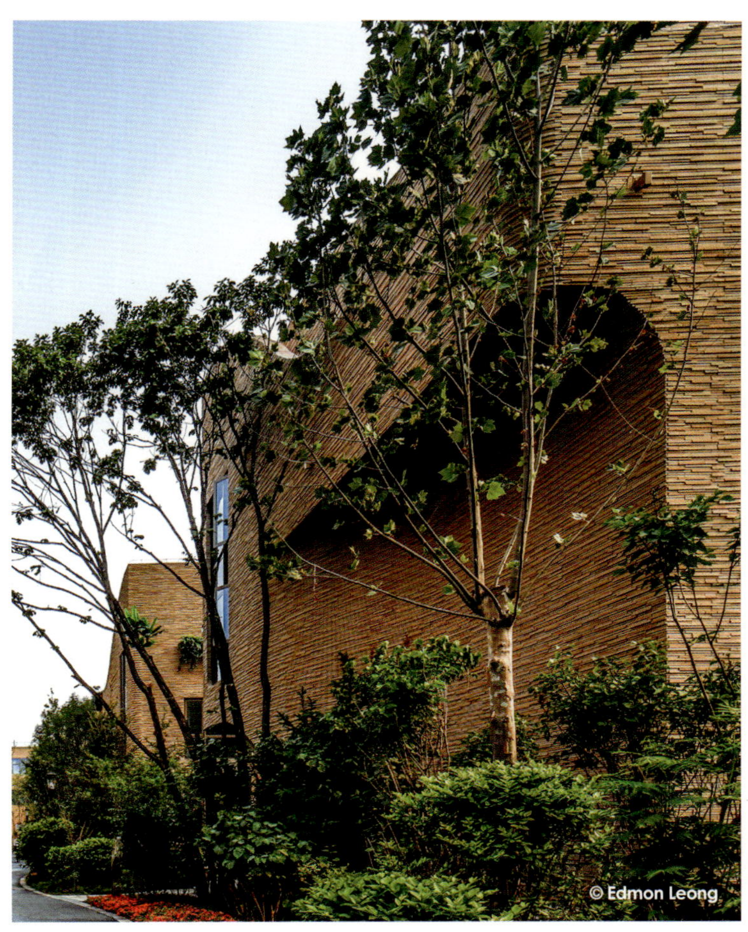

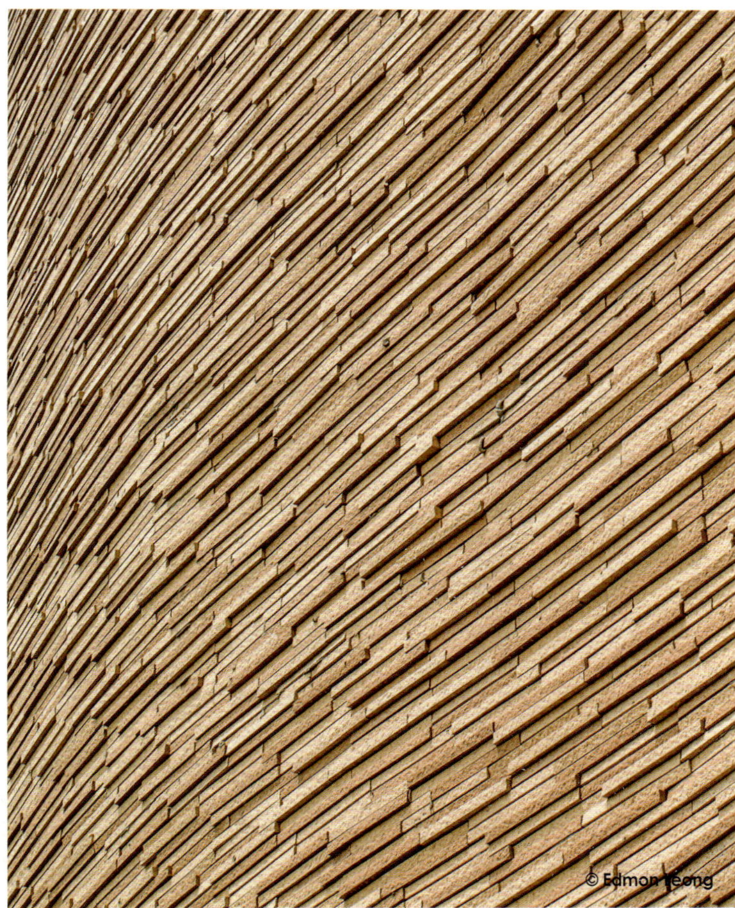

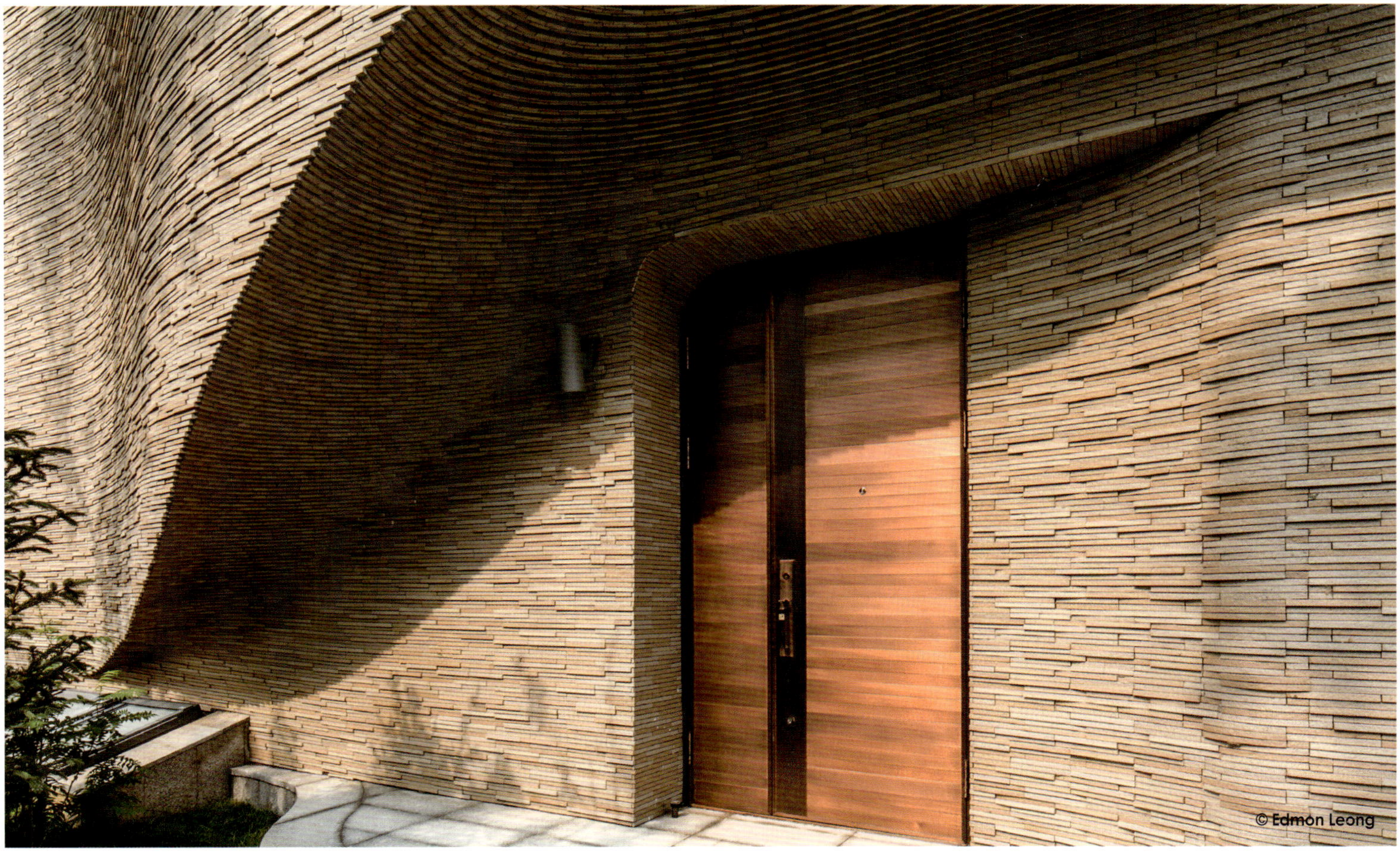
© Edmon Leong

Gapyong Space Invader, Gyeonggi-do, Korea
NL Architects

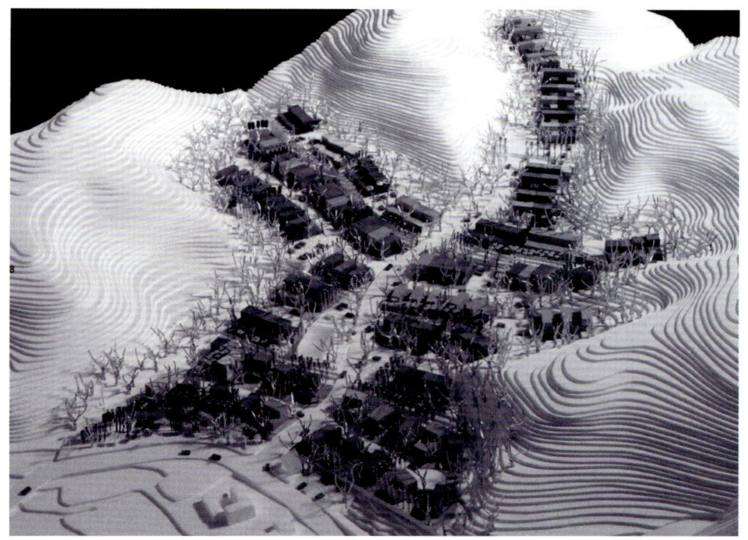

Study Modeling

Section

Kameleon, Amsterdam, the Natherlands
NL Architects

© Luuk Kramer and Marcel van der Burg

© Luuk Kramer and Marcel van der Burg

© Luuk Kramer and Marcel van der Burg

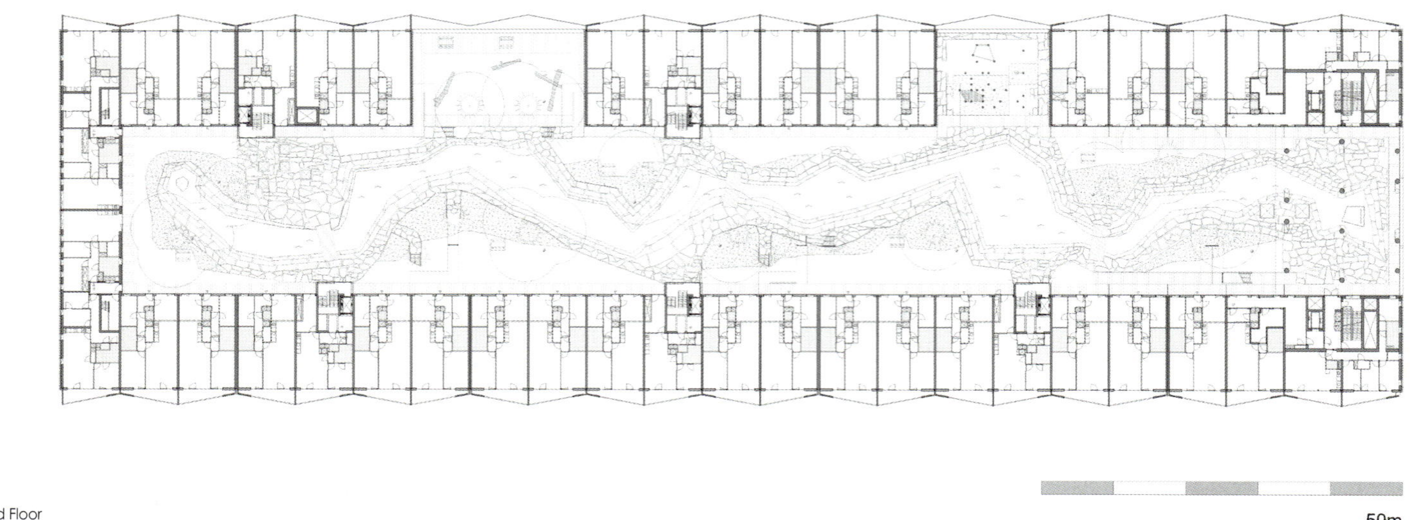

3rd Floor

50m

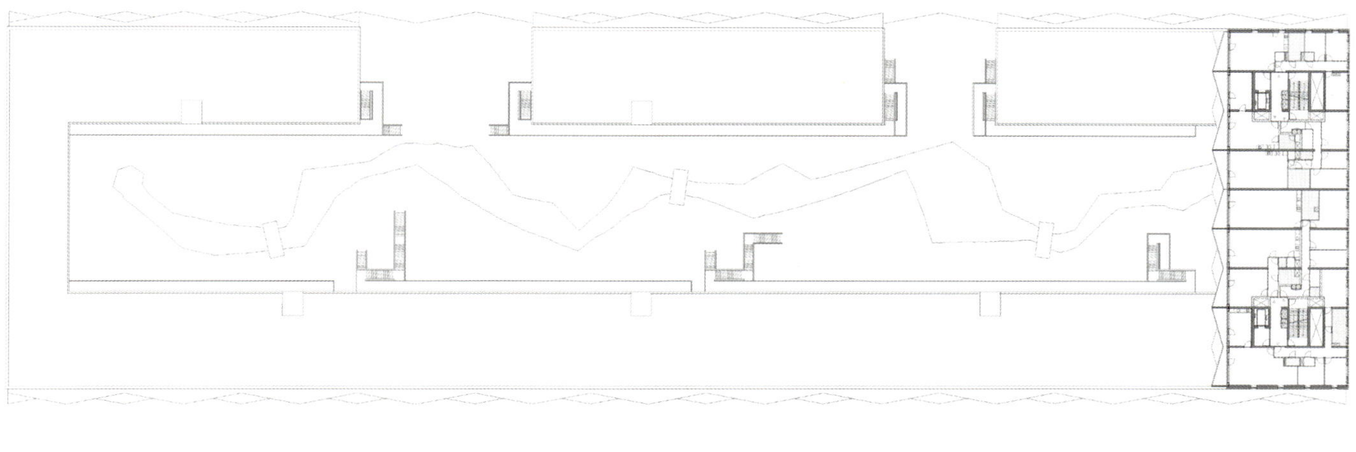

9th, 11th, 13th, 15th Floor

50m

© Luuk Kramer and Marcel van der Burg

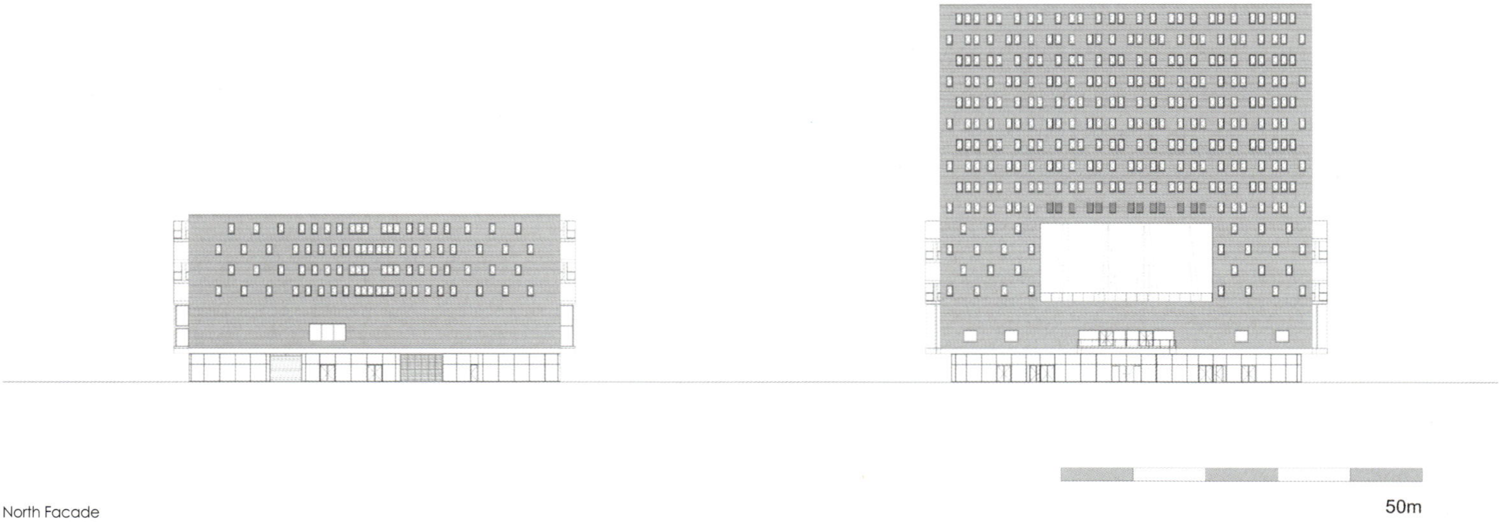

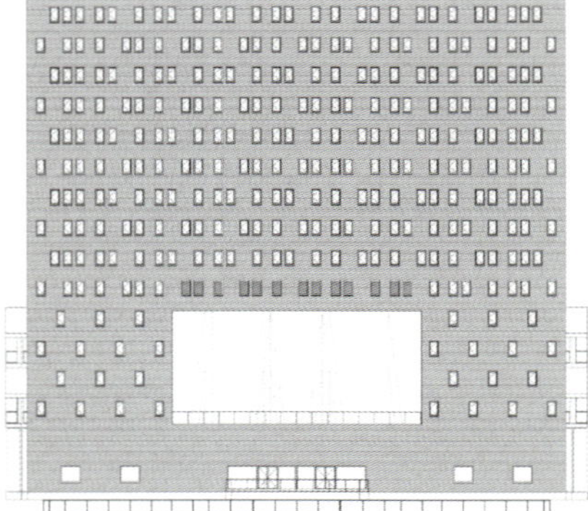

North Facade

© Luuk Kramer and Marcel van der Burg

VILLA TORPED, Saint Denis, France

PERIPHERIQUES

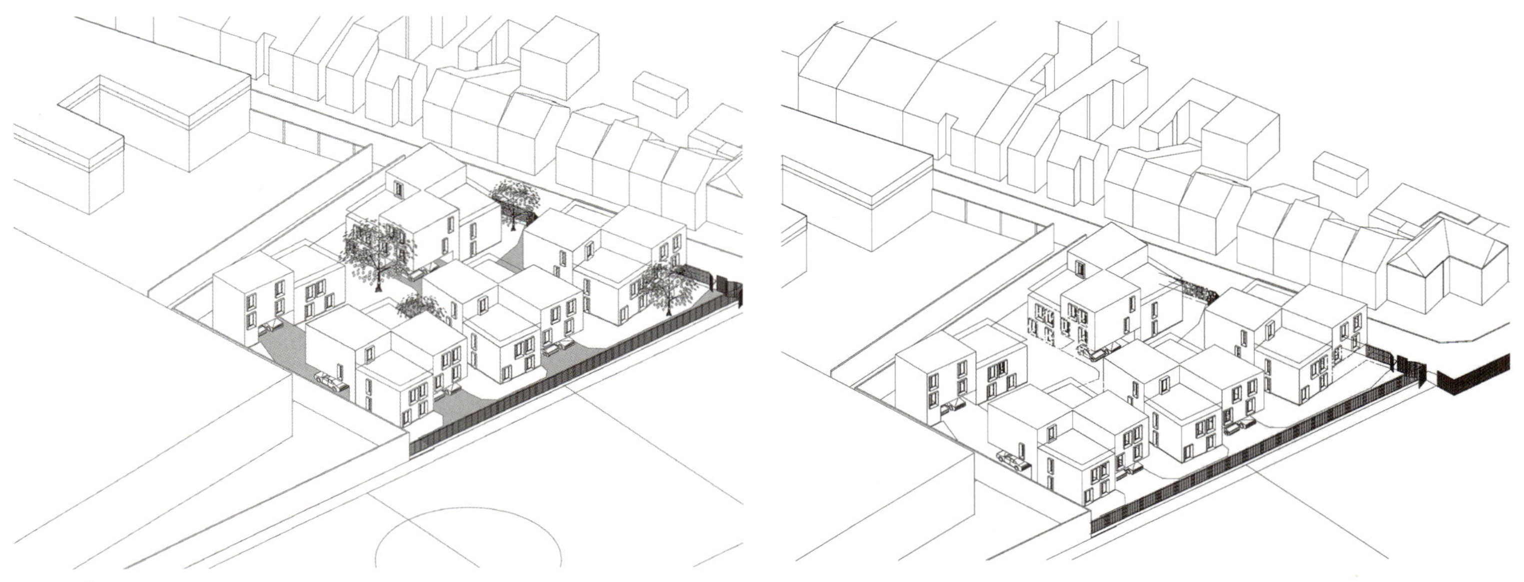

Axonometric

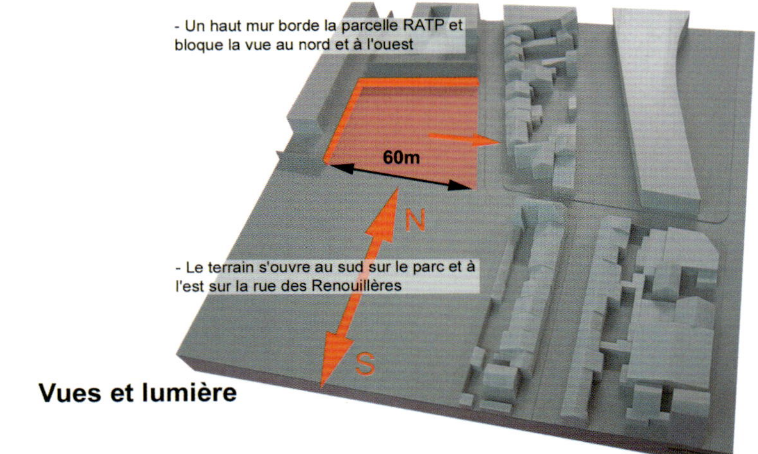

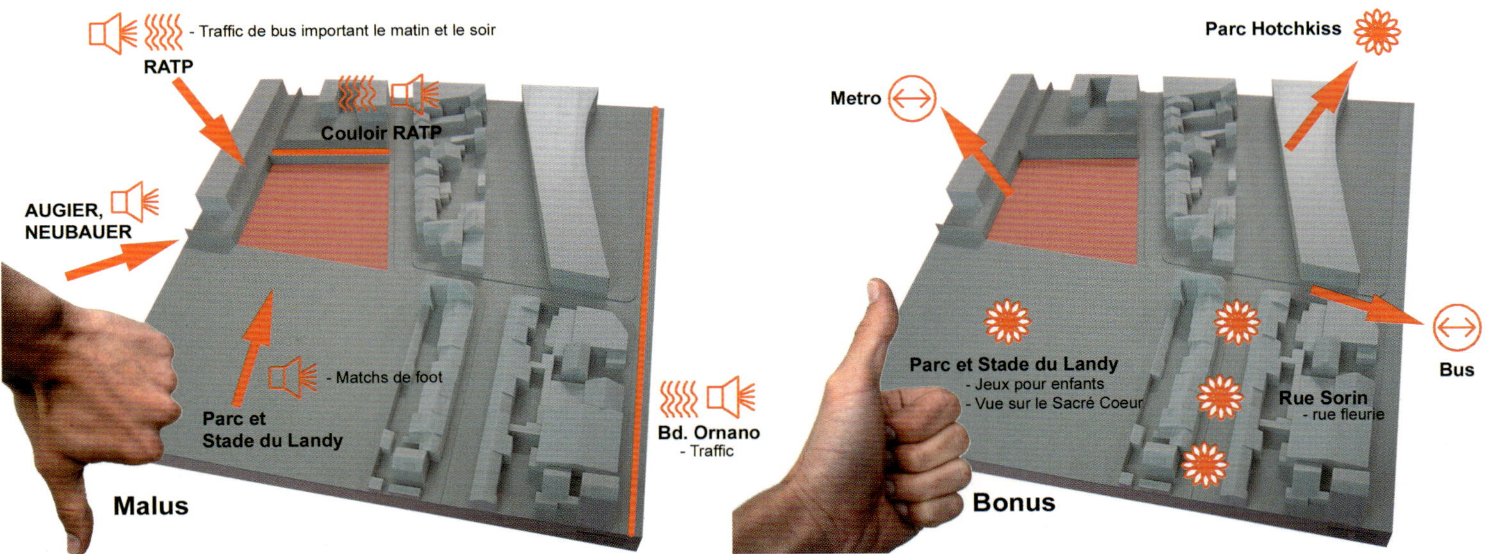

Site Analysis

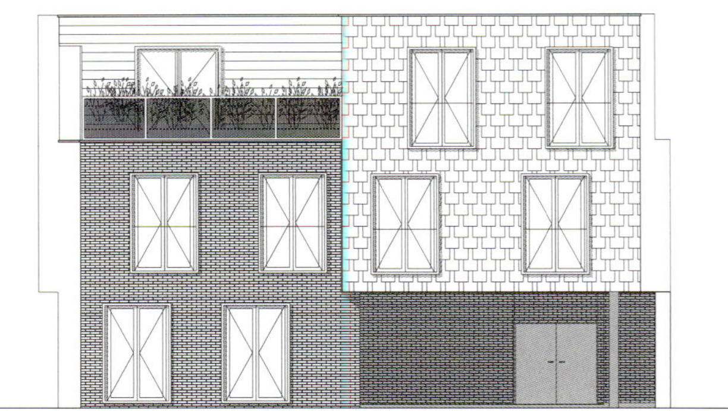

Elevation A3

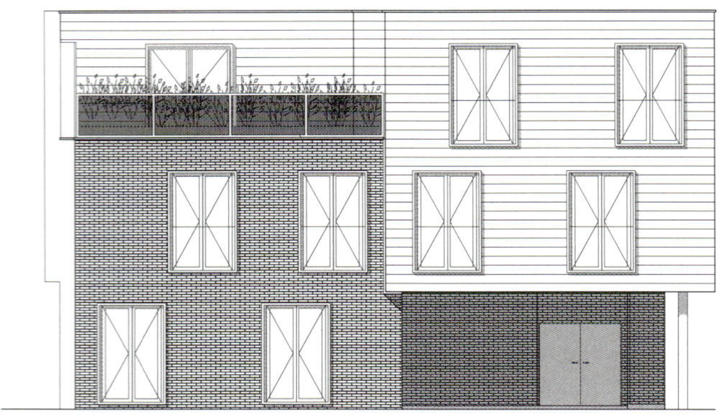

Elevation A1

Elevation A2

Elevation A4

Section

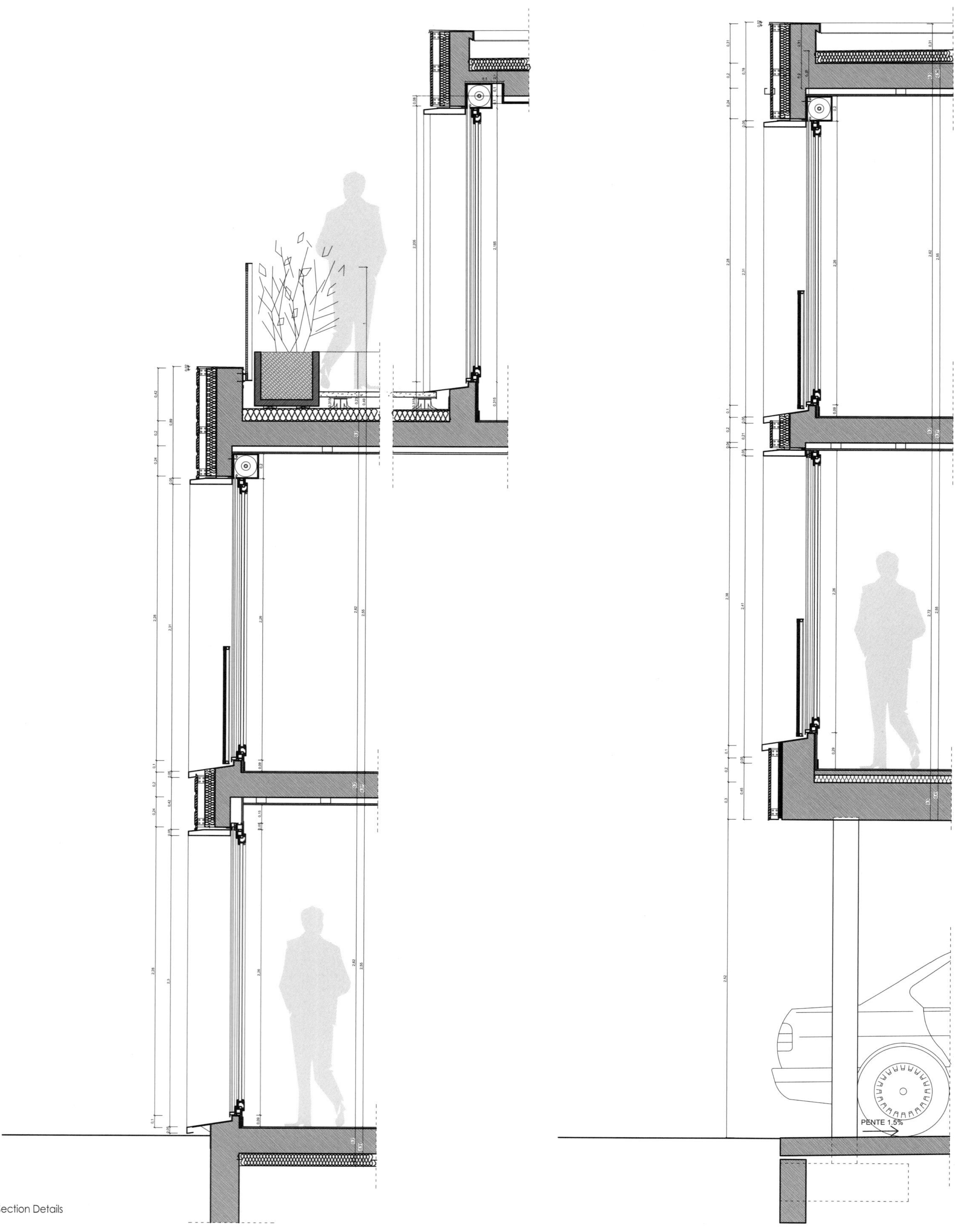

Section Details

COLÉGIO DOS PLÁTANOS, Sintra, Portugal

murmuro

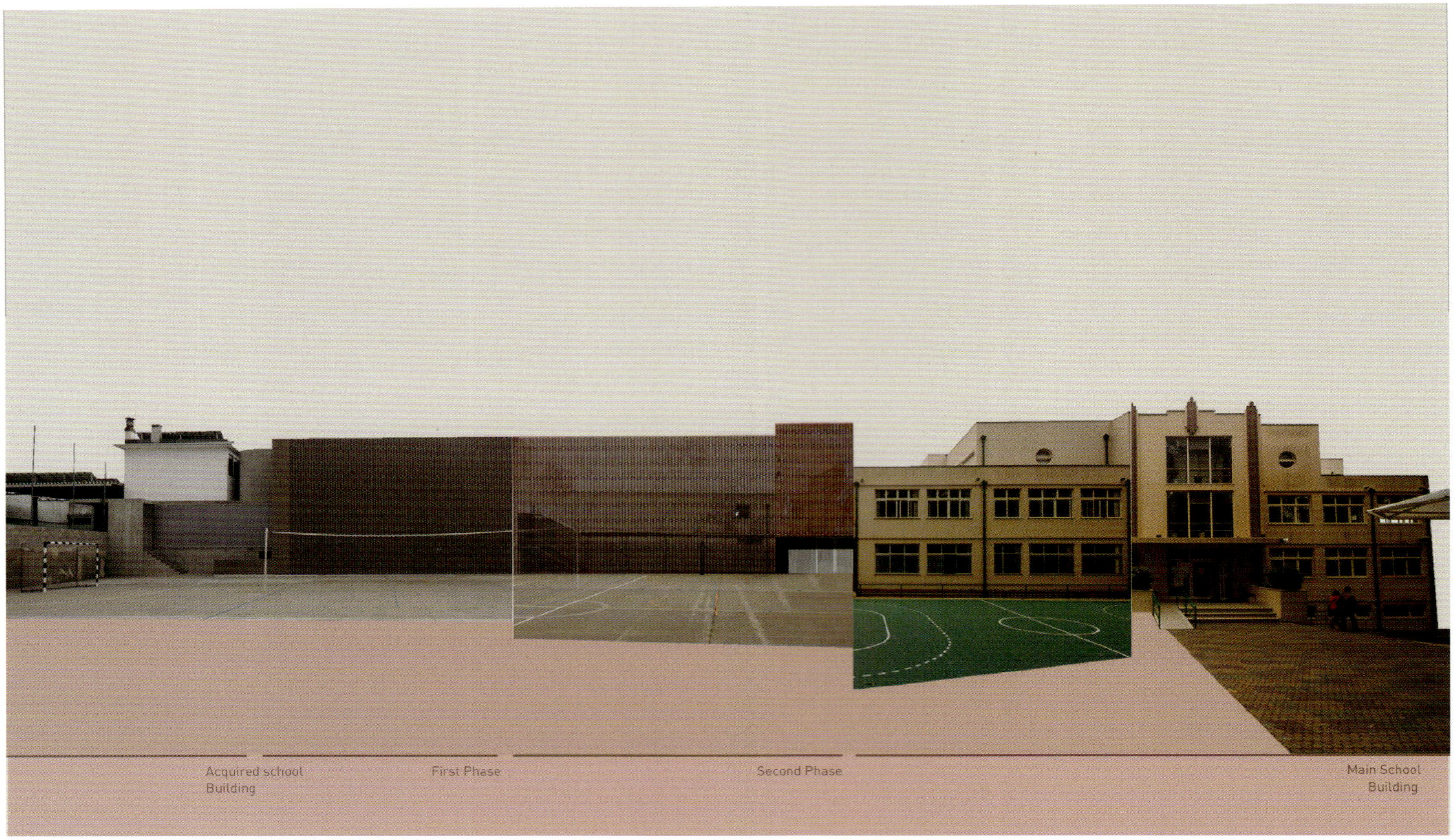

Acquired school
Building
First Phase
Second Phase
Main School
Building

Colagem

1. Classroom
2. WC
3. Office
4. Teacher's WC
5. Entrance

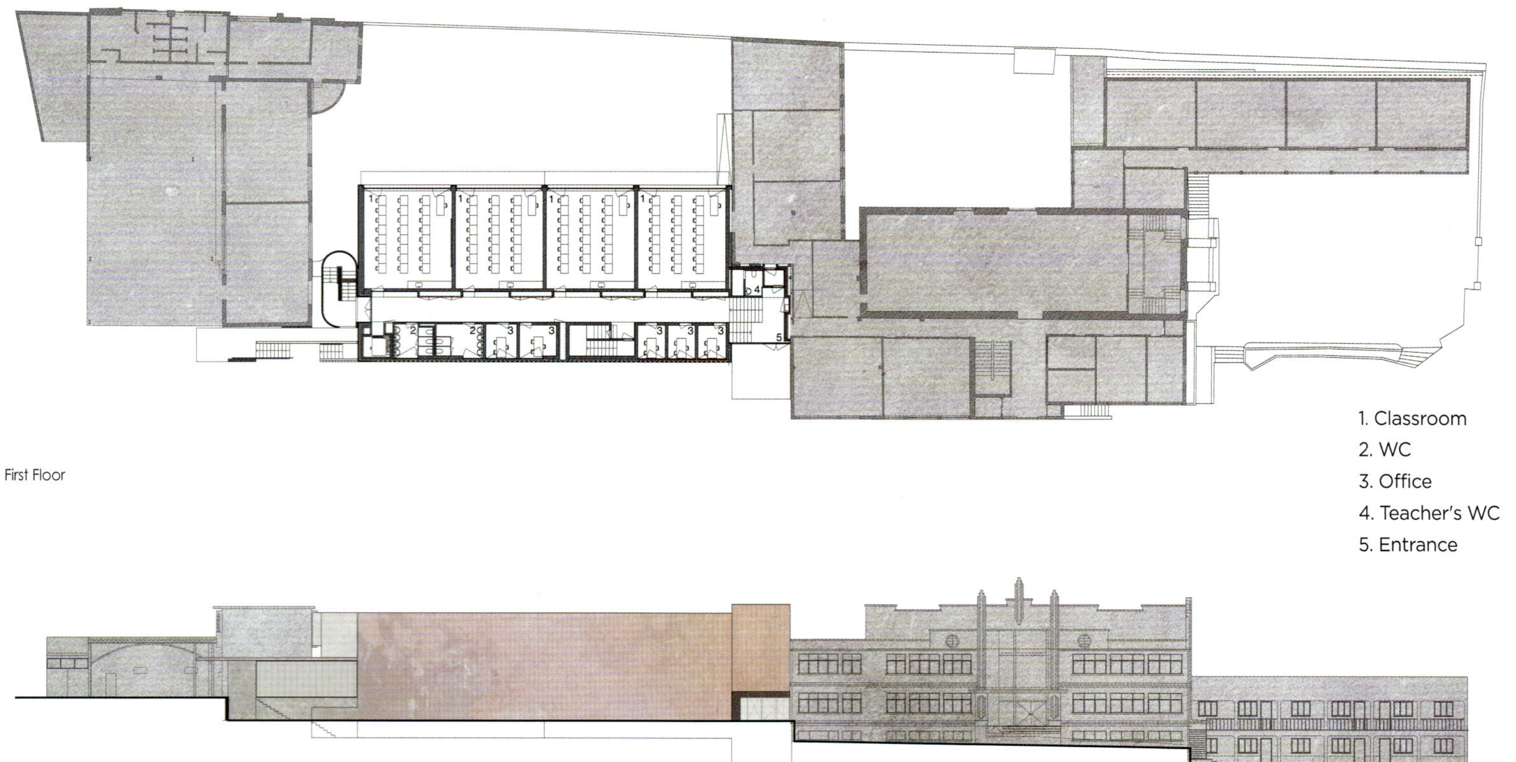

First Floor

North Elevation

© Pedro Nuno Pacheco

© Pedro Nuno Pacheco

© Pedro Nuno Pacheco

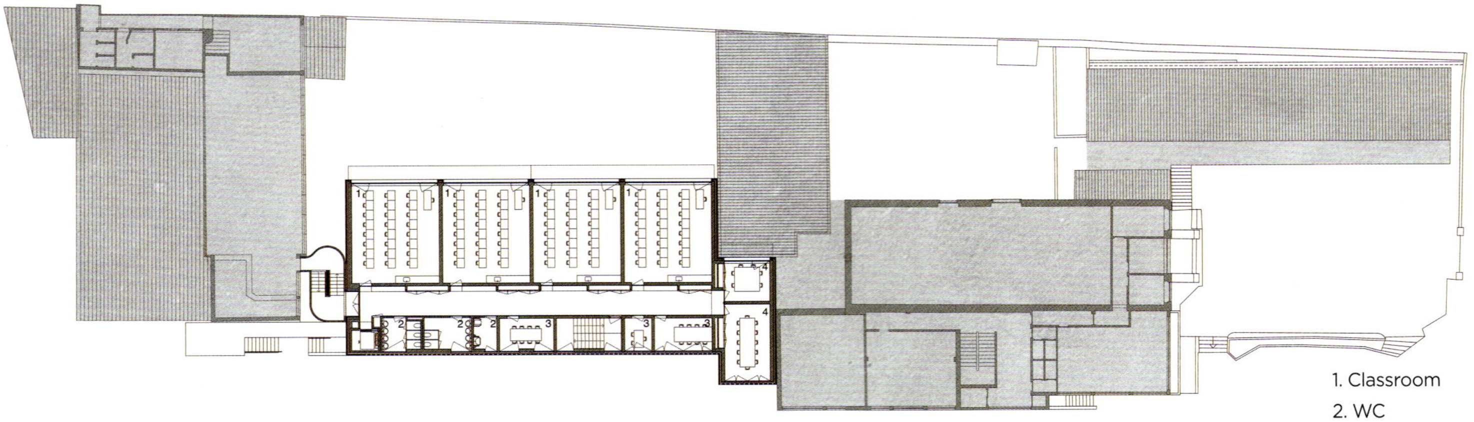

1. Classroom
2. WC
3. Office
4. Meeting Room

Second Floor

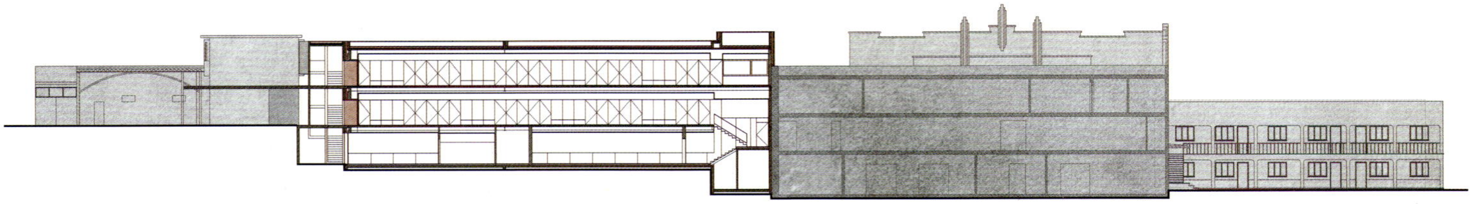

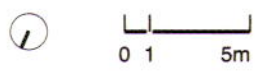

Section A

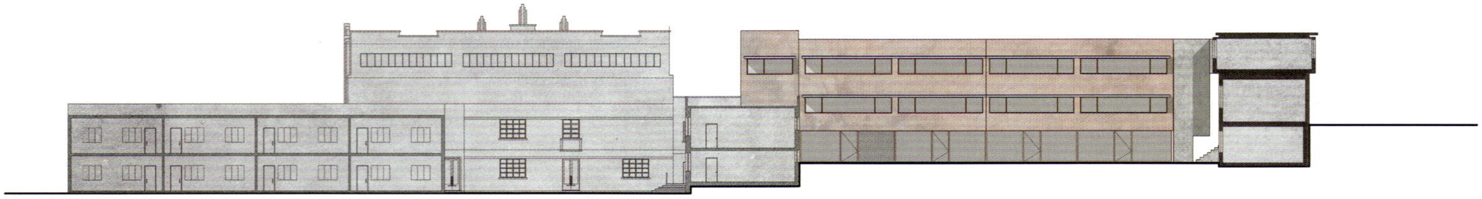

Section B

Section C/D

© Pedro Nuno Pacheco

© Pedro Nuno Pacheco

© Pedro Nuno Pacheco

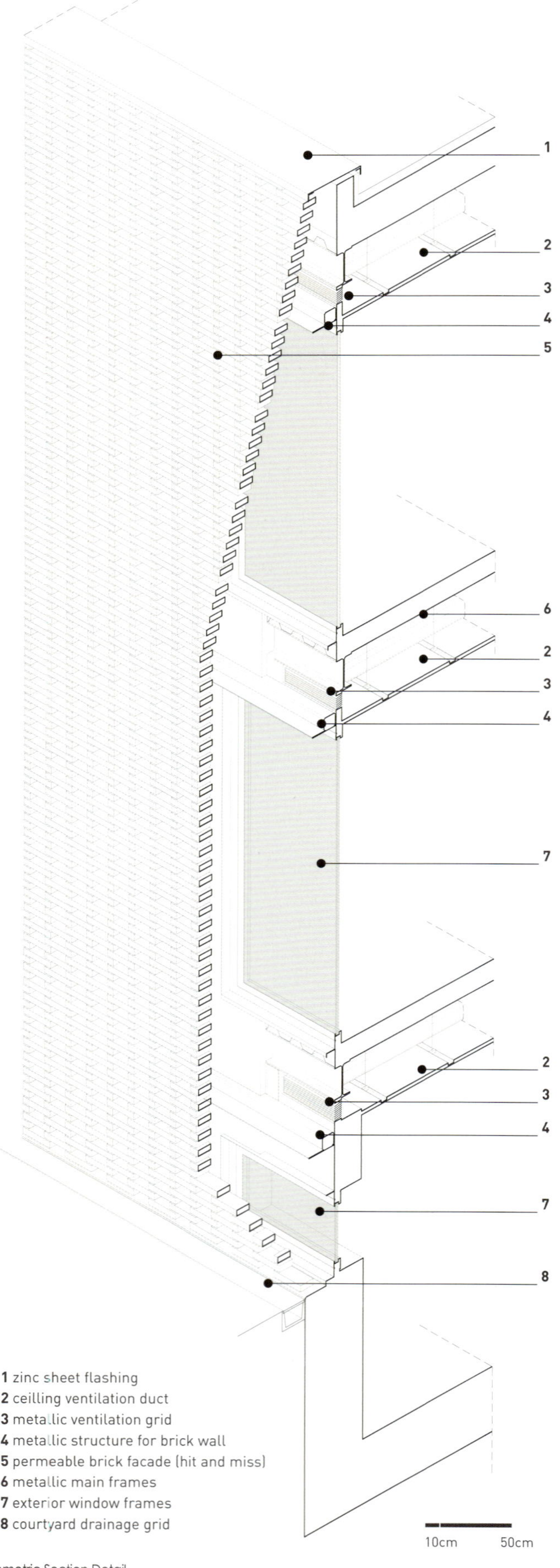

Axonometric Section Detail

Bricks Detail

© Pedro Nuno Pacheco

Library and Cafeteria for the Technical University of Applied Sciences in Wildau, Wildau, Germany

Rebecca Chestnutt, Robert Niess

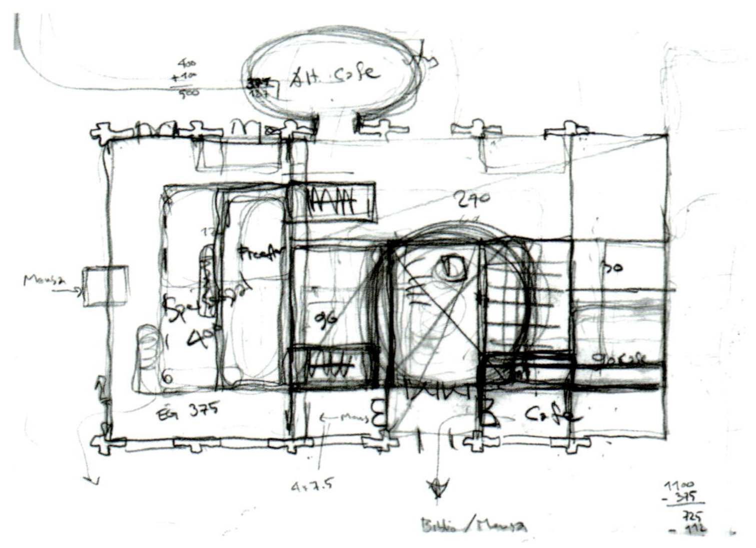

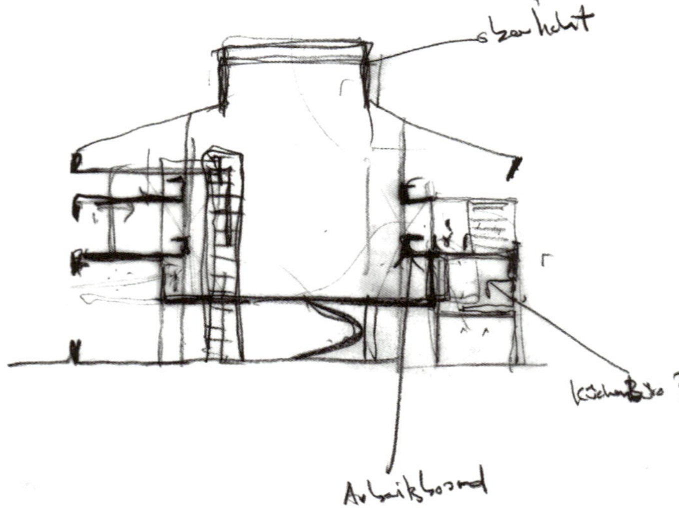

Idea Sketch

South Elevation

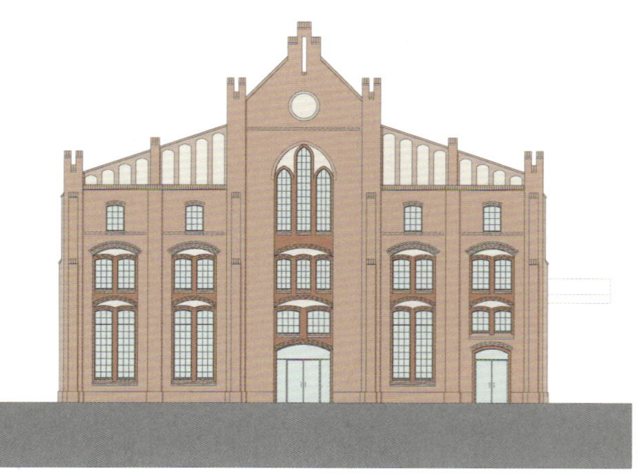

West Elevation

© Chestnutt Niess

© Chestnutt Niess

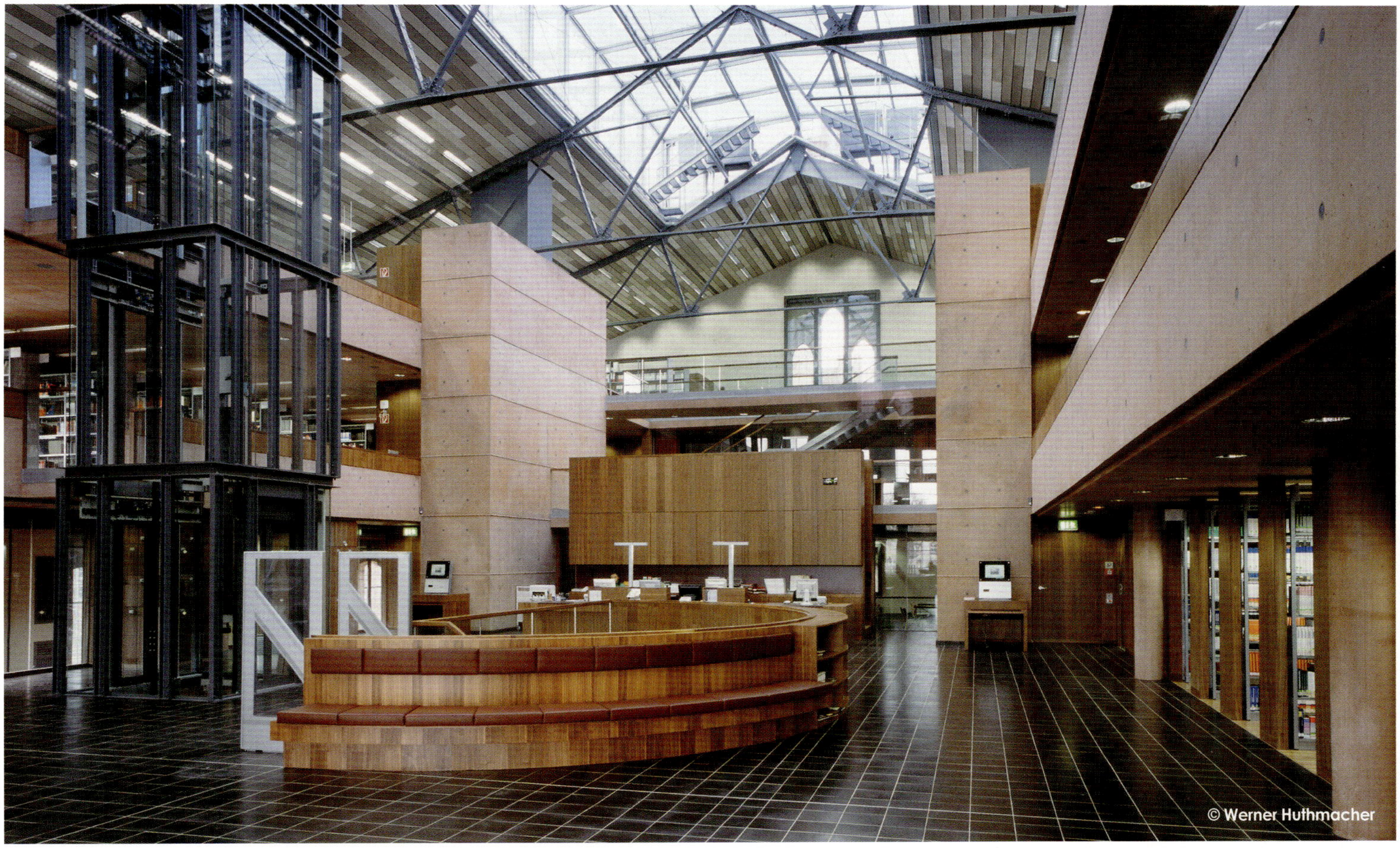

© Werner Huthmacher

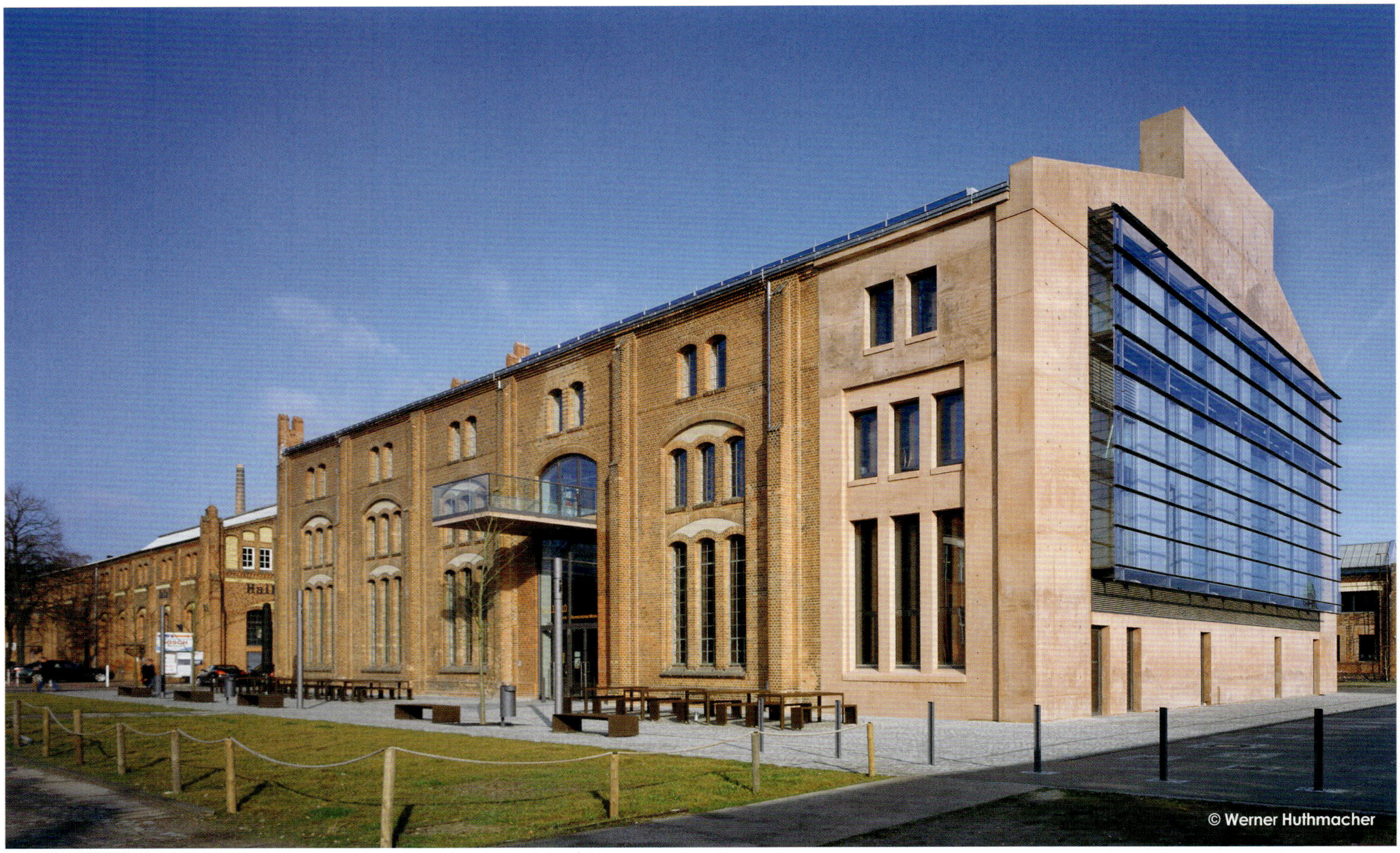

© Werner Huthmacher

Halle
FACHHOCHSCHULE

MARIA GRAZIA CUTULI SCHOOL, Herat, Afghanistan

2A+P/A, IaN+, ma0, Mario Cutuli

Idea Sketch

Study Modeling

1.

2.

3.

4.

5.

6.

7.

8.

Construction Process

Binet Business Center, Paris, France

AZC

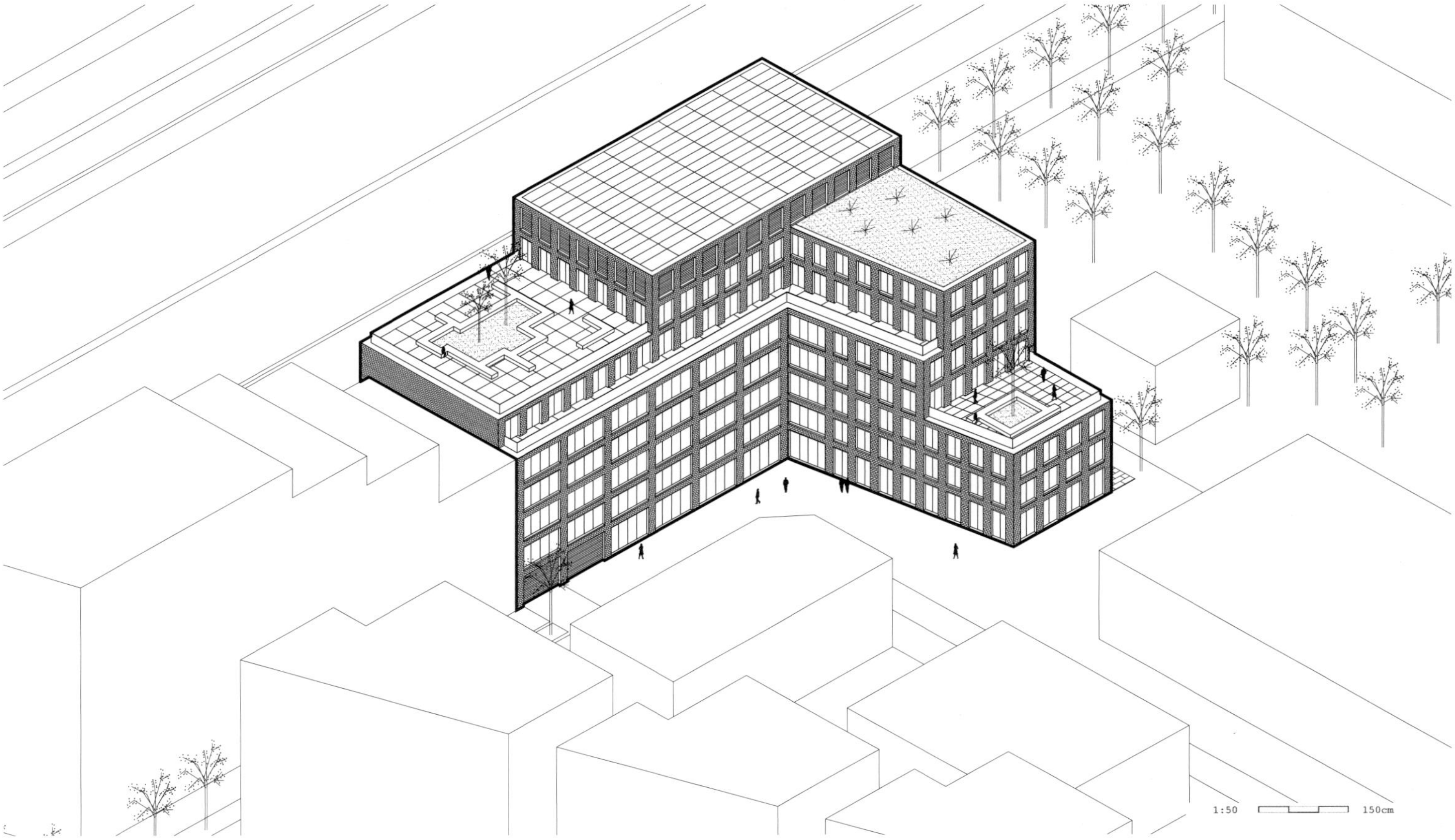

1:50 150cm

Axonometric

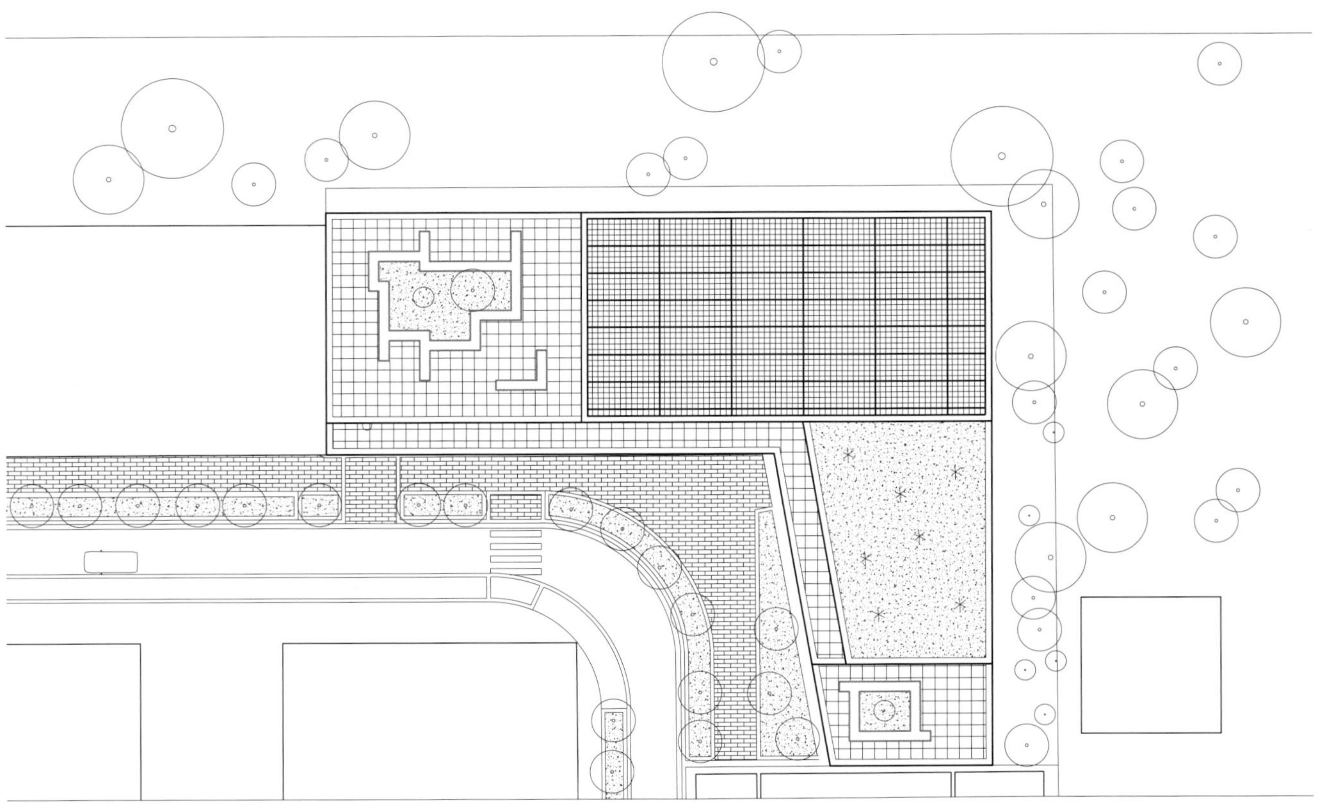

Ground Floor

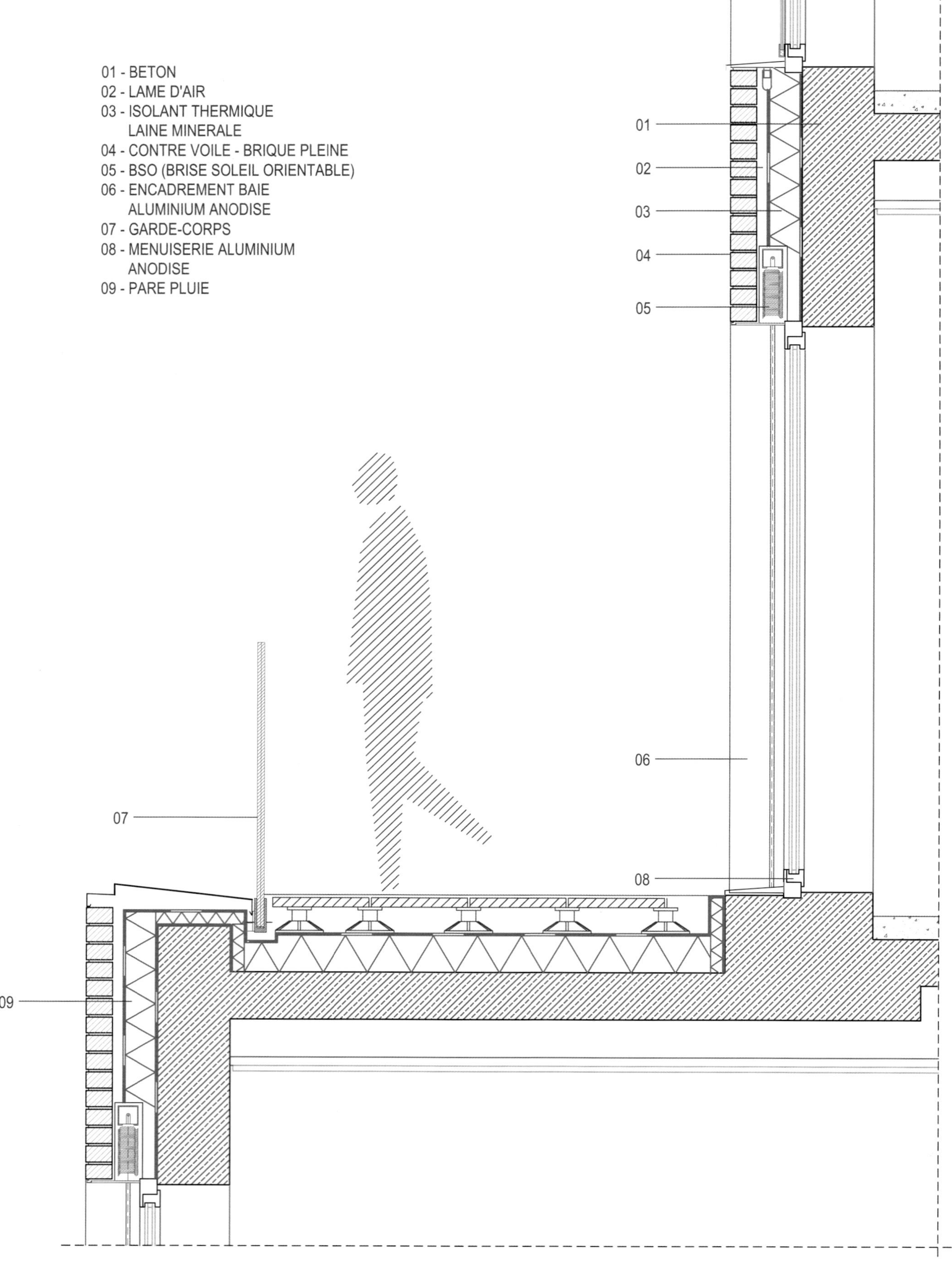
01 - BETON
02 - LAME D'AIR
03 - ISOLANT THERMIQUE
LAINE MINERALE
04 - CONTRE VOILE - BRIQUE PLEINE
05 - BSO (BRISE SOLEIL ORIENTABLE)
06 - ENCADREMENT BAIE
ALUMINIUM ANODISE
07 - GARDE-CORPS
08 - MENUISERIE ALUMINIUM
ANODISE
09 - PARE PLUIE
01
02
03
04
05
06
07
08
09

Detail

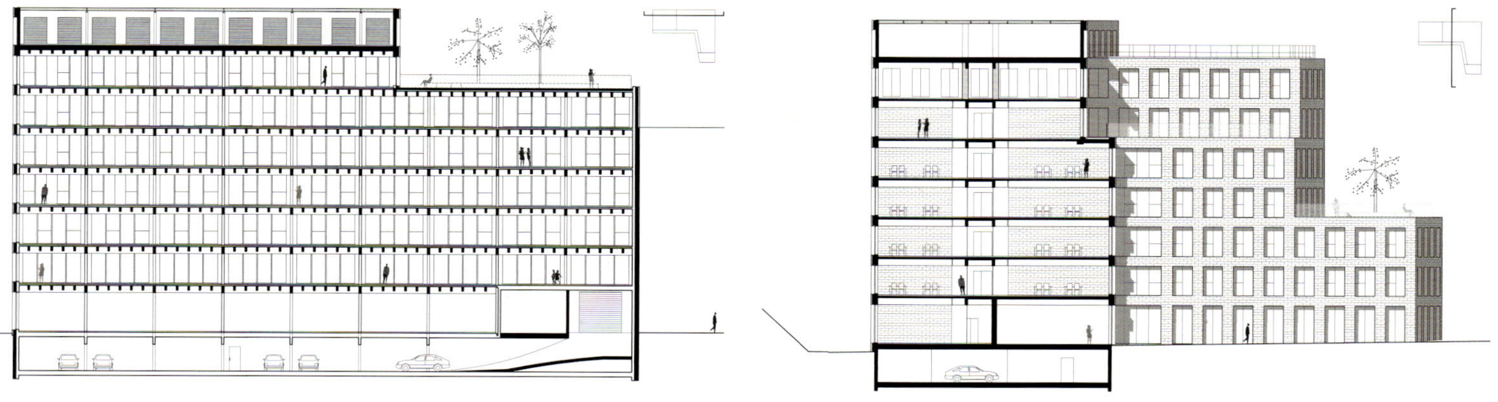

Longitudinal Elevation

Transverse Elevation

TOI TOI
ensemble
La ville

Labels Clothes Shop, Sittard, Sittard, the Netherlands

Maurice Mentjens Design

© Leon Abraas, Geleen

© Leon Abraas, Geleen

© Leon Abraas, Geleen

National and University Library II, Ljubljana, Slovenia

NL Architects

KNJIŽNICA NUK II

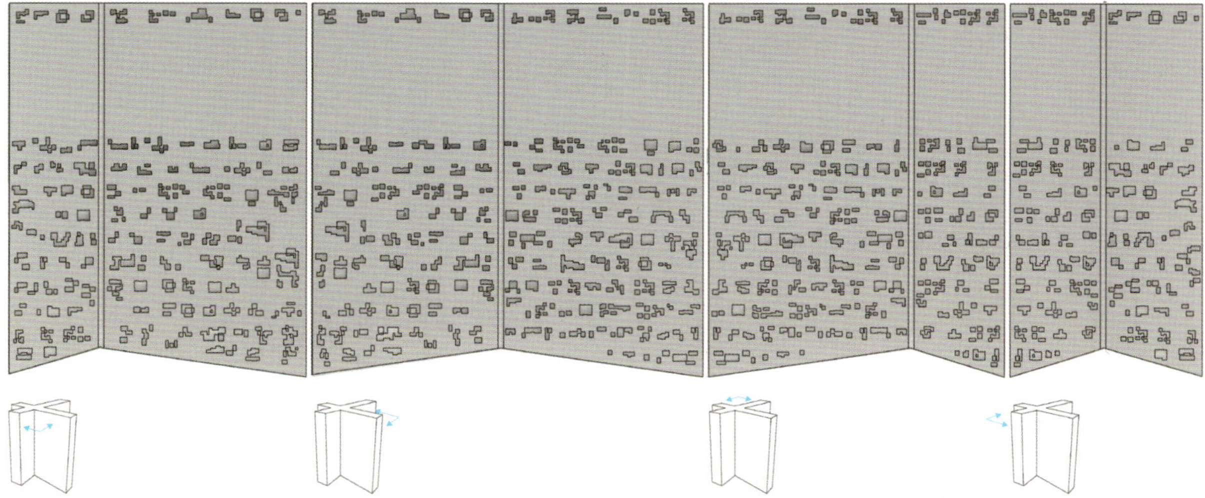

Diagram

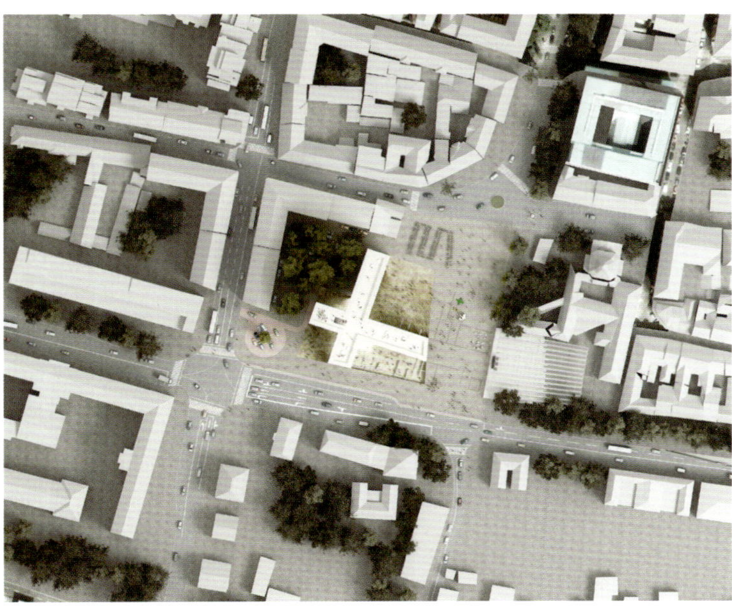

Site Plan

Axonometric

TECHNICAL
STORAGE
STORAGE
STORAGE
STORAGE
STORAGE
STORAGE
STORAGE
OFFICE
OFFICE
OFFICE
LIBRARY
LIBRARY
LIBRARY
LIBRARY
LIBRARY
LIBRARY
LIBRARY
PUBLIC
PUBLIC
PUBLIC

observation deck roof terrace
observation room cafe service/lavatories NUK cafe/restaurant
computer centre technical workshops and storage failities
special collections storage facility
storage facilities layout classical closed storage facility
storage facilities layout
storage facilities layout
computer and informatics d. shipping office and mailroom conference room special collection d.
user services division conservation and preservation conference room
reading room free access area info centre for library science storage of info centre
free access area info desk3 lavatories map and pictorial collection
music collection free access area group workrooms semi silent reading room study cells
semi silent reading room info desk2 free access area lavatories silent reading room
free access transl. booth free access newspapers lavatories
equipment room multipurpose hall
entrance reception lavatories cloakroom night reading room

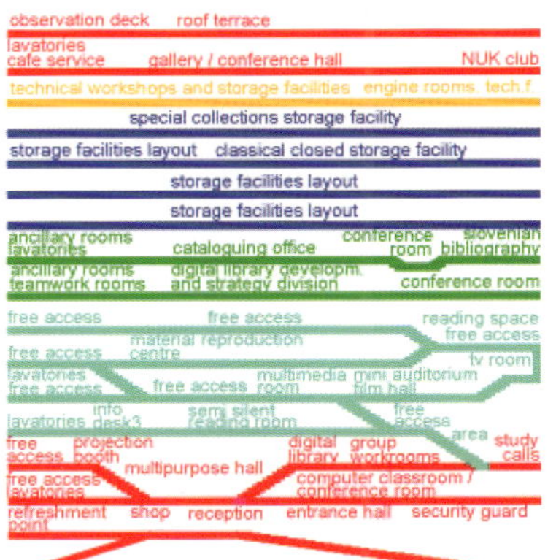

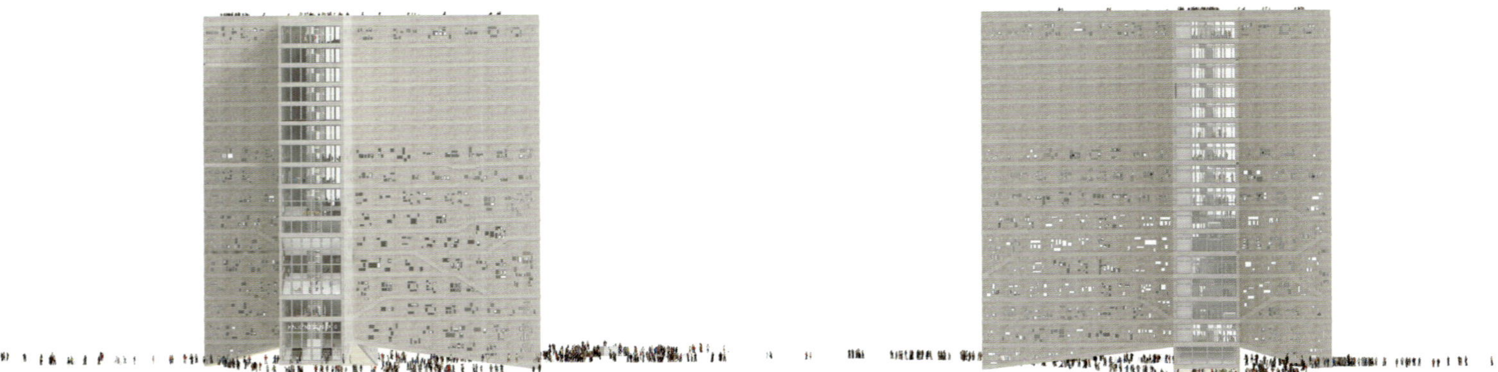

Elevation

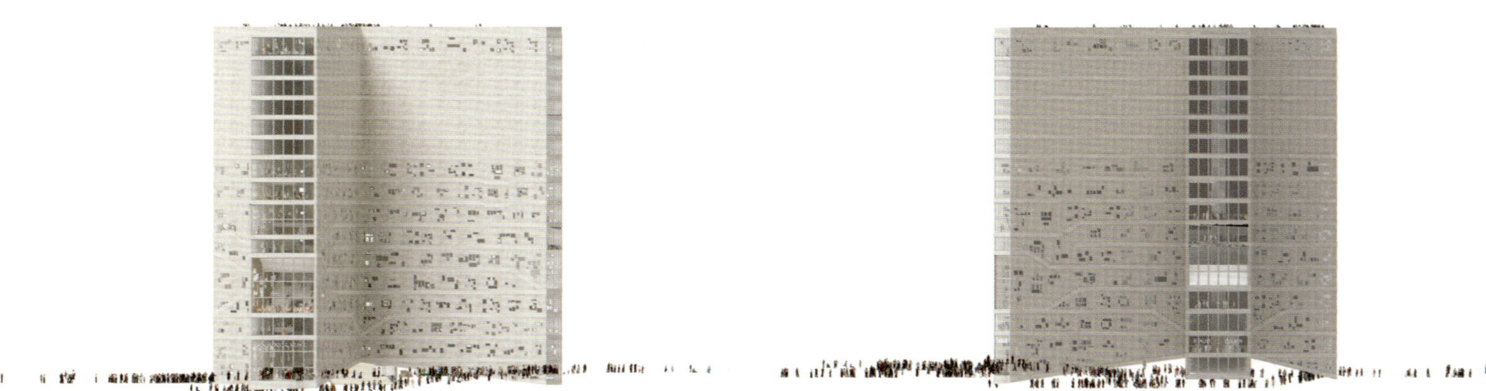

Medi Flower Oriental Hospital, Seoul, South Korea

jay is working

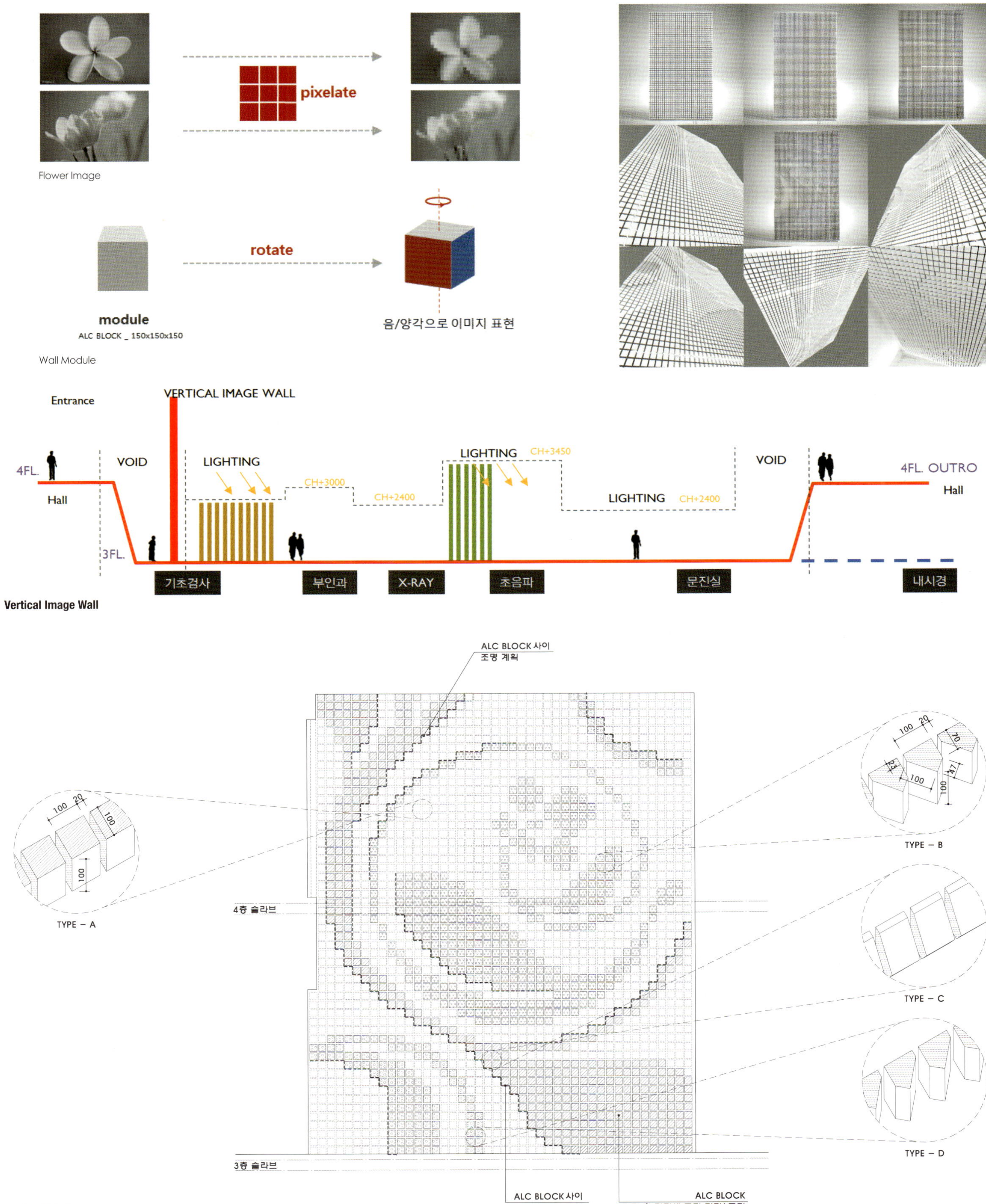
pixelate
Flower Image
rotate
module
ALC BLOCK _ 150x150x150
음/양각으로 이미지 표현
Wall Module
Entrance
VERTICAL IMAGE WALL
4FL.
VOID
LIGHTING
CH+3000
CH+2400
LIGHTING
CH+3450
LIGHTING
CH+2400
VOID
4FL. OUTRO
Hall
Hall
3FL.
기초검사
부인과
X-RAY
초음파
문진실
내시경
Vertical Image Wall
ALC BLOCK 사이
조명 계획
4층 슬라브
3층 슬라브
TYPE – A
TYPE – B
TYPE – C
TYPE – D
ALC BLOCK 사이
조명 계획
ALC BLOCK
커팅 후 지정색 도장 마감/ 코팅
section Detail

1.

2.

3.

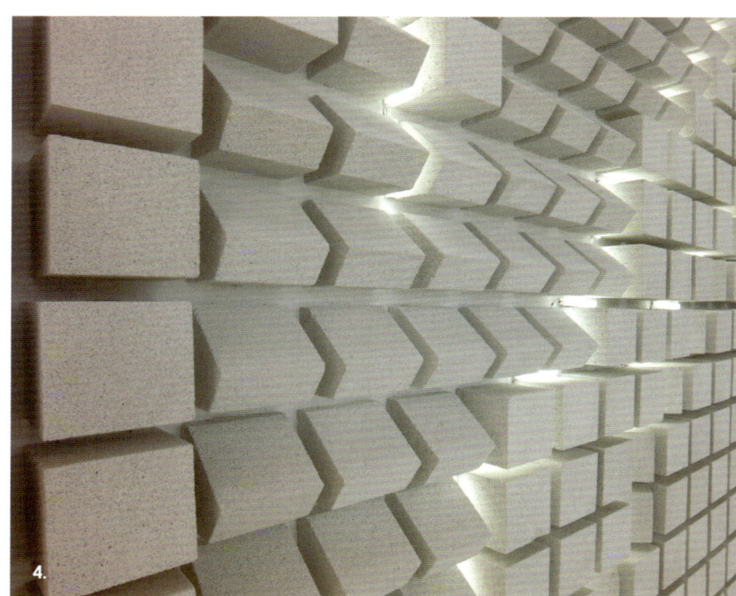
4.

5.

6.

Construction Process

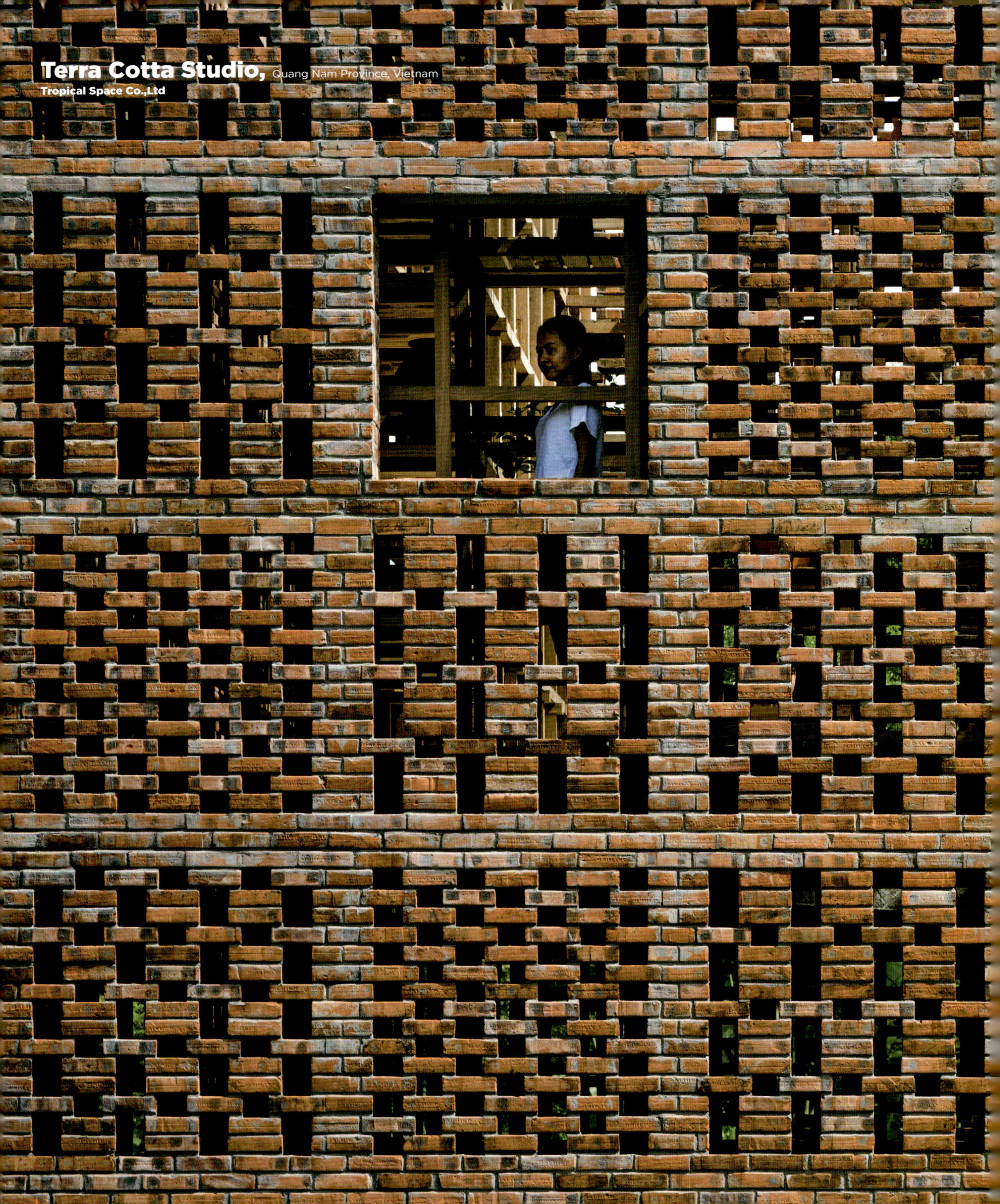

Terra Cotta Studio, Quang Nam Province, Vietnam
Tropical Space Co.,Ltd

© Oki Hiroyuki

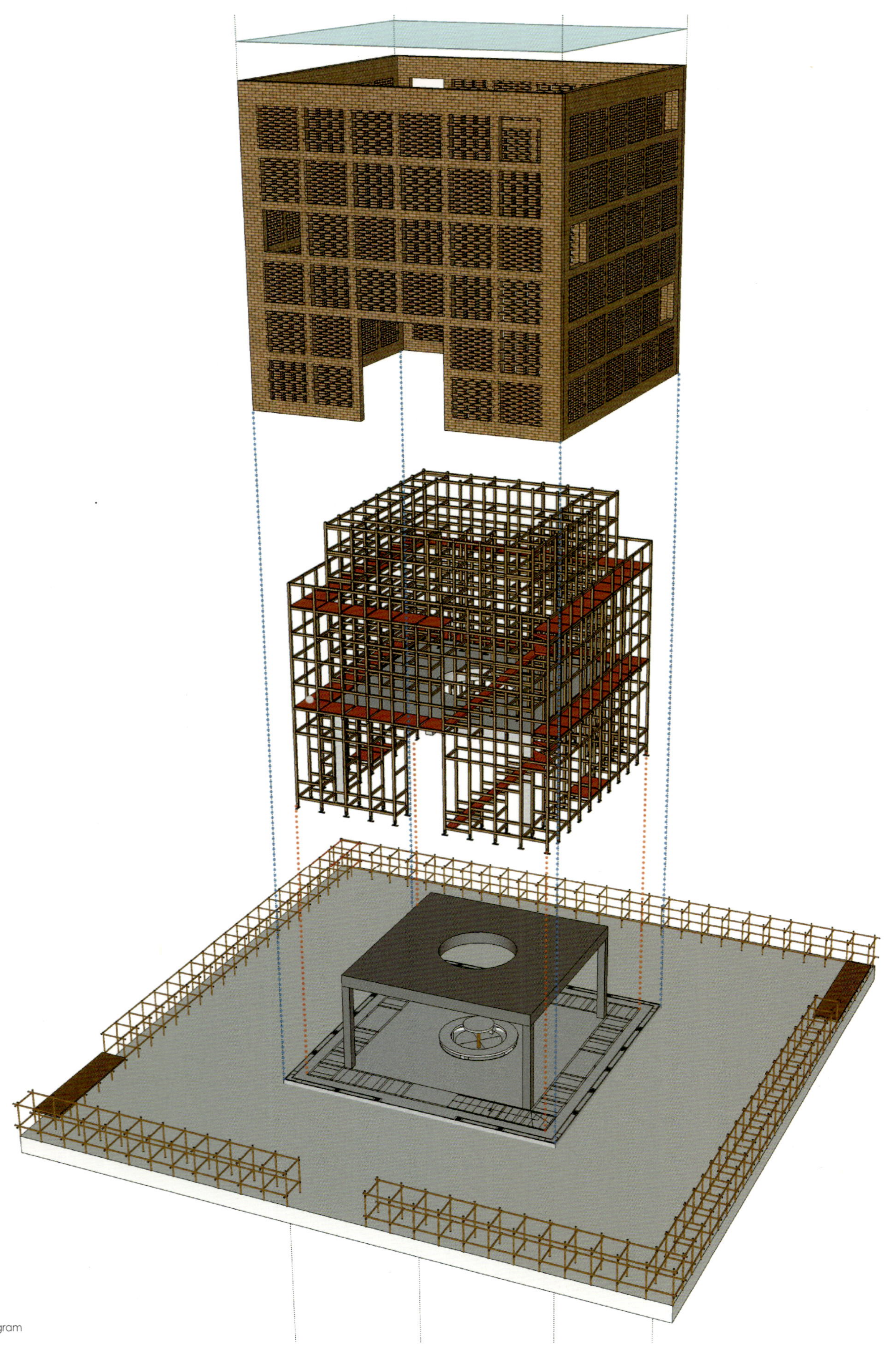

Architectural Diagram

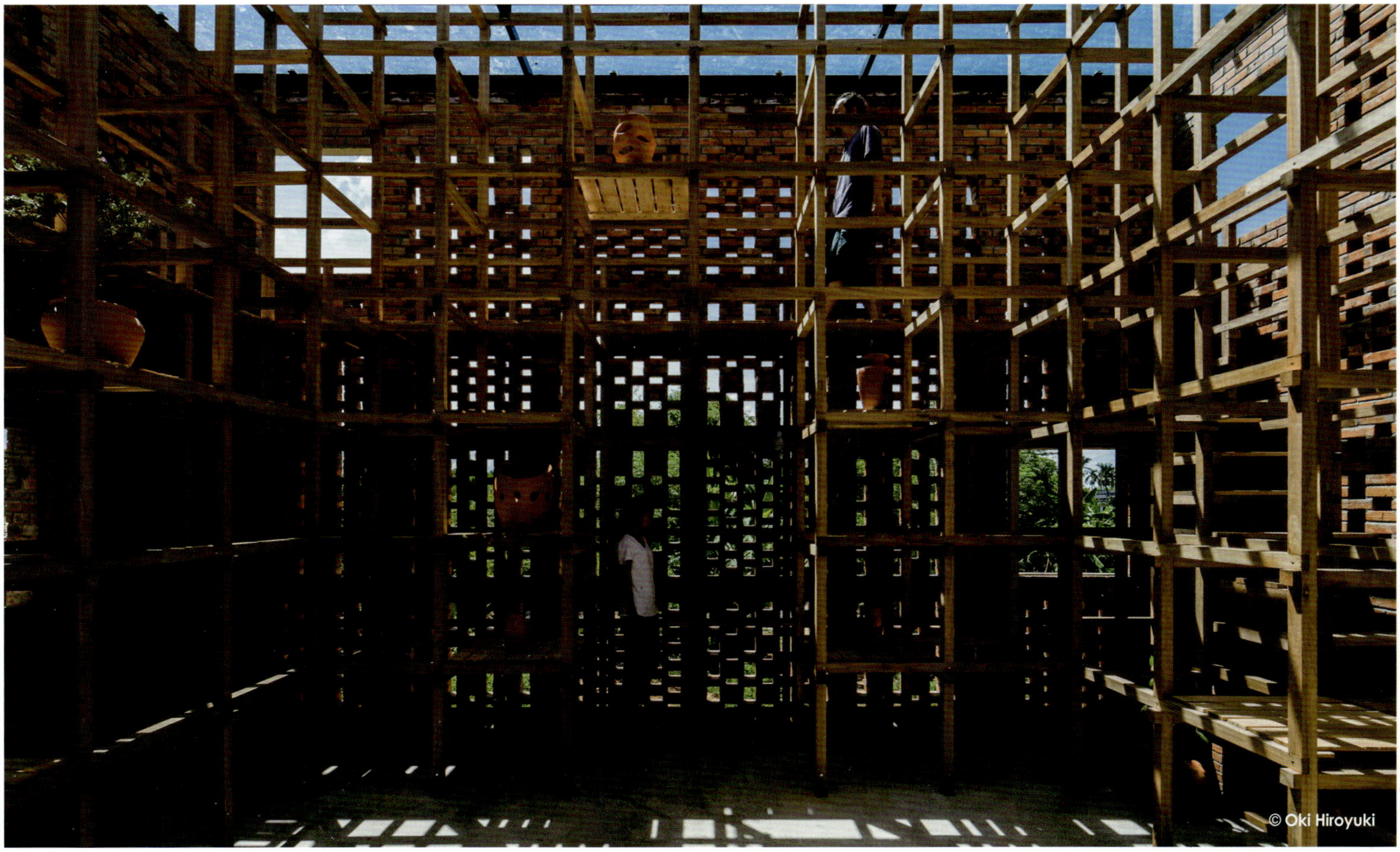

© Oki Hiroyuki

© Oki Hiroyuki

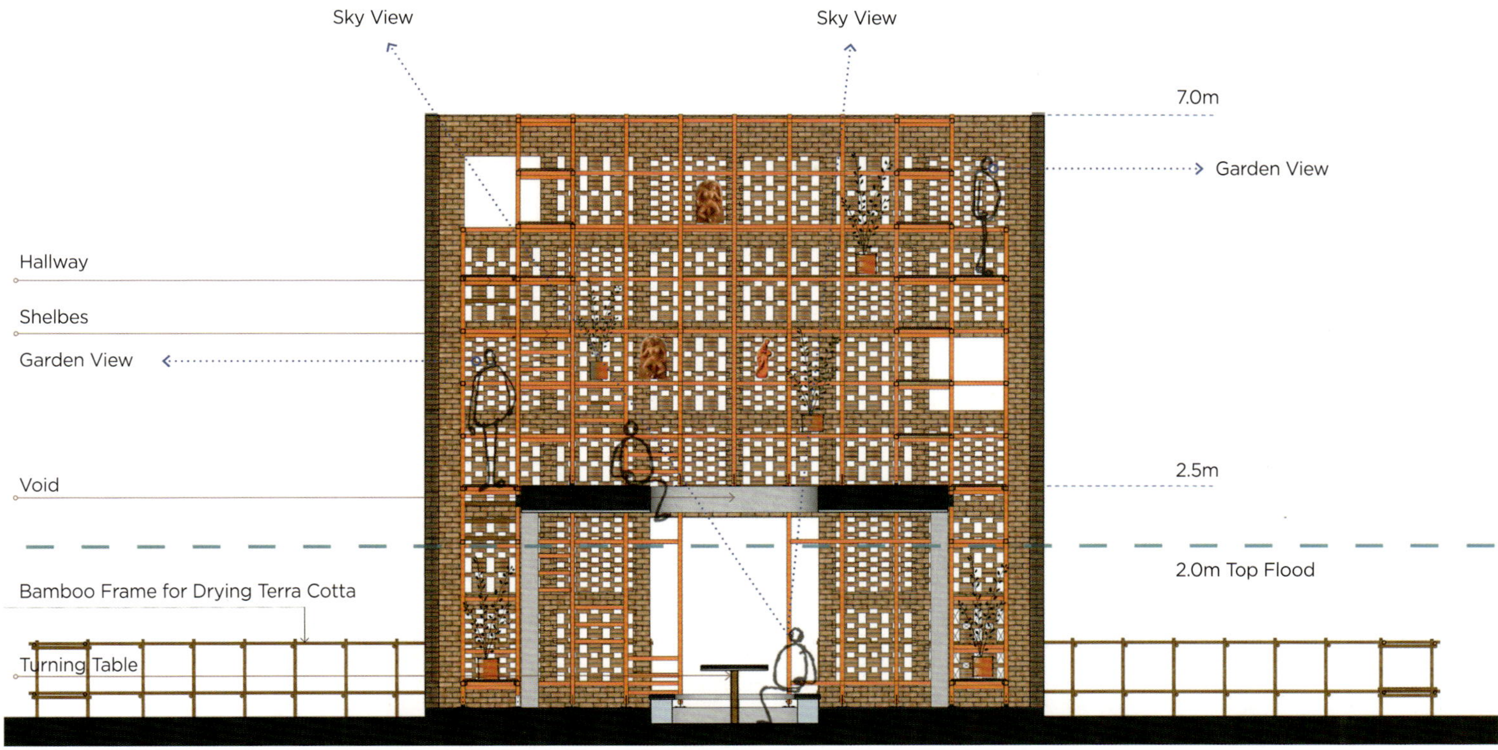

Section

© Oki Hiroyuki

WAP Art Space, Seoul, Korea

DAVIDE MACULLO ARCHITECTS

© Yousub Song

Idea Sketch

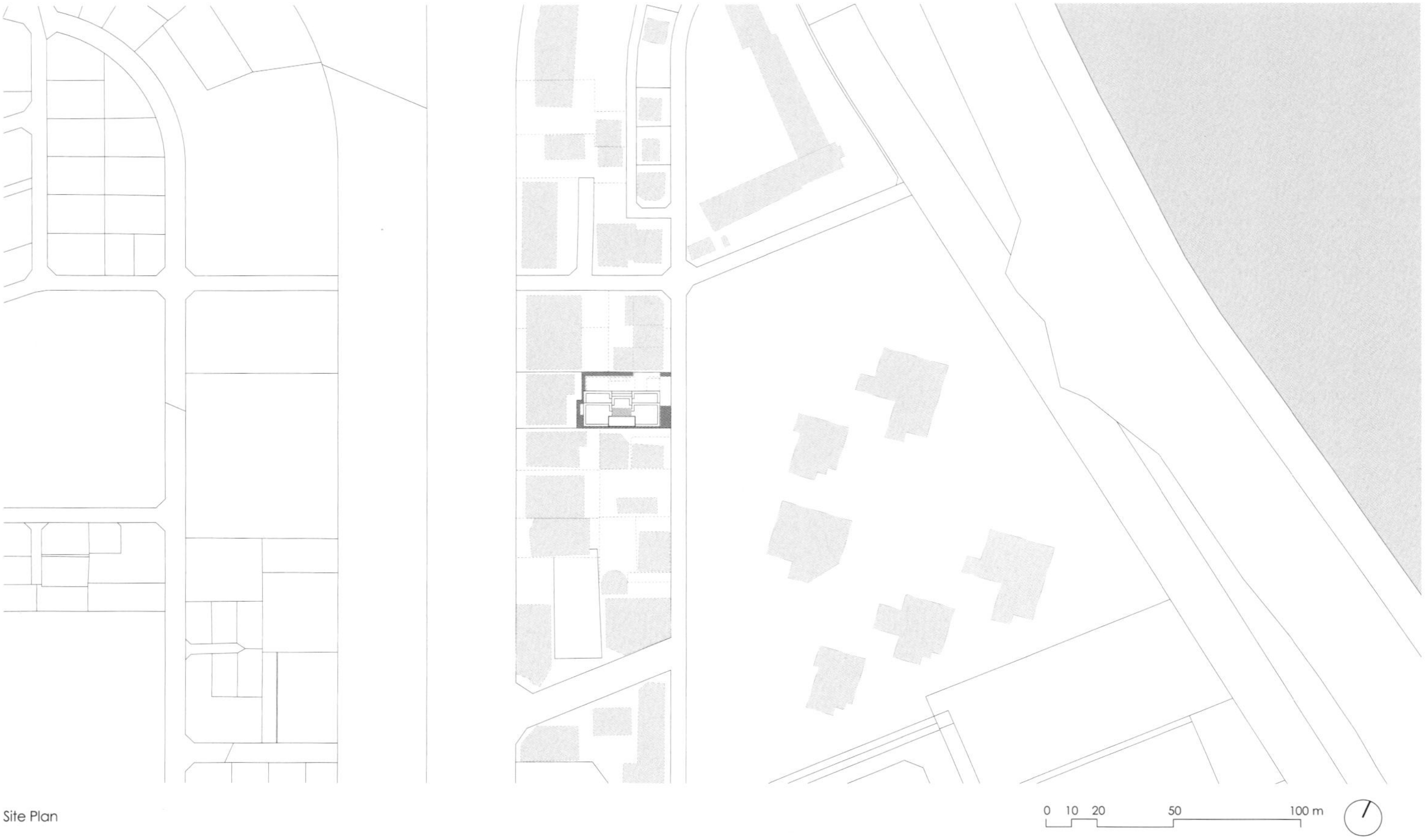

Site Plan

0 10 20 50 100 m

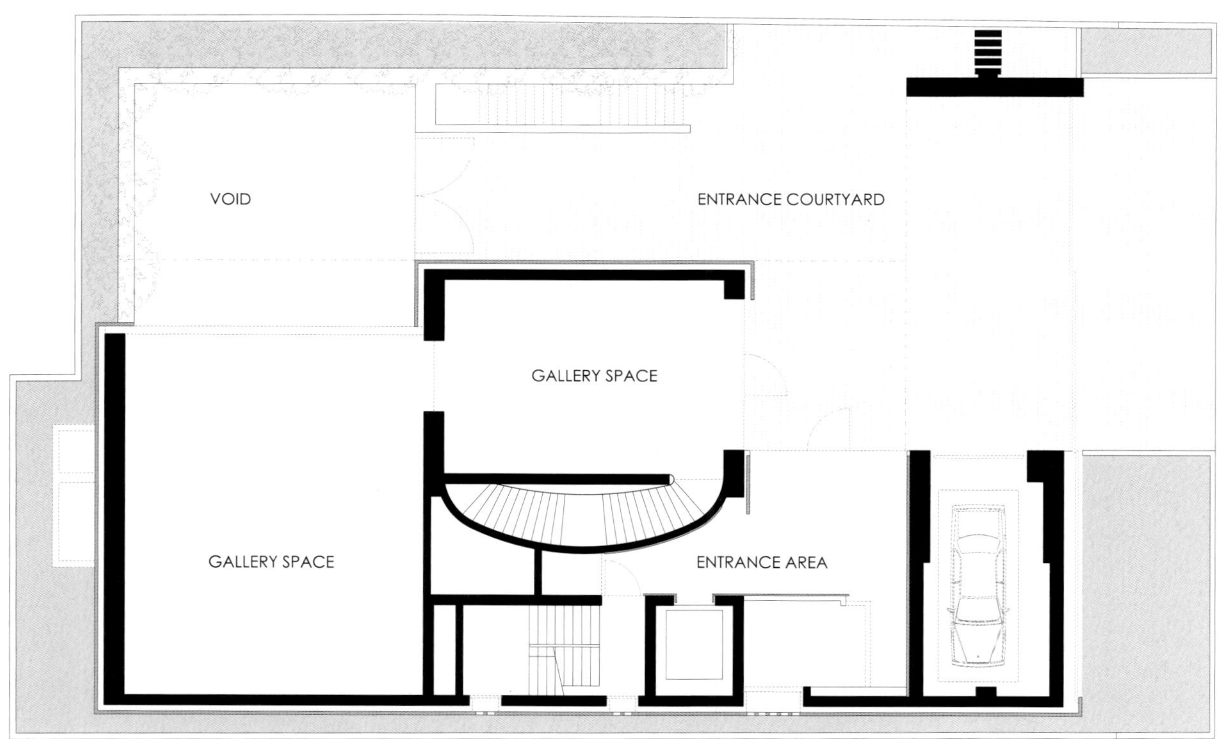

Ground Floor

0 1 2 10 m

© Yousub Song

© Yousub Song

© Yousub Song

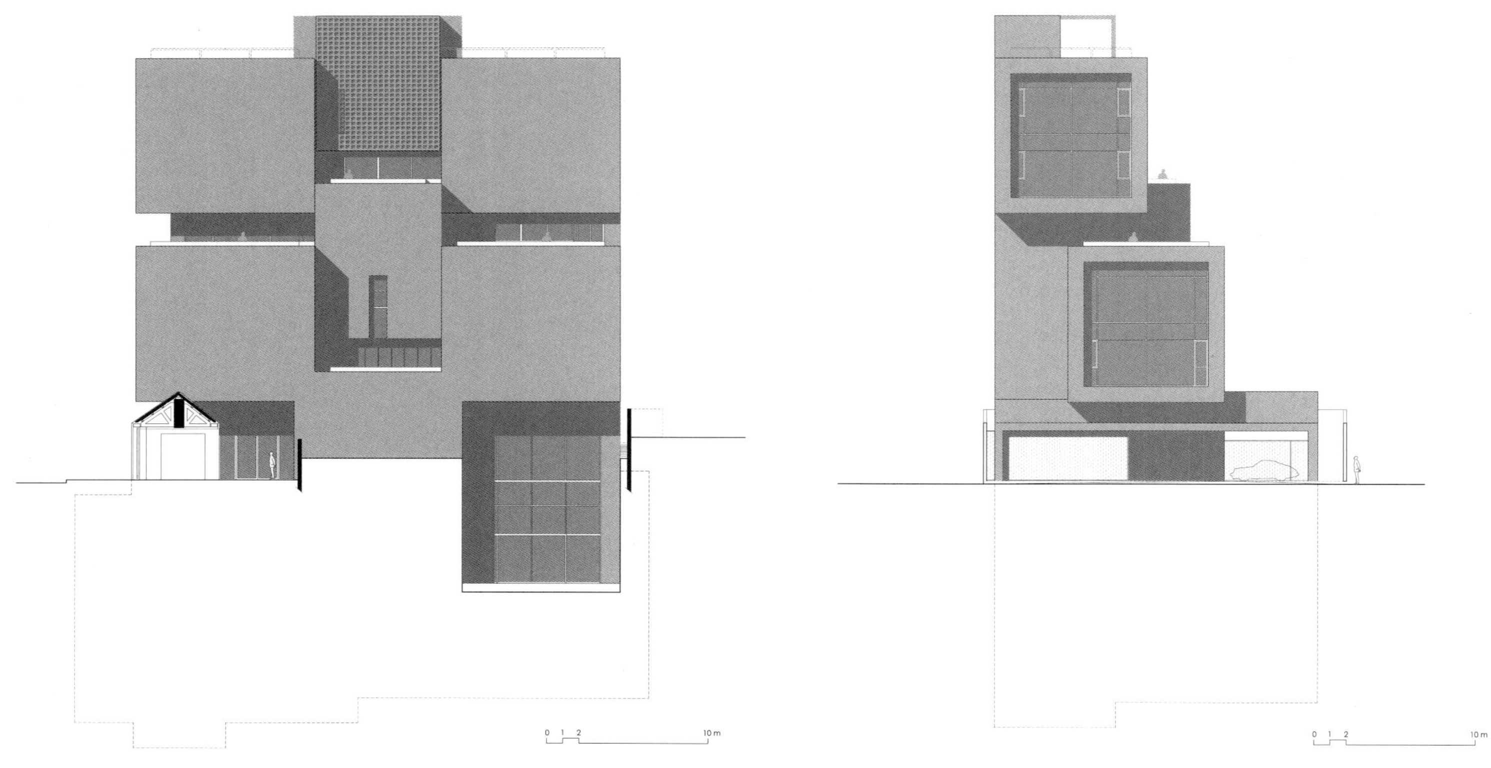

North Elevation

East Elevation

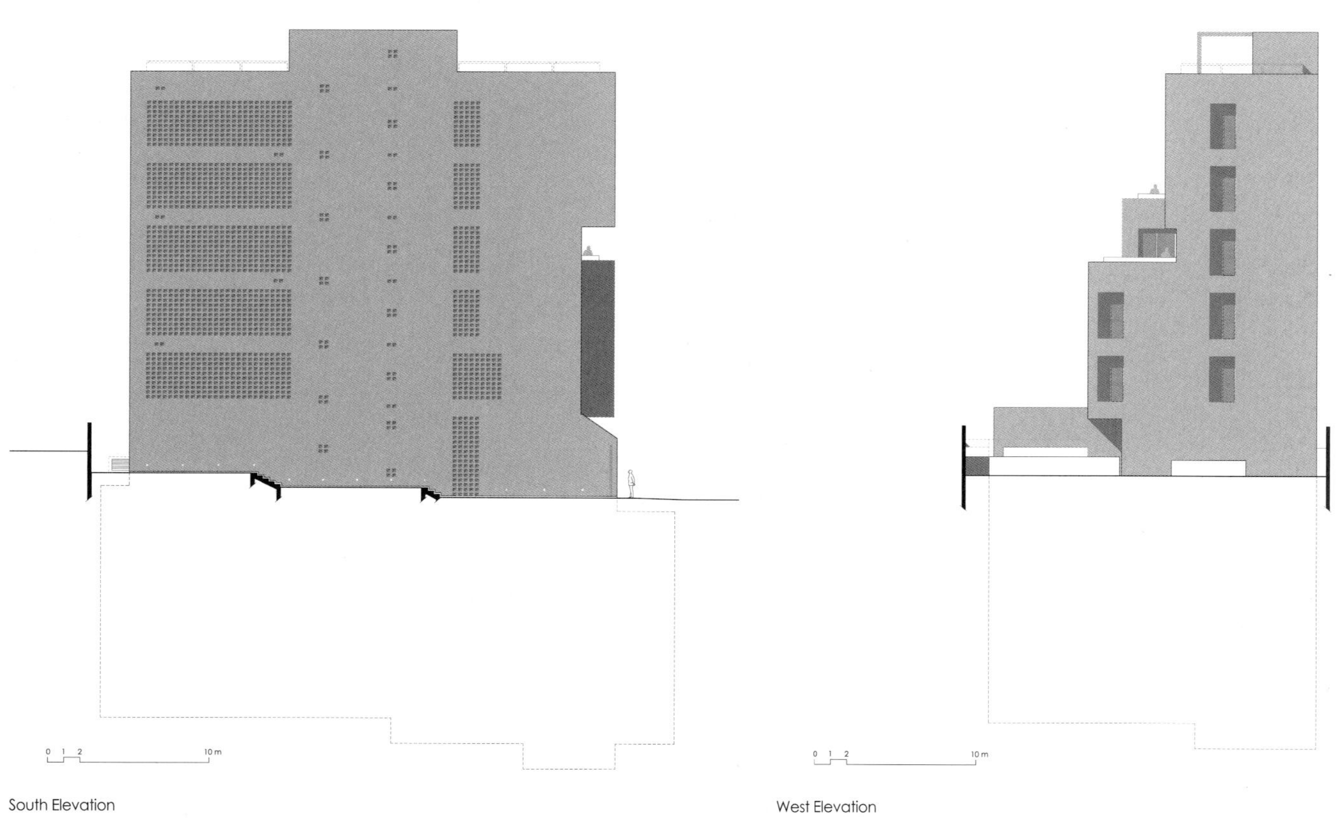

South Elevation

West Elevation

© Yousub Song

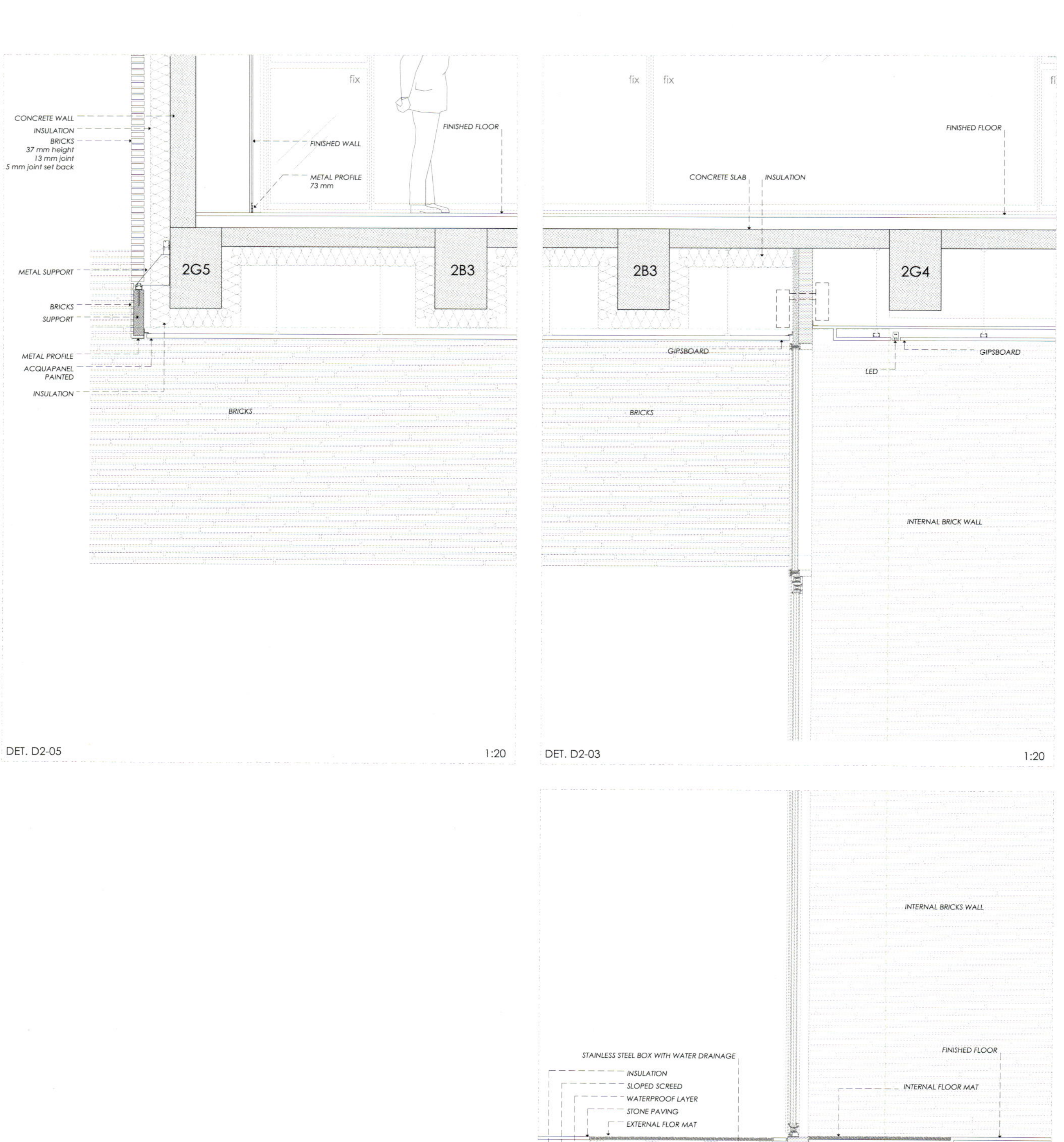

INTERNAL BRICKS WALL
STAINLESS STEEL BOX WITH WATER DRAINAGE
INSULATION
SLOPED SCREED
WATERPROOF LAYER
STONE PAVING
EXTERNAL FLOR MAT
FINISHED FLOOR
INTERNAL FLOOR MAT
1G4
CONCRETE
SLAB
GIPSBOARD
FINISHED WALL
LIGHT + VENT.
GIPSBOARD
FINISHED WALL
DET. D2-01
1:20

Section Detail

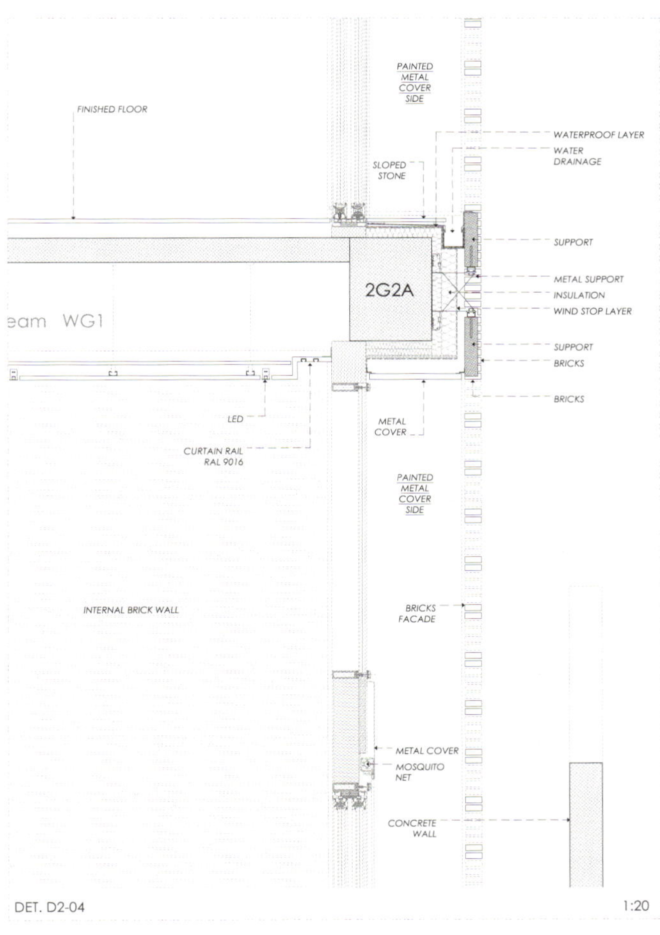

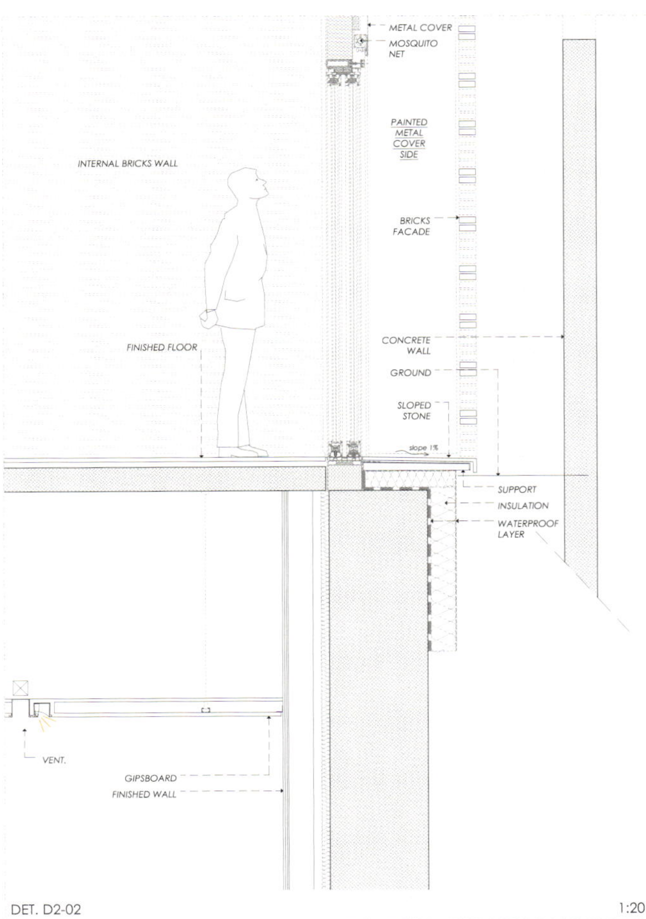

Construction Process

제일정
소나무아
COFFEE &
PIZZA

© Yousub Song

Warehouse 17C, Madrid, Spain

Arturo Franco, Fabrice van Teslaar

© Carlos Fernandez Piñar

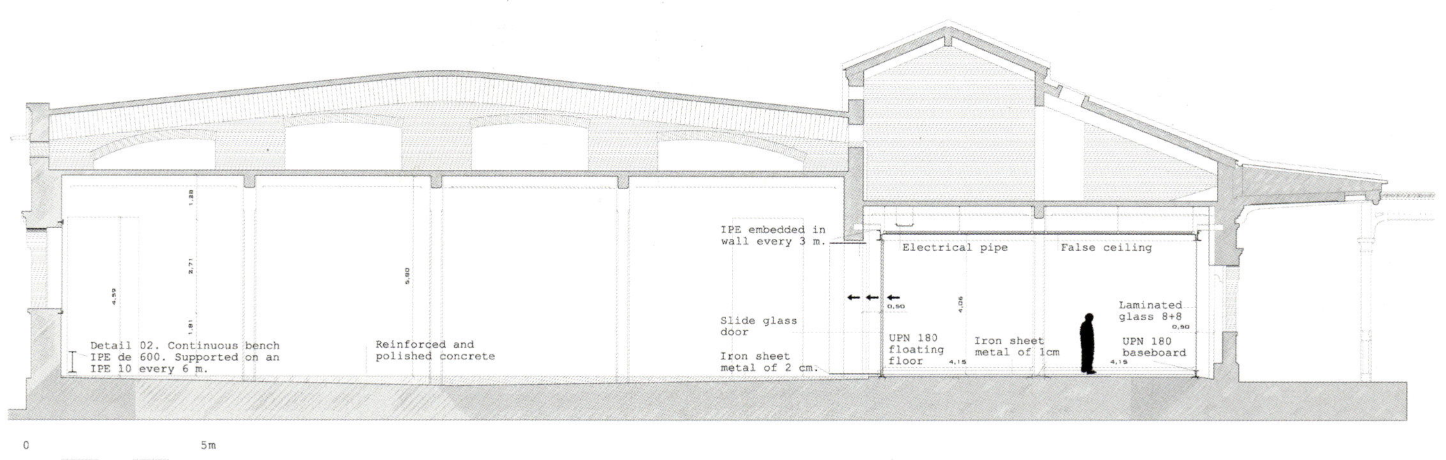

© Carlos Fernandez Piñar

© Carlos Fernandez Piñar

Entrepreneur Offices. Madridejos, Castilla La Mancha, Spain

OOIIO Architecture

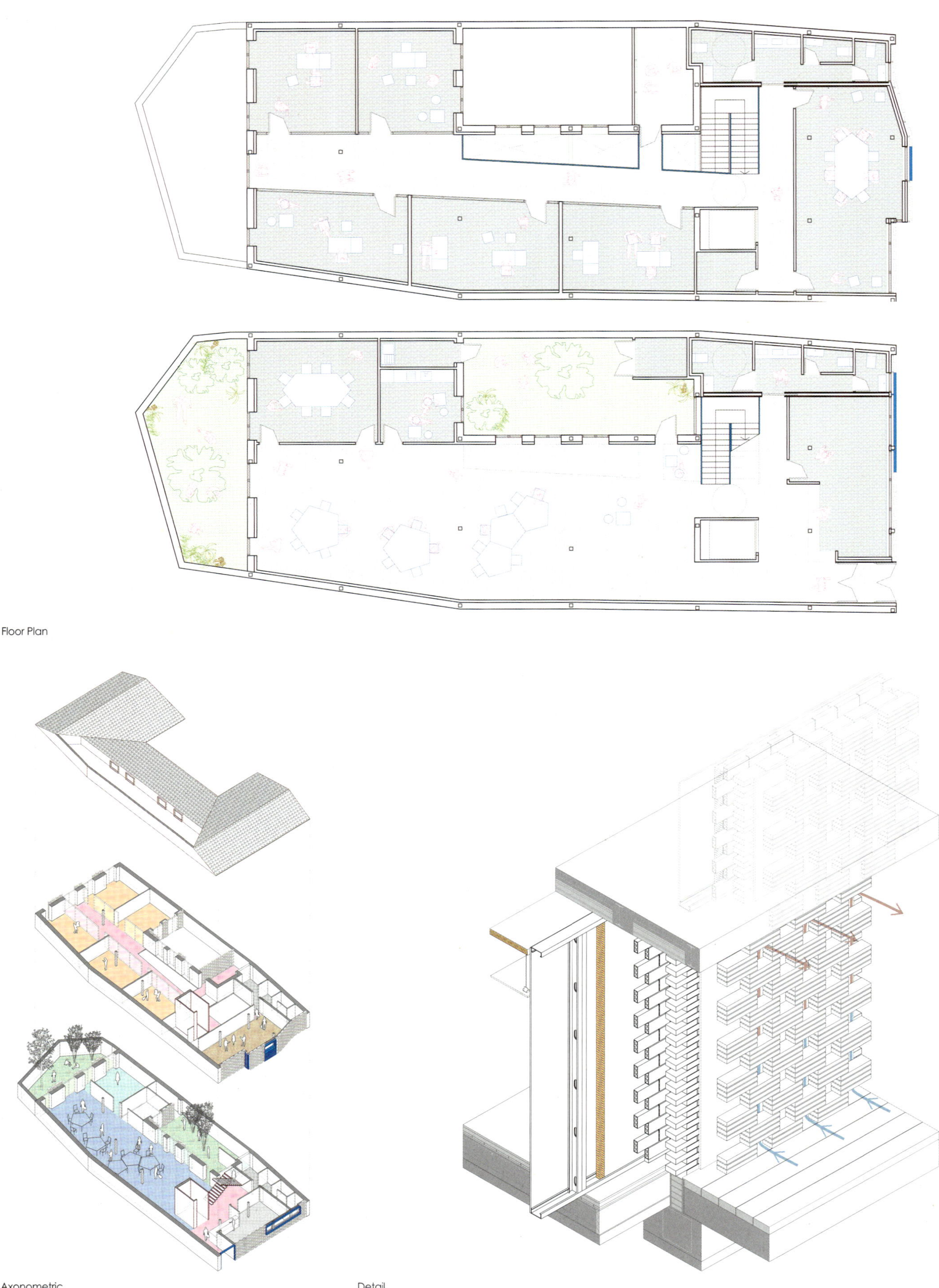

Floor Plan

Axonometric

Detail

Concept

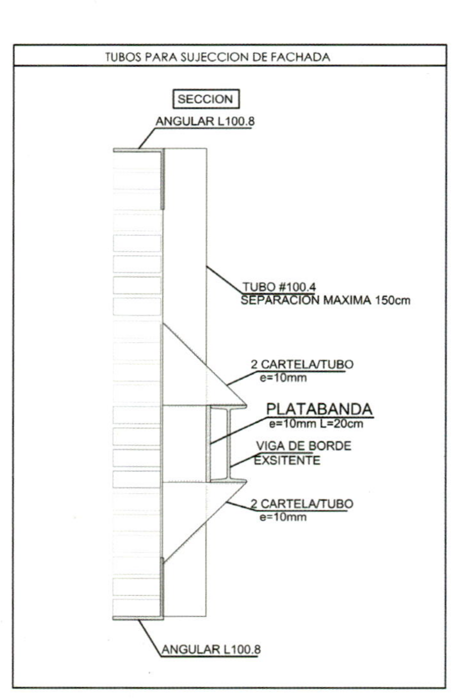

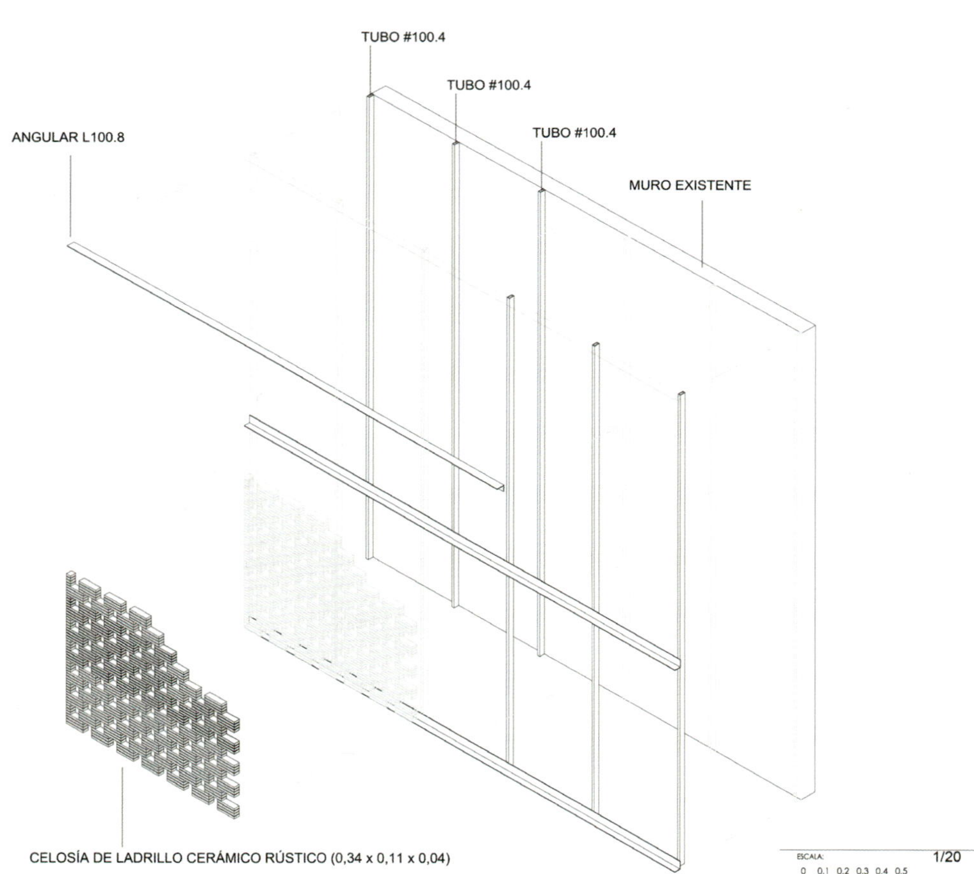

Detail

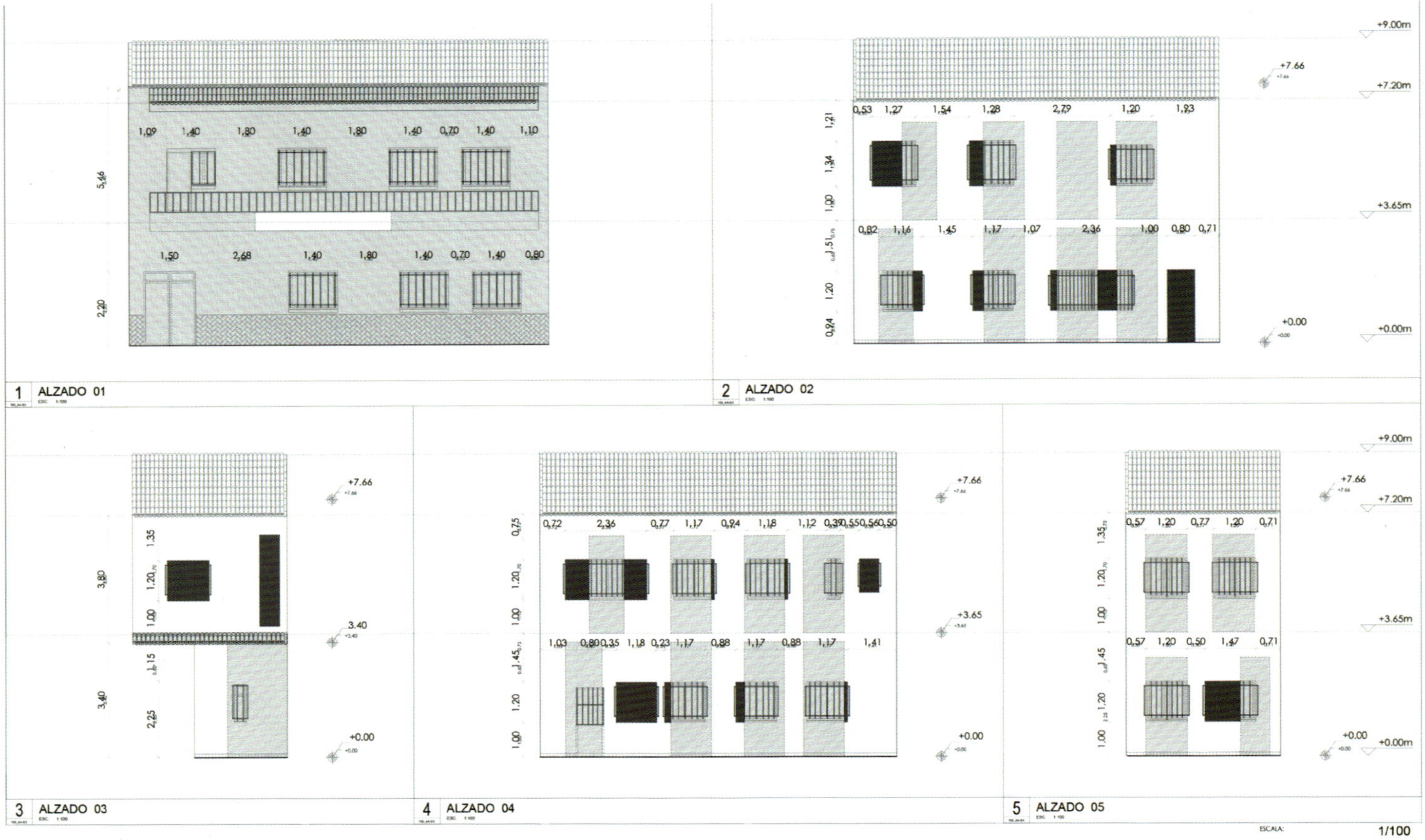

1 ALZADO 01

2 ALZADO 02

3 ALZADO 03

4 ALZADO 04

5 ALZADO 05

ESCALA: 1/100

Section Detail

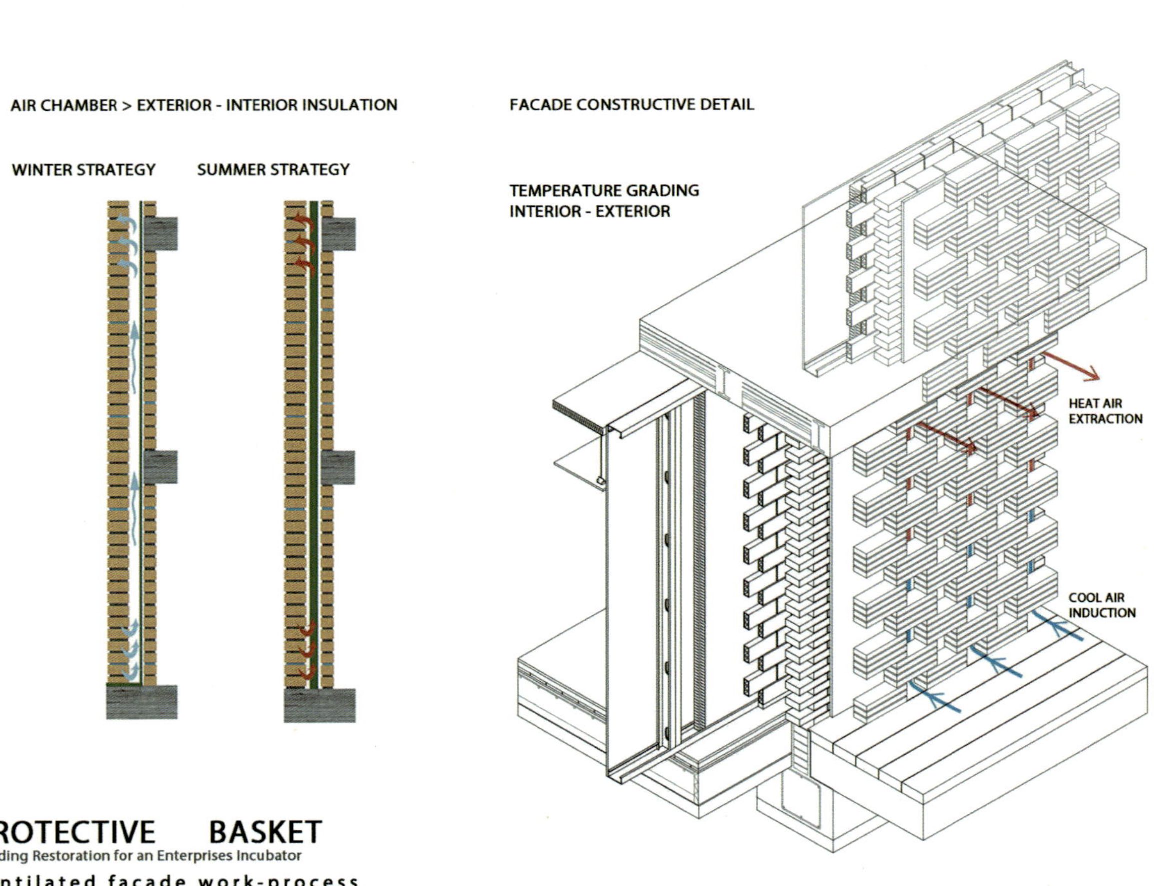

Section Detail2

FACHADA PATIOS
D1

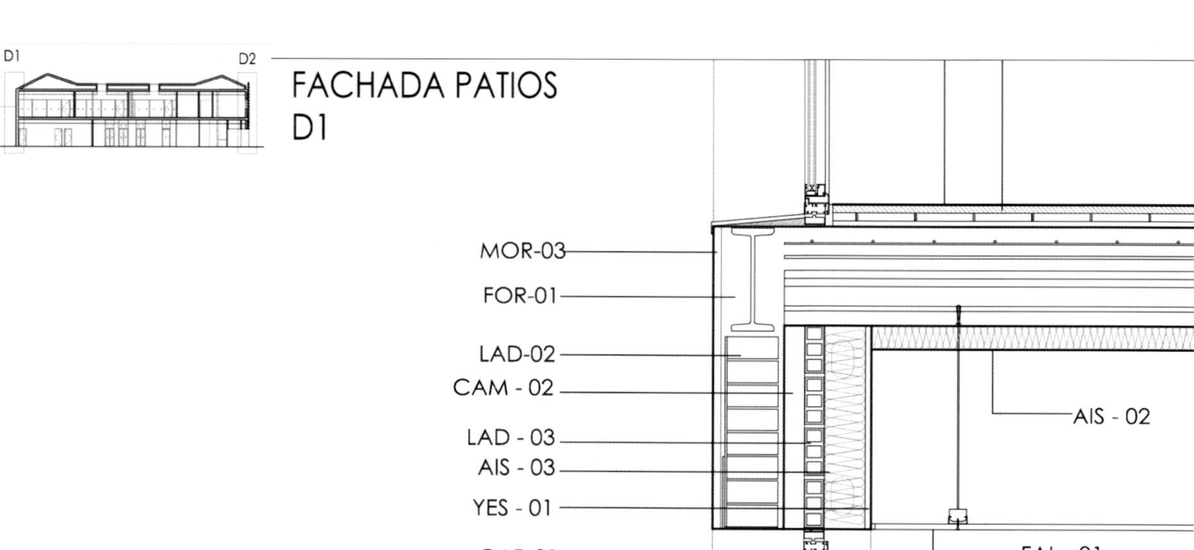

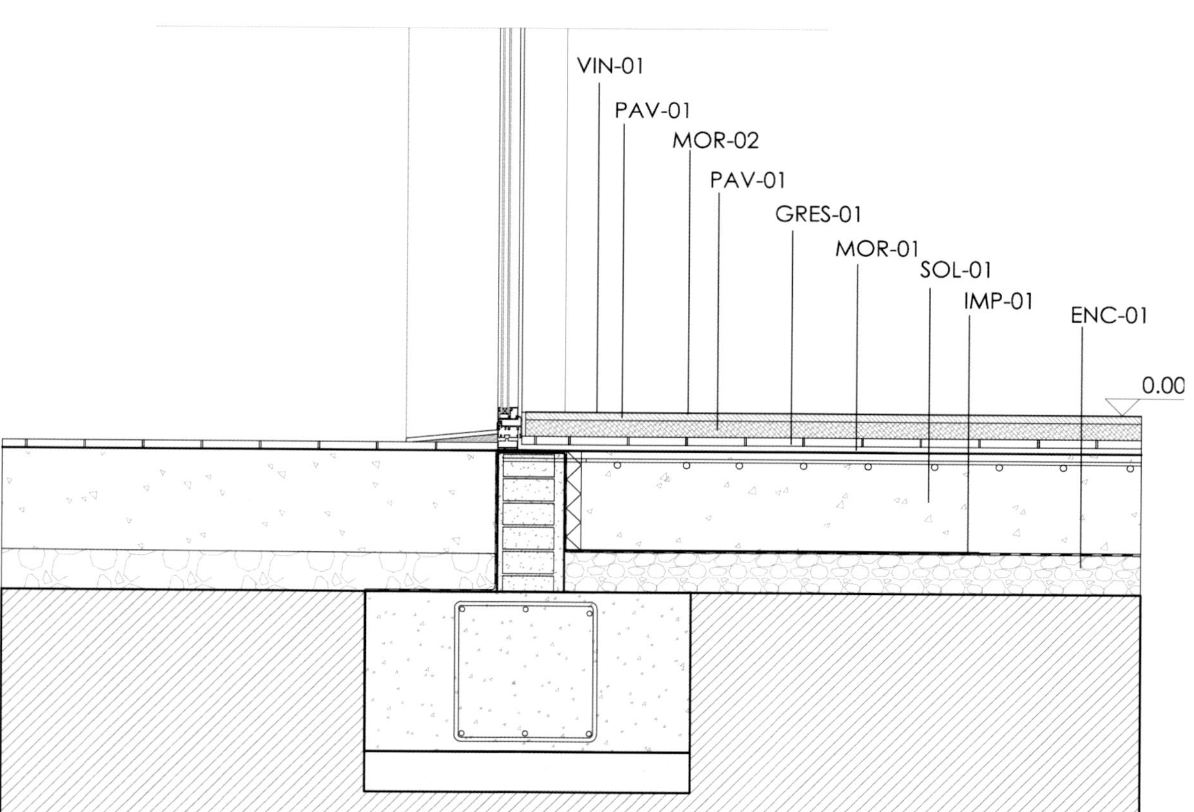

Section Detail

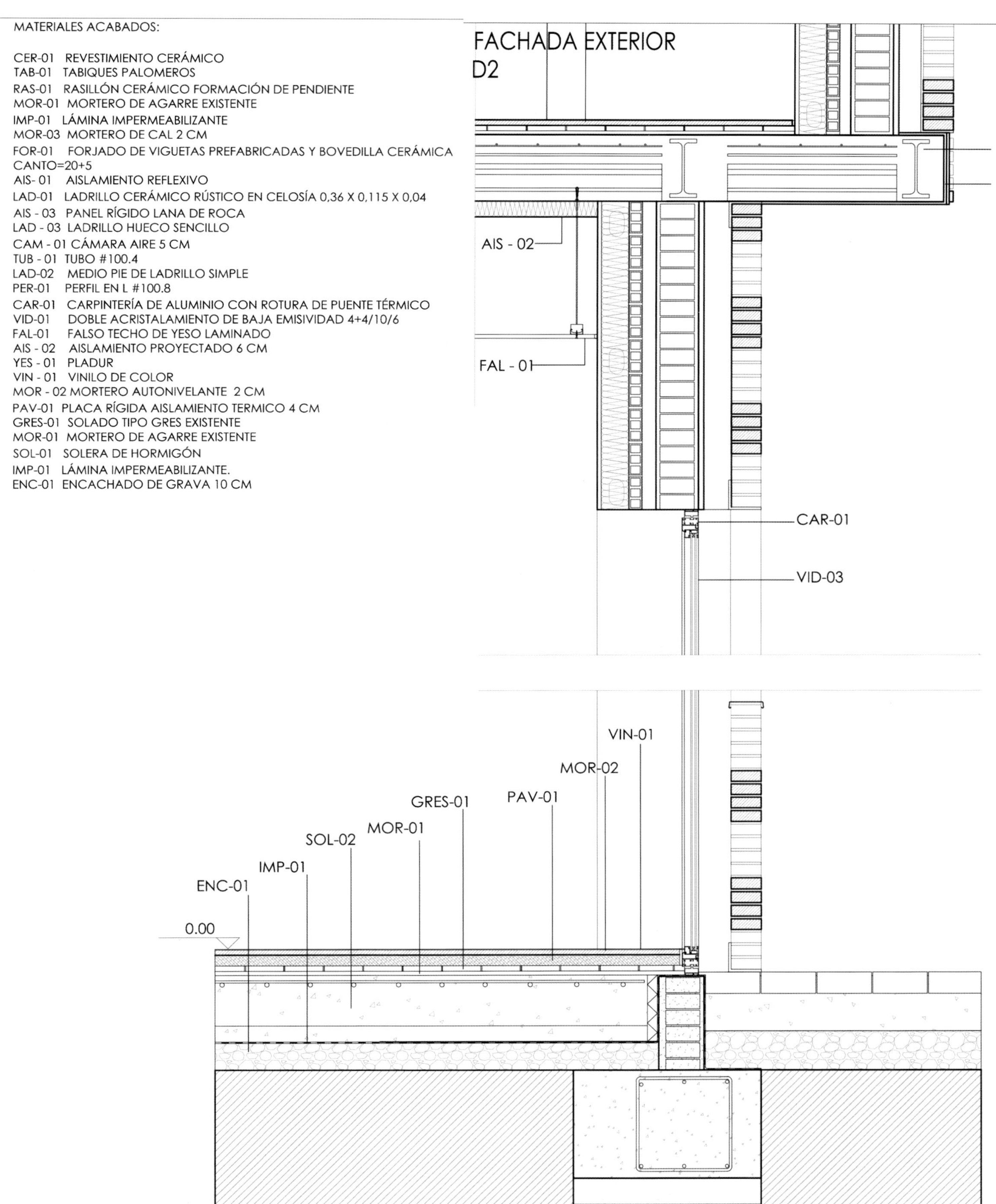
MATERIALES ACABADOS:
CER-01 REVESTIMIENTO CERÁMICO
TAB-01 TABIQUES PALOMEROS
RAS-01 RASILLÓN CERÁMICO FORMACIÓN DE PENDIENTE
MOR-01 MORTERO DE AGARRE EXISTENTE
IMP-01 LÁMINA IMPERMEABILIZANTE
MOR-03 MORTERO DE CAL 2 CM
FOR-01 FORJADO DE VIGUETAS PREFABRICADAS Y BOVEDILLA CERÁMICA CANTO=20+5
AIS- 01 AISLAMIENTO REFLEXIVO
LAD-01 LADRILLO CERÁMICO RÚSTICO EN CELOSÍA 0,36 X 0,115 X 0,04
AIS - 03 PANEL RÍGIDO LANA DE ROCA
LAD - 03 LADRILLO HUECO SENCILLO
CAM - 01 CÁMARA AIRE 5 CM
TUB - 01 TUBO #100.4
LAD-02 MEDIO PIE DE LADRILLO SIMPLE
PER-01 PERFIL EN L #100.8
CAR-01 CARPINTERÍA DE ALUMINIO CON ROTURA DE PUENTE TÉRMICO
VID-01 DOBLE ACRISTALAMIENTO DE BAJA EMISIVIDAD 4+4/10/6
FAL-01 FALSO TECHO DE YESO LAMINADO
AIS - 02 AISLAMIENTO PROYECTADO 6 CM
YES - 01 PLADUR
VIN - 01 VINILO DE COLOR
MOR - 02 MORTERO AUTONIVELANTE 2 CM
PAV-01 PLACA RÍGIDA AISLAMIENTO TERMICO 4 CM
GRES-01 SOLADO TIPO GRES EXISTENTE
MOR-01 MORTERO DE AGARRE EXISTENTE
SOL-01 SOLERA DE HORMIGÓN
IMP-01 LÁMINA IMPERMEABILIZANTE.
ENC-01 ENCACHADO DE GRAVA 10 CM
FACHADA EXTERIOR
D2
AIS - 02
FAL - 01
CAR-01
VID-03
VIN-01
MOR-02
PAV-01
GRES-01
MOR-01
SOL-02
IMP-01
ENC-01
0.00

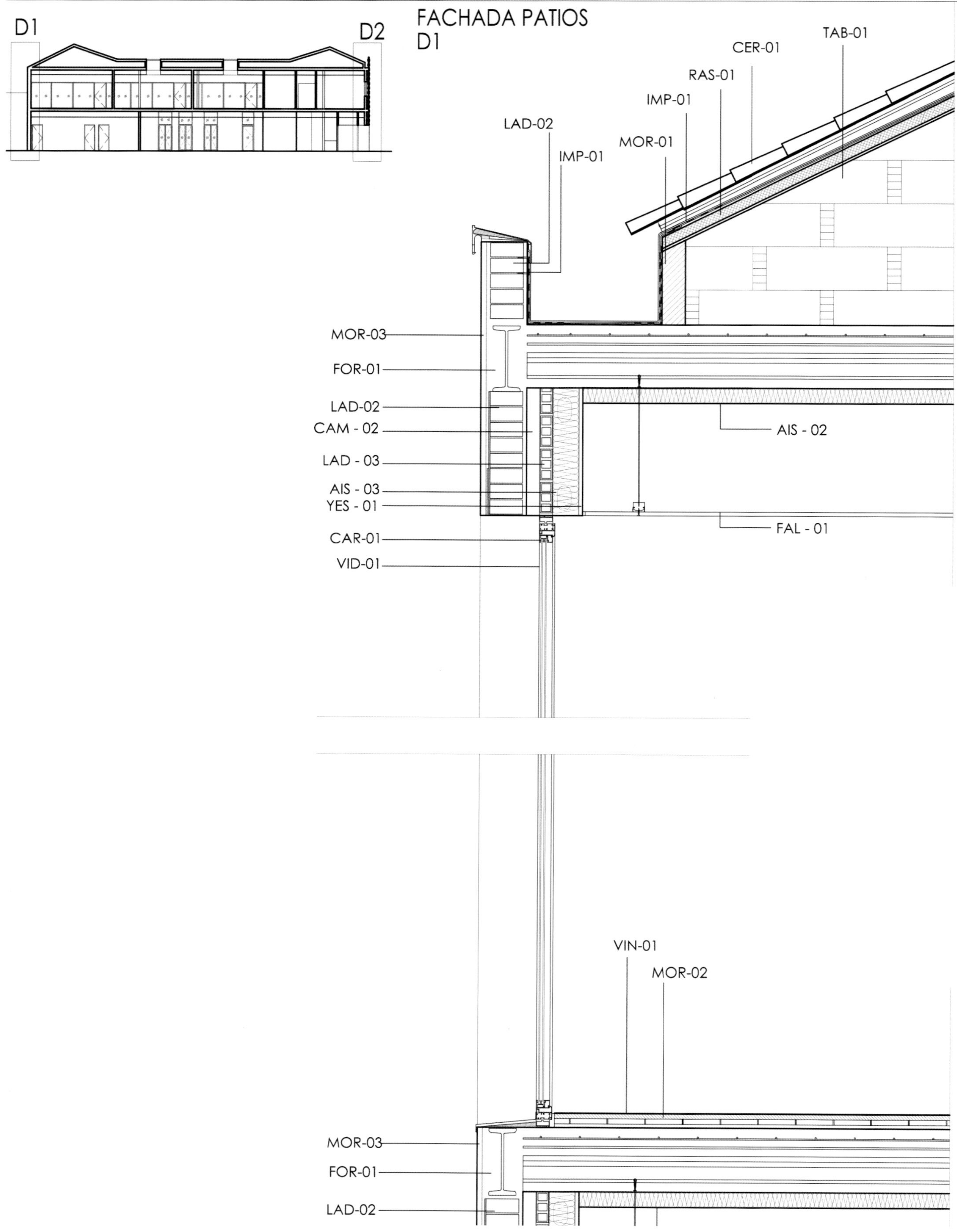

Section Detail

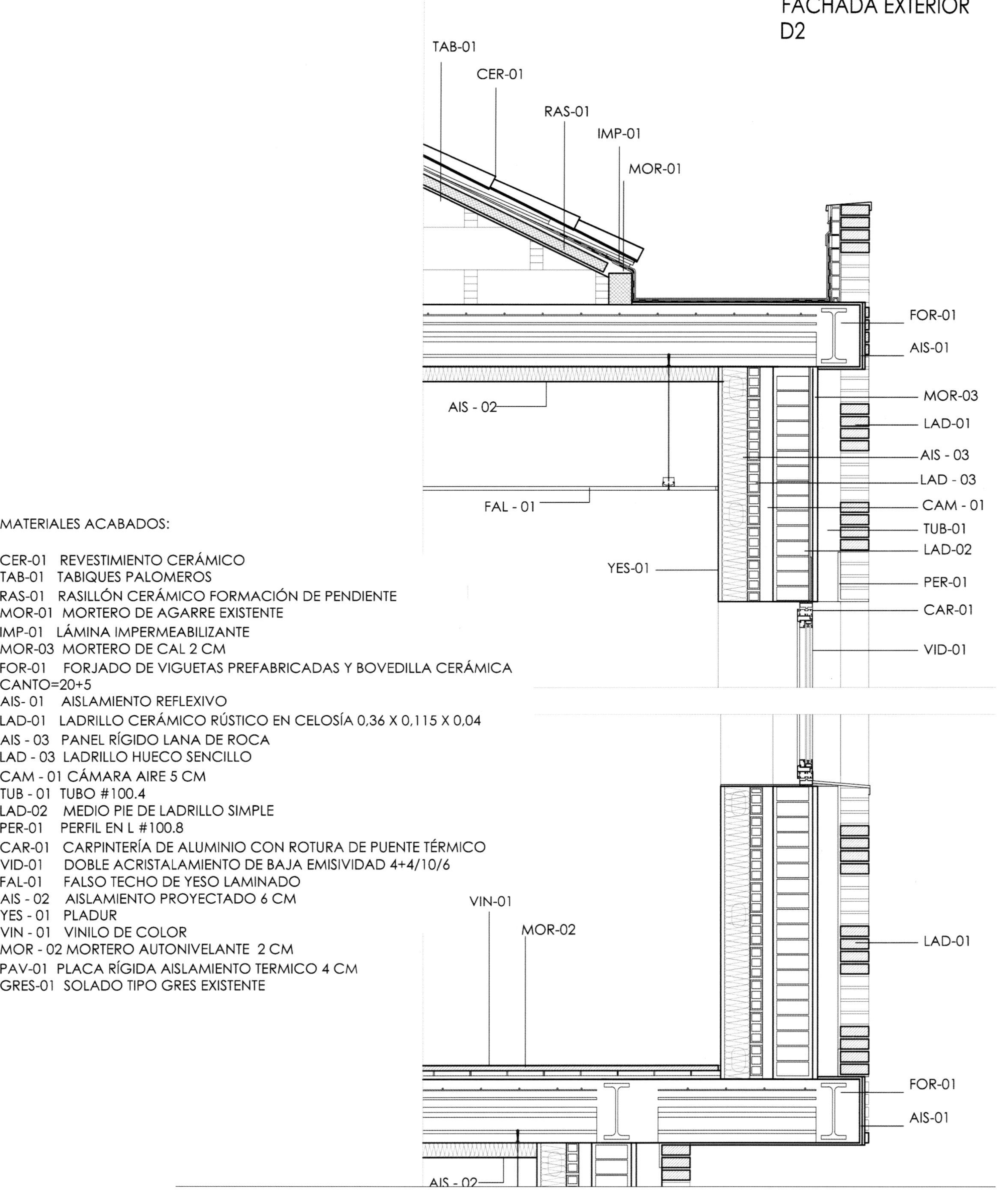
FACHADA EXTERIOR
D2
TAB-01
CER-01
RAS-01
IMP-01
MOR-01
FOR-01
AIS-01
AIS - 02
MOR-03
LAD-01
AIS - 03
LAD - 03
CAM - 01
FAL - 01
TUB-01
LAD-02
YES-01
PER-01
CAR-01
VID-01
MATERIALES ACABADOS:
CER-01 REVESTIMIENTO CERÁMICO
TAB-01 TABIQUES PALOMEROS
RAS-01 RASILLÓN CERÁMICO FORMACIÓN DE PENDIENTE
MOR-01 MORTERO DE AGARRE EXISTENTE
IMP-01 LÁMINA IMPERMEABILIZANTE
MOR-03 MORTERO DE CAL 2 CM
FOR-01 FORJADO DE VIGUETAS PREFABRICADAS Y BOVEDILLA CERÁMICA CANTO=20+5
AIS- 01 AISLAMIENTO REFLEXIVO
LAD-01 LADRILLO CERÁMICO RÚSTICO EN CELOSÍA 0,36 X 0,115 X 0,04
AIS - 03 PANEL RÍGIDO LANA DE ROCA
LAD - 03 LADRILLO HUECO SENCILLO
CAM - 01 CÁMARA AIRE 5 CM
TUB - 01 TUBO #100.4
LAD-02 MEDIO PIE DE LADRILLO SIMPLE
PER-01 PERFIL EN L #100.8
CAR-01 CARPINTERÍA DE ALUMINIO CON ROTURA DE PUENTE TÉRMICO
VID-01 DOBLE ACRISTALAMIENTO DE BAJA EMISIVIDAD 4+4/10/6
FAL-01 FALSO TECHO DE YESO LAMINADO
AIS - 02 AISLAMIENTO PROYECTADO 6 CM
YES - 01 PLADUR
VIN - 01 VINILO DE COLOR
MOR - 02 MORTERO AUTONIVELANTE 2 CM
PAV-01 PLACA RÍGIDA AISLAMIENTO TERMICO 4 CM
GRES-01 SOLADO TIPO GRES EXISTENTE
VIN-01
MOR-02
LAD-01
FOR-01
AIS-01
AIS - 02

VIVERO
MADRIDEJOS

FACTORY IN BINH DUONG, Binh Duong city, Vietnam

NISHIZAWA ARCHITECTS

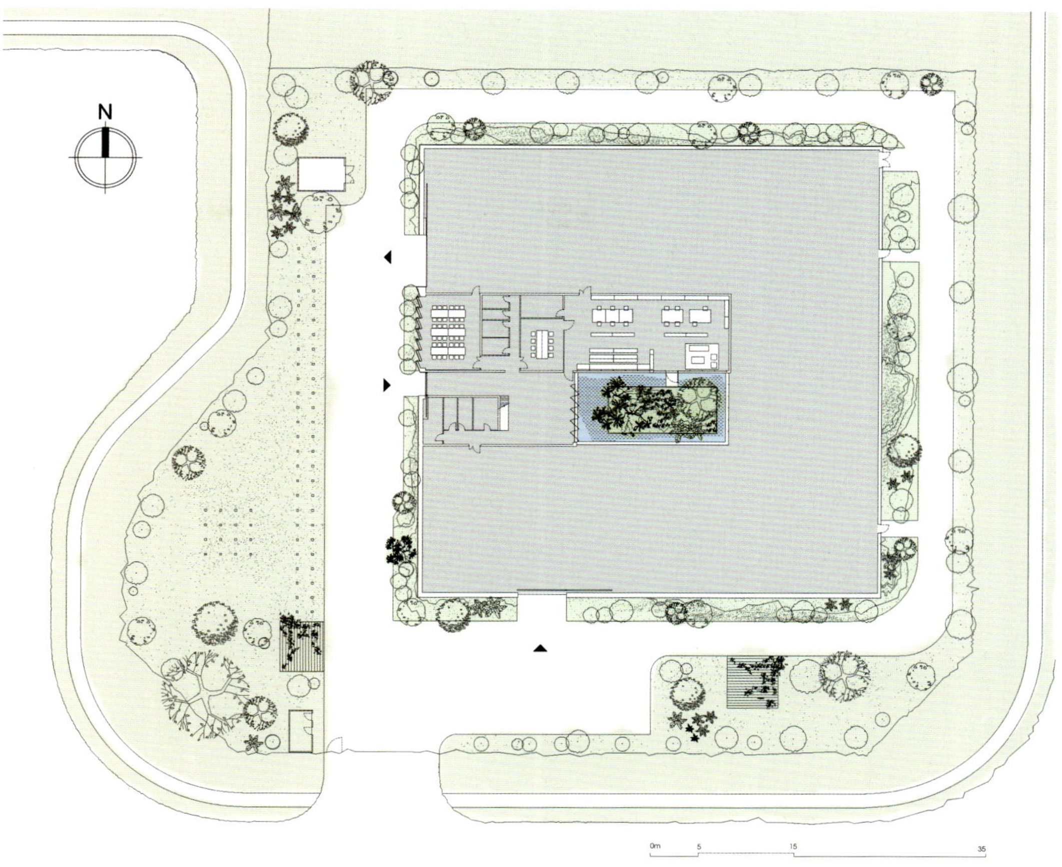

Site Plan

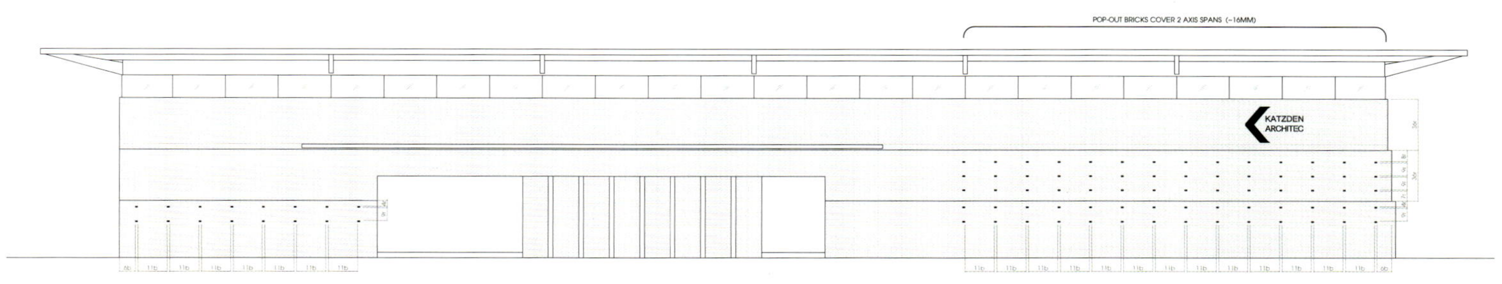

West Elevation

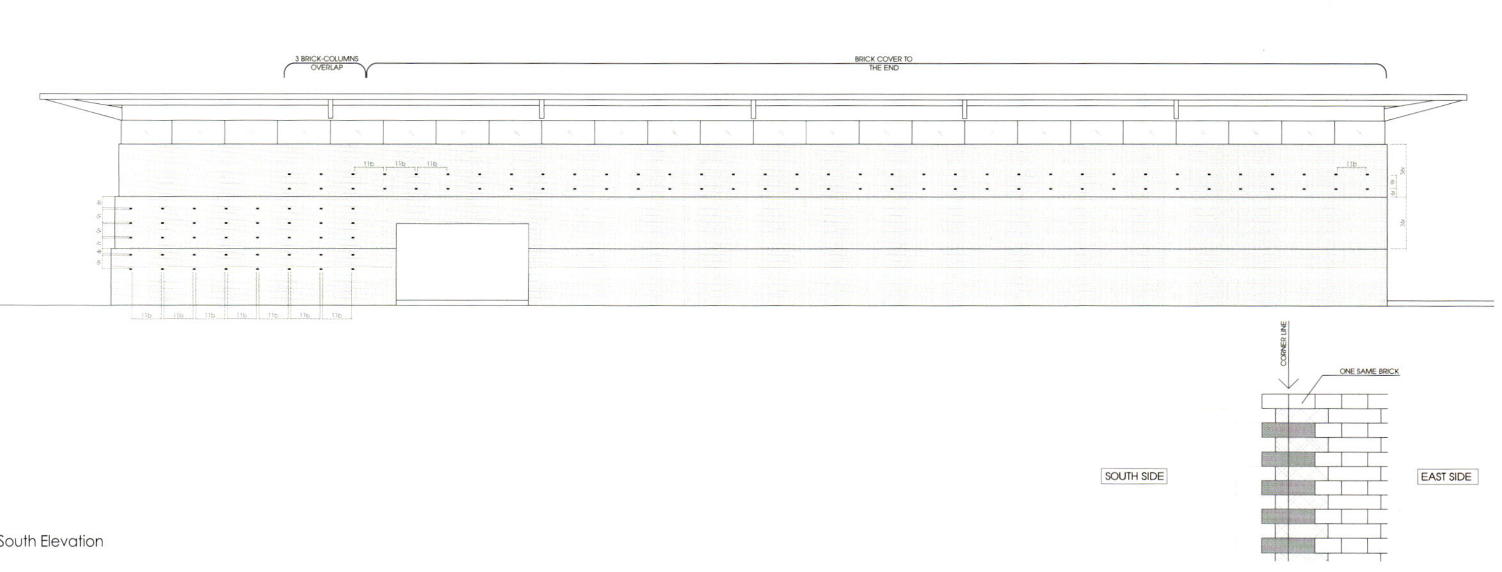

South Elevation

© Hiroyuki Oki

© Hiroyuki Oki

Section A

© Hiroyuki Oki

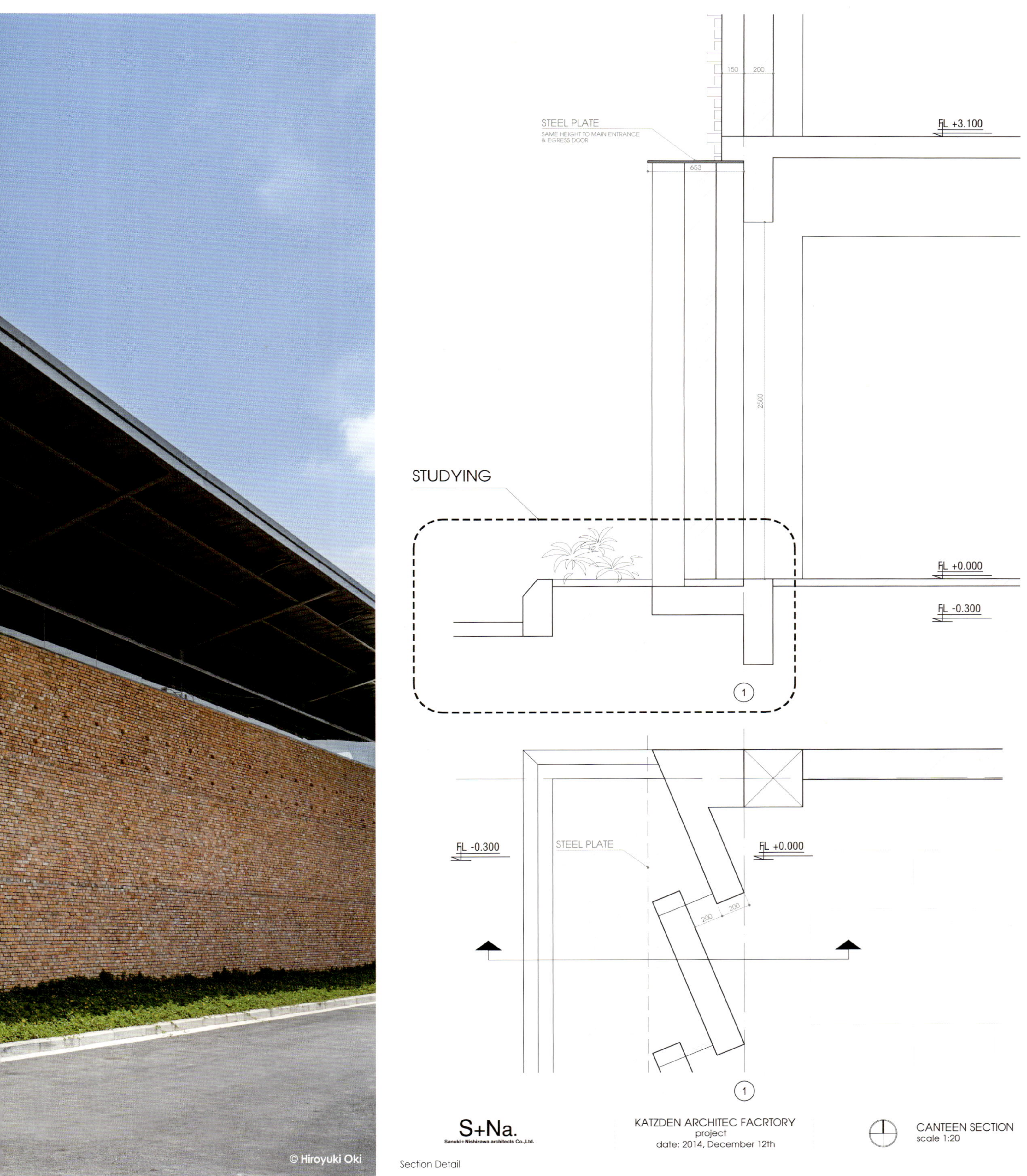

Section Detail

Construction Process

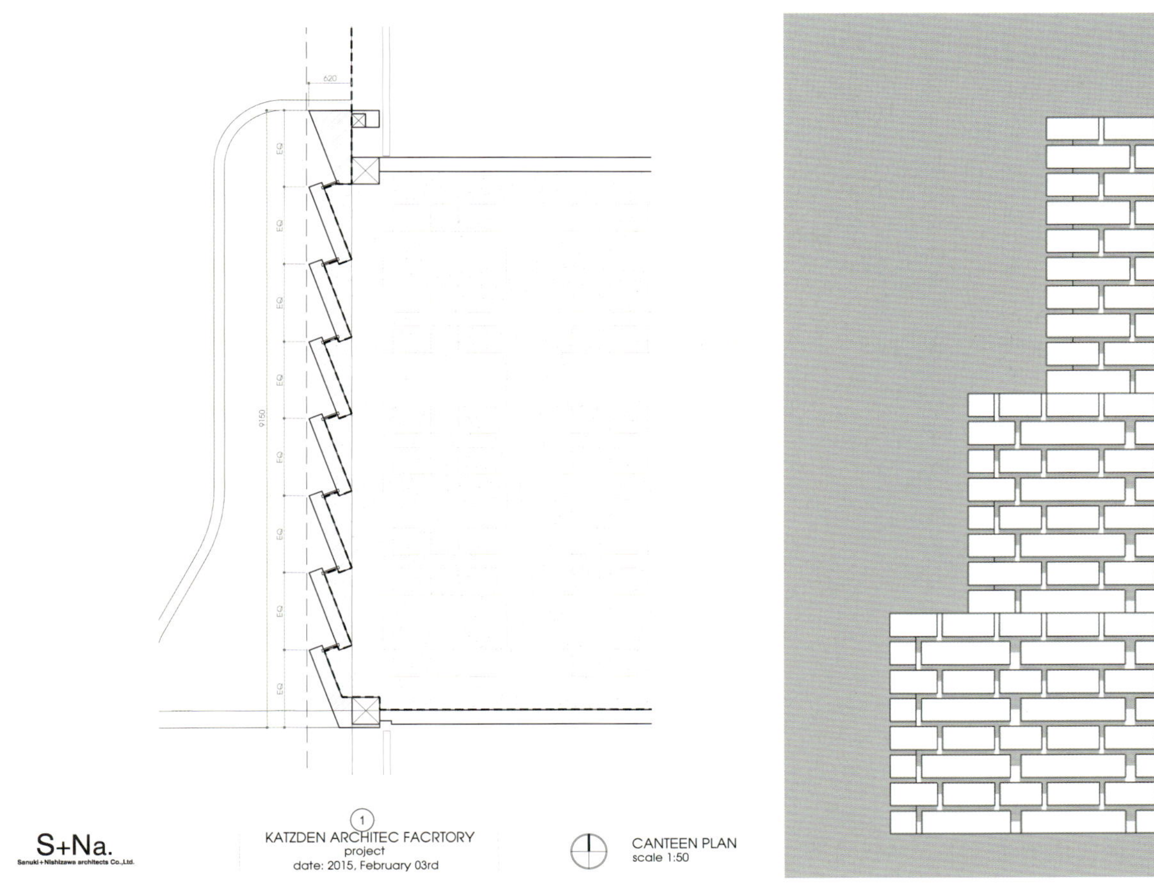

Section Detail

© Hiroyuki Oki

© Hiroyuki Oki

Namsan Patio, Busan, South Korea

Architects Group Raum

© Joonhwan Yoon

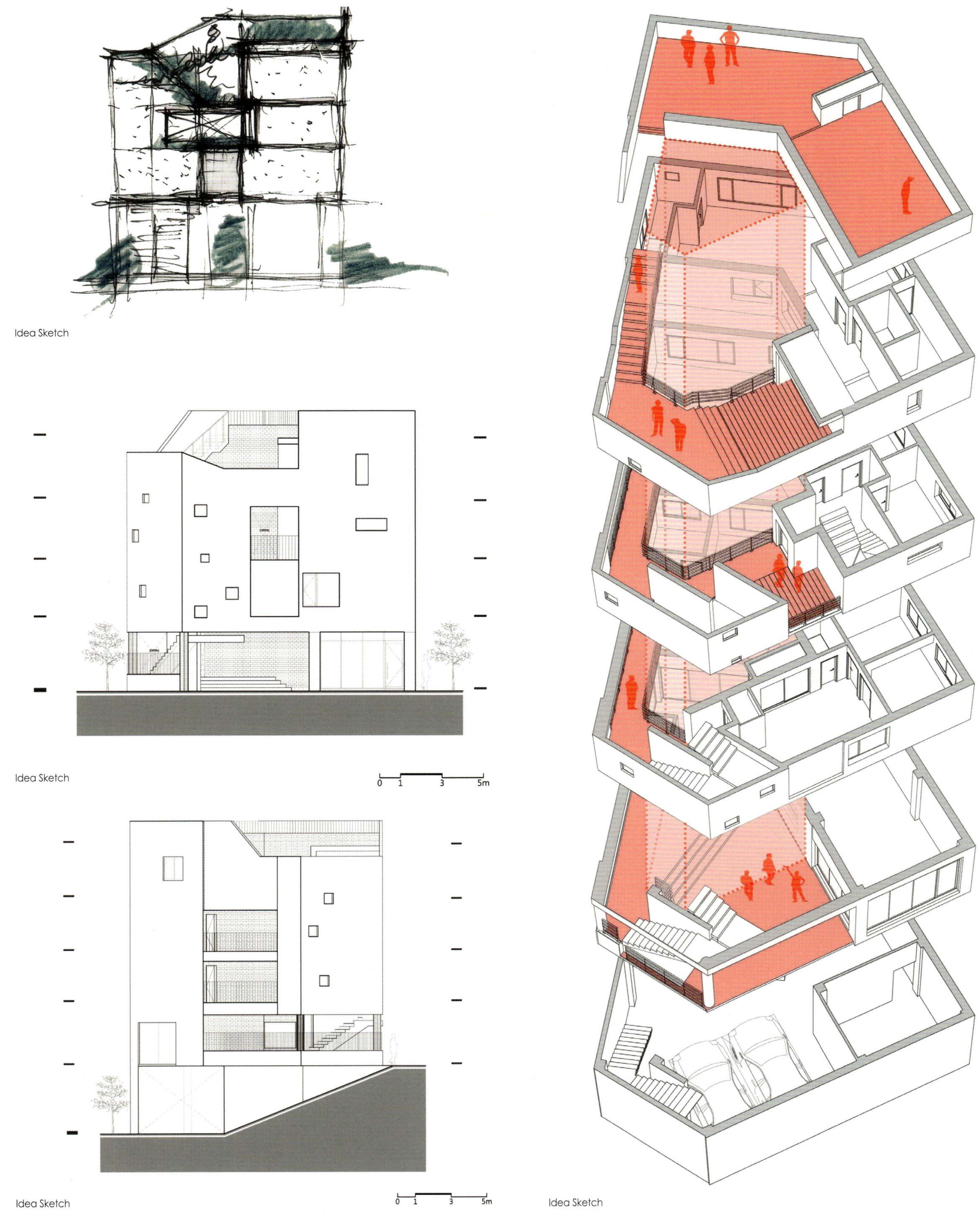

Idea Sketch

Idea Sketch

Idea Sketch

Idea Sketch

© Joonhwan Yoon

© Joonhwan Yoon

© Joonhwan Yoon

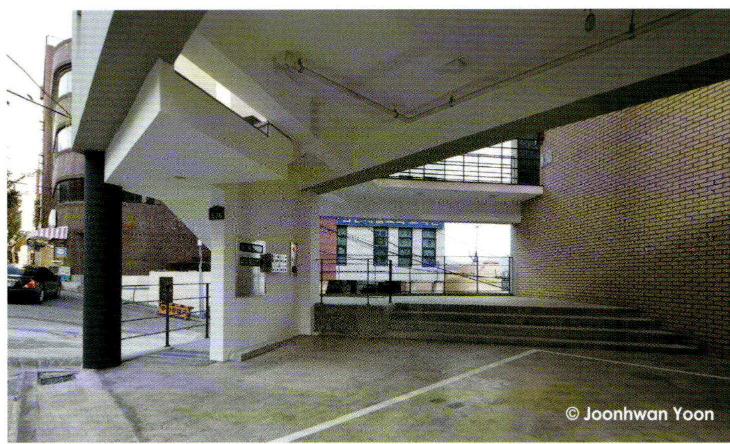

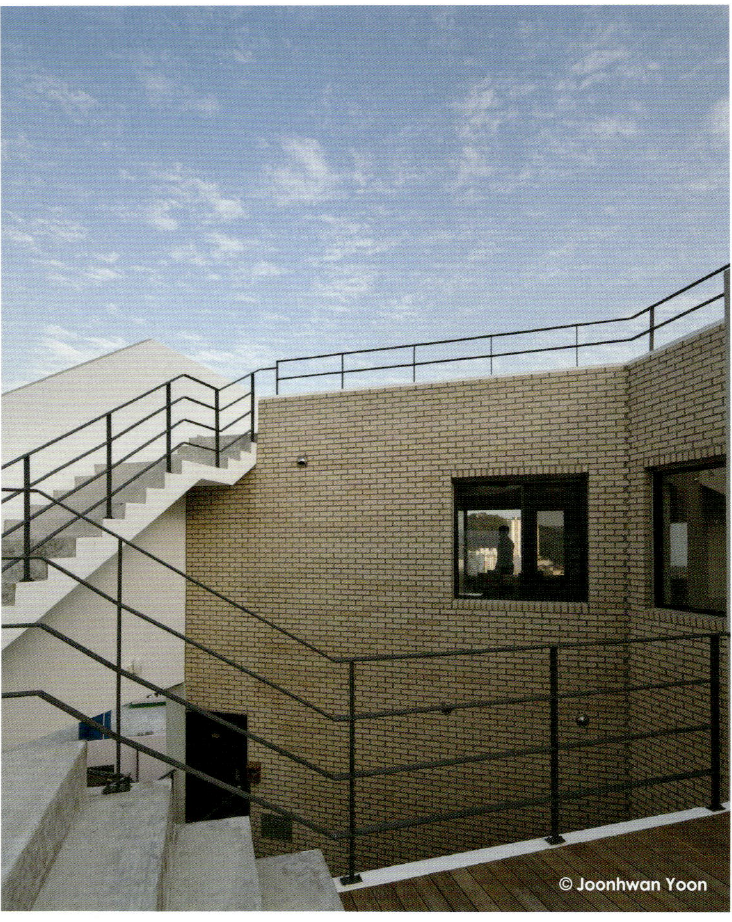

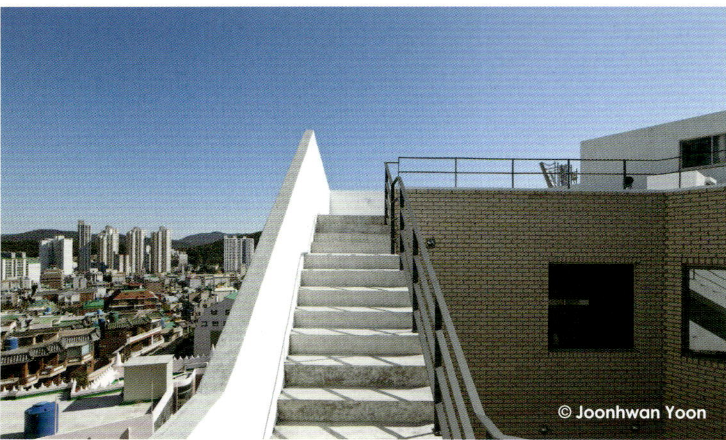

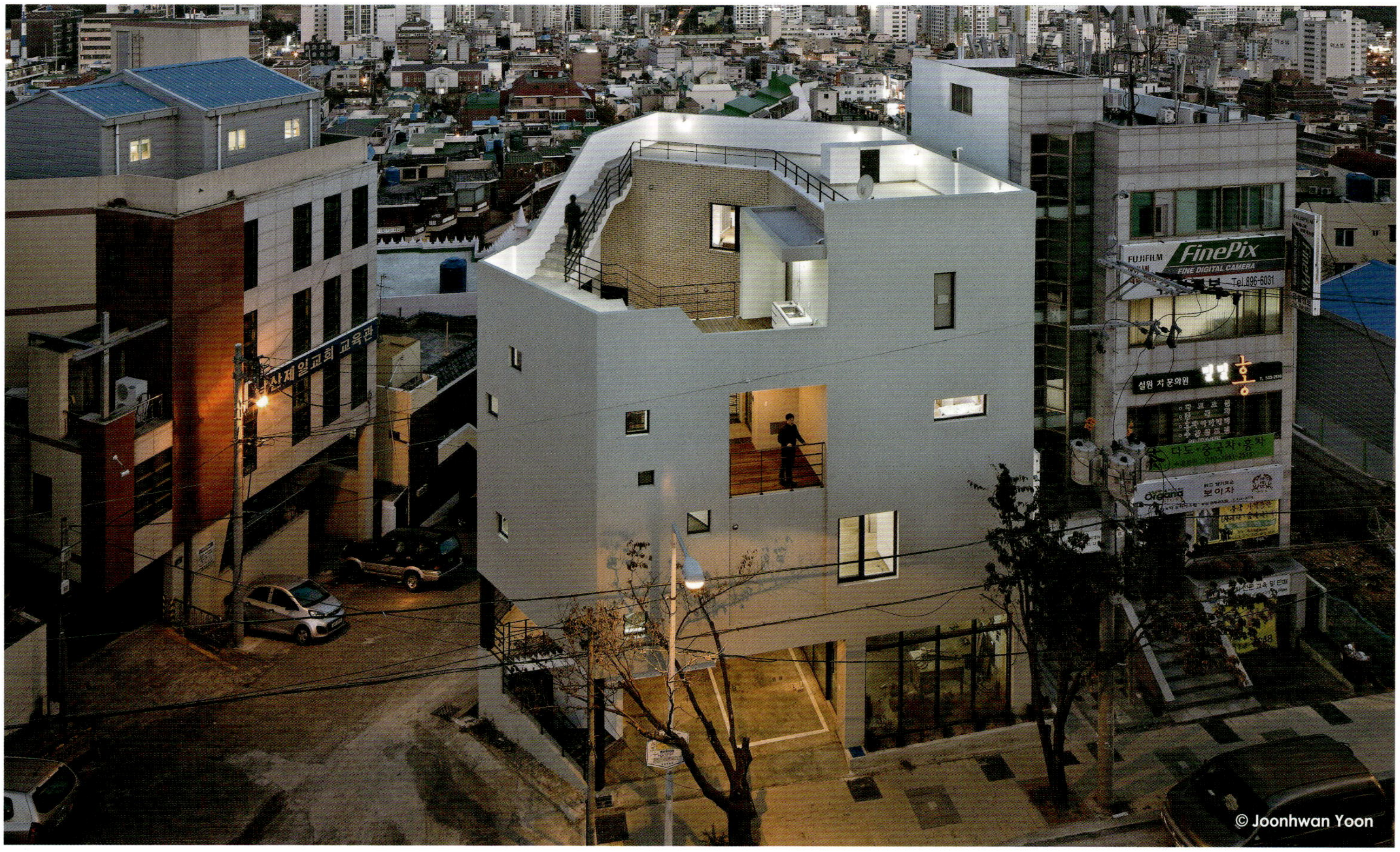
FUJIFILM
FinePix
FINE DIGITAL CAMERA
© Joonhwan Yoon

© Joonhwan Yoon

Dissolving Arch, jeju, South Korea

stpmj Architecture

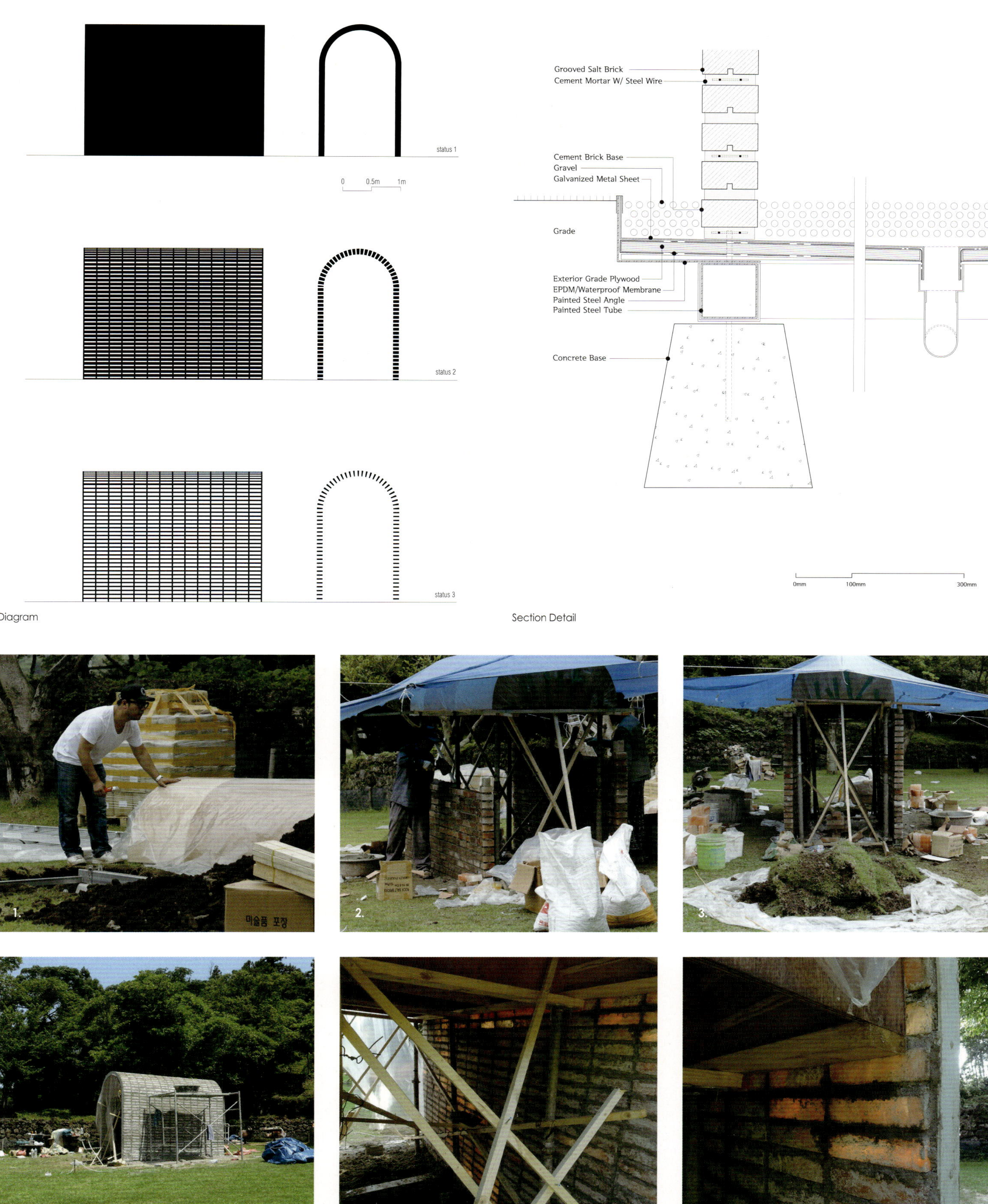

Diagram

Section Detail

Construction Process

© stpmj

© Kyoungtae Kim

© Kyoungtae Kim

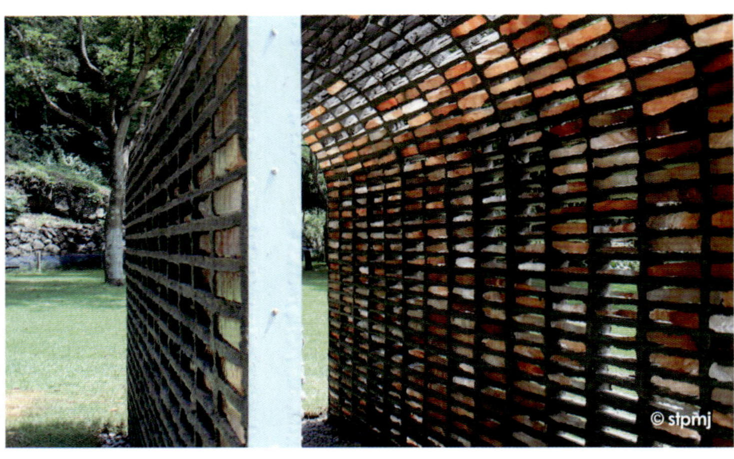

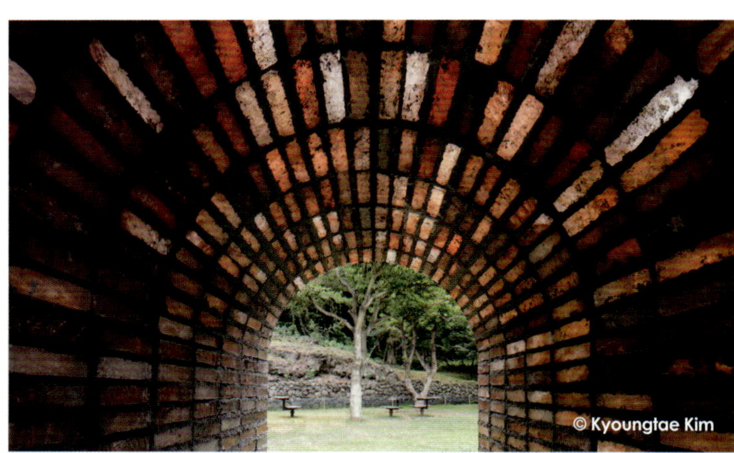

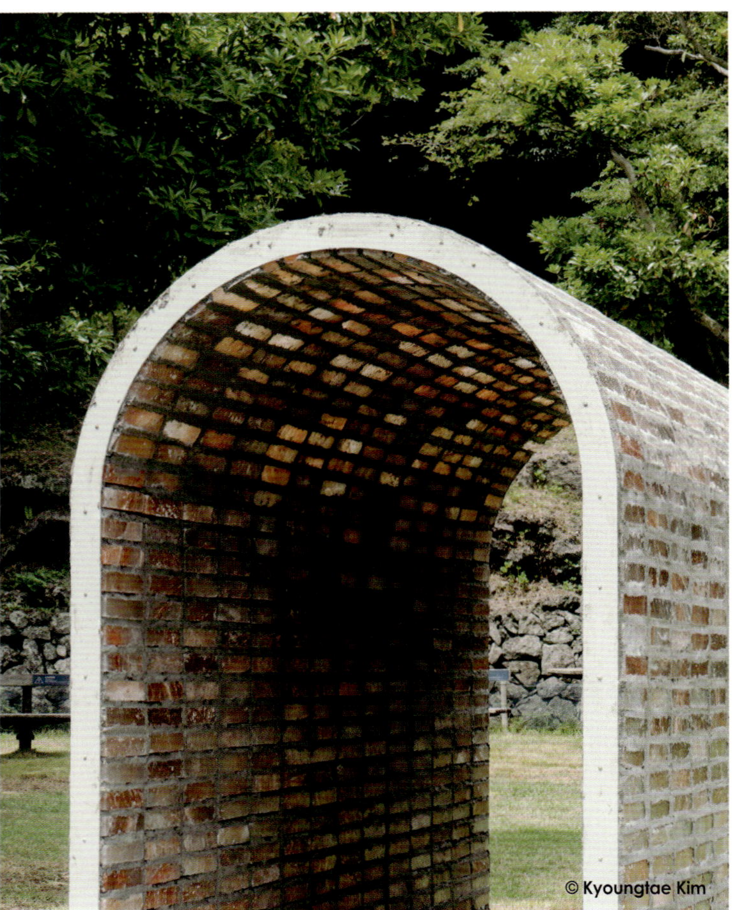

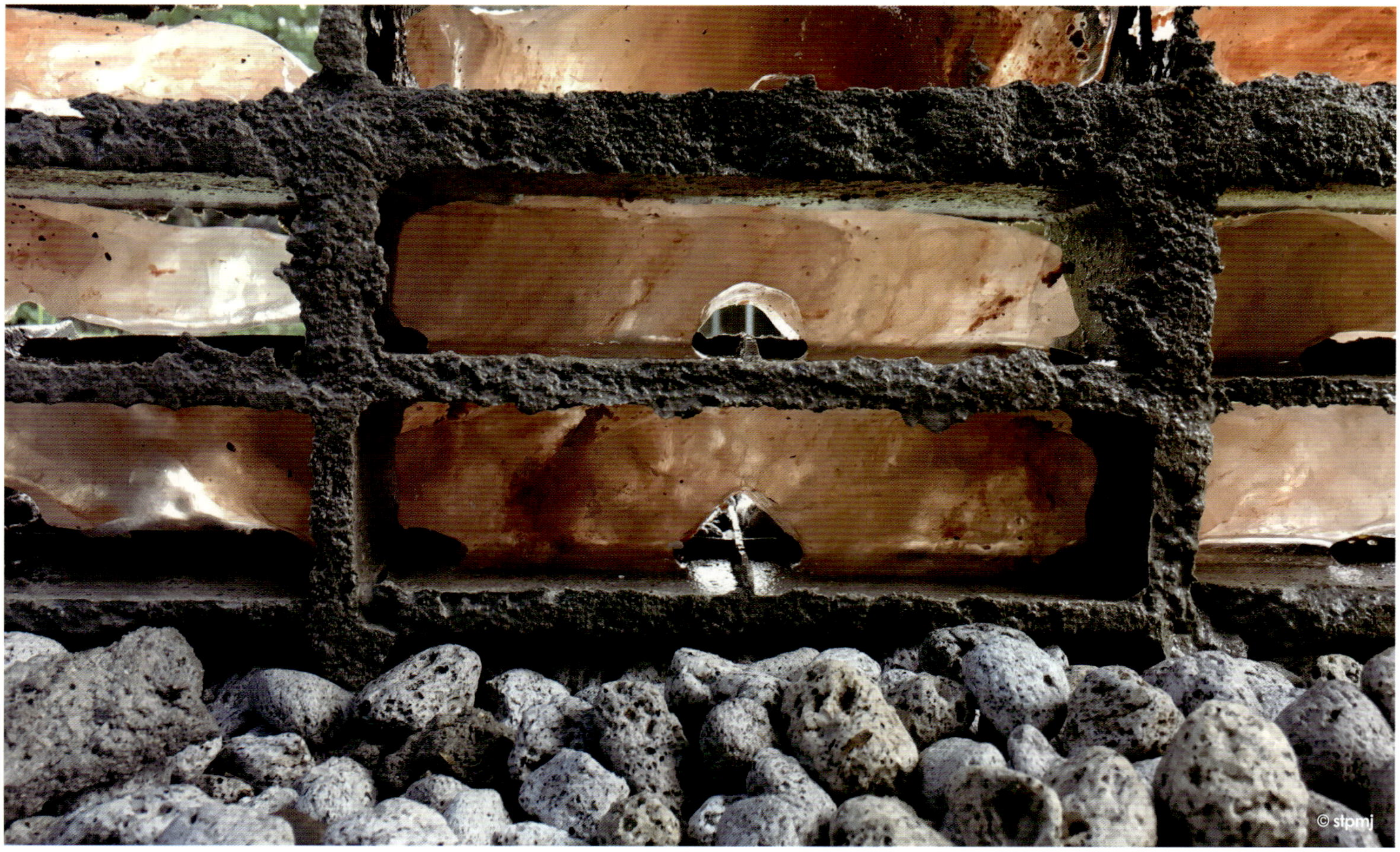
© stpmj

Antwerp Zoo, Antwerp, Belgium
Studio Farris Architects

© Koen Van Damme

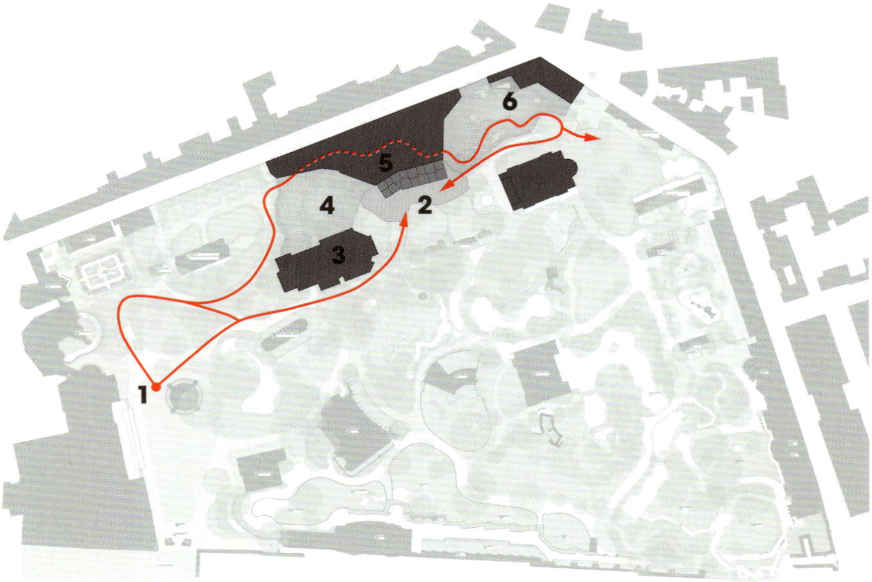

MASTERPLAN OF THE ANTWERP ZOO

1 New entrance
2 Plaza
3 Existing apes building
4 Extension of the apes shelter, connected to the restaurant
5 Restaurant
6 Buffalo shelter and birds volière, connected to the restaurant

Master Plan

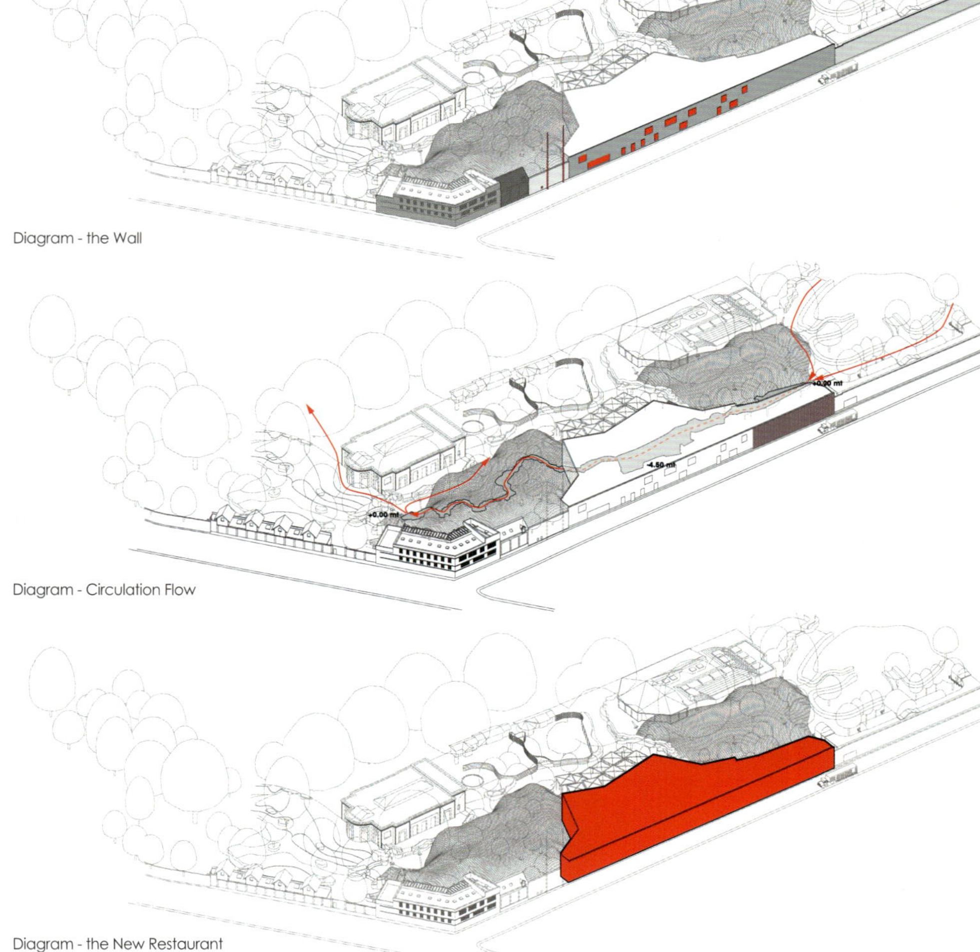

Diagram - the Wall

Diagram - Circulation Flow

Diagram - the New Restaurant

© Koen Van Damme

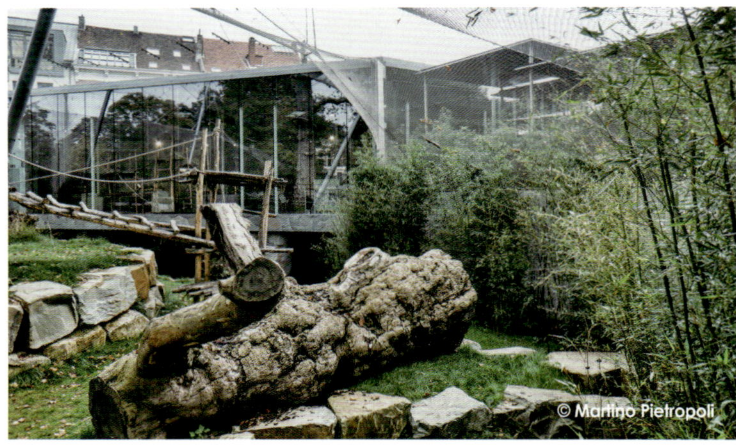

© Martino Pietropoli

GIRAFFEN
© Martino Pietropoli

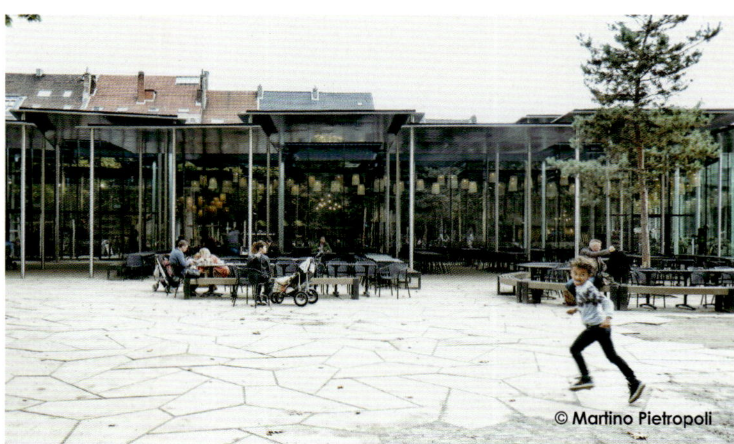

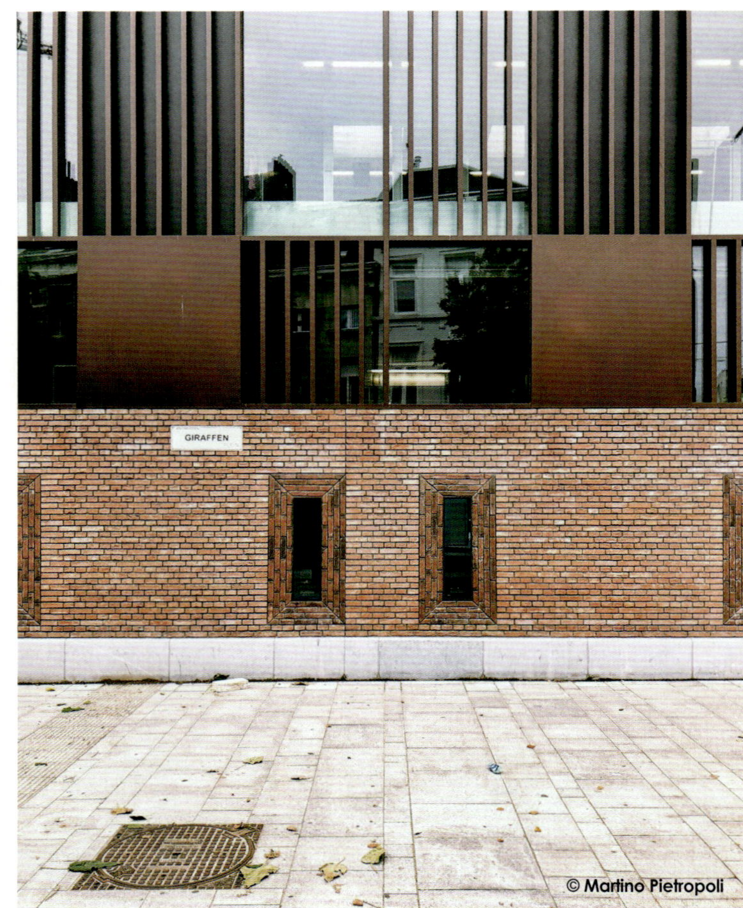

© Martino Pietropoli

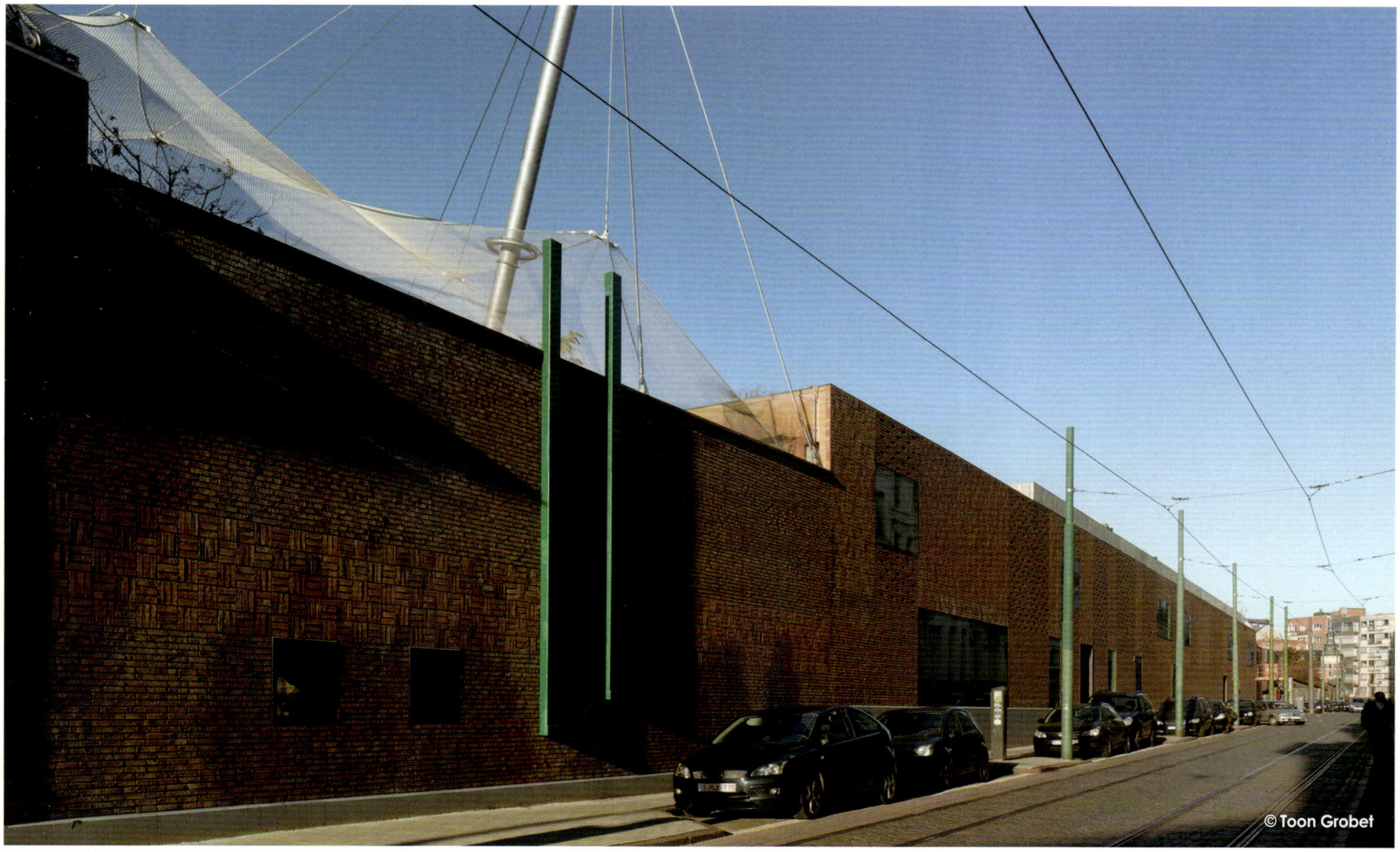
© Toon Grobet

Artena Archaeological Park, Rome, Italy

2TR ARCHITETTURA

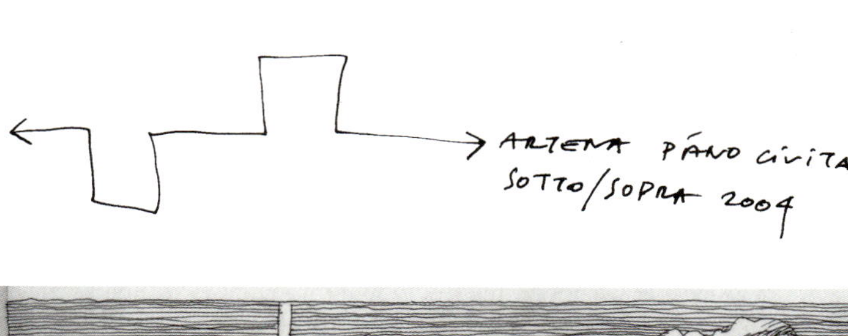
ARTENA PIANO CIVITA
SOTTO/SOPRA 2004

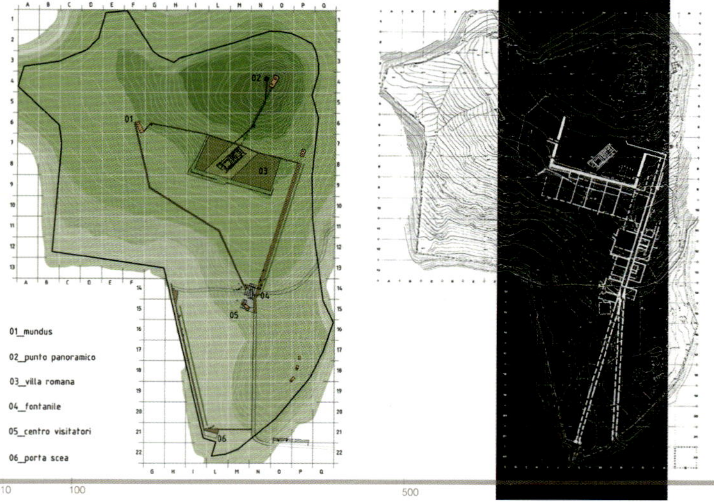
01_mundus
02_punto panoramico
03_villa romana
04_fontanile
05_centro visitatori
06_porta scea
0 10
100
500

Section

Community Centre Aussersihl Part2, Zurich, Switzerland

EM2N

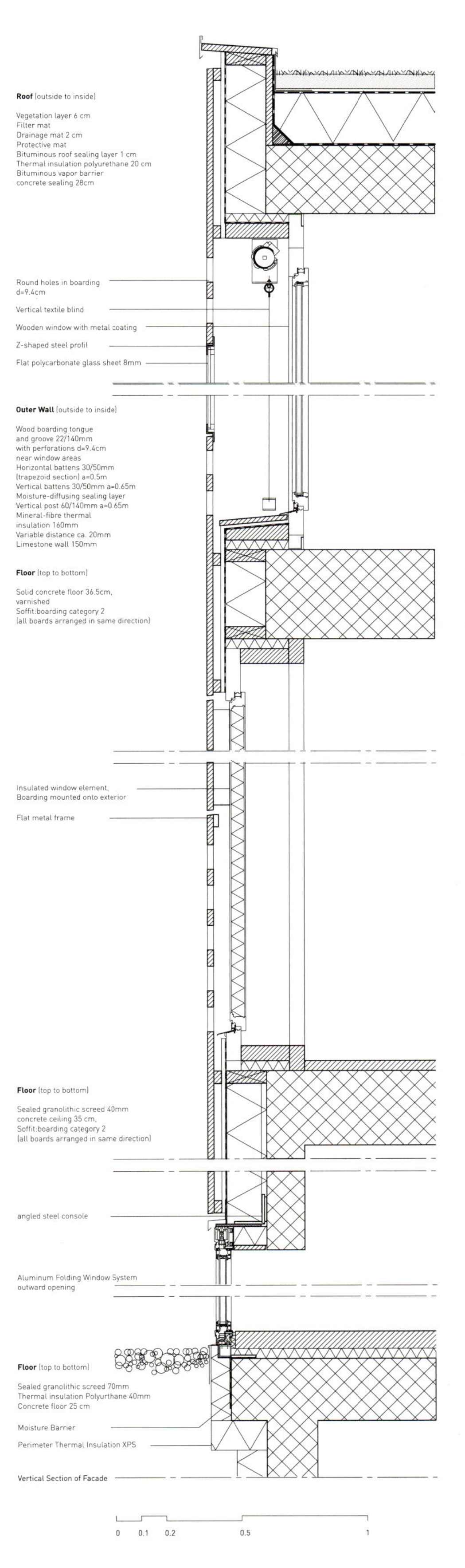
Roof (outside to inside)
Vegetation layer 6 cm
Filter mat
Drainage mat 2 cm
Protective mat
Bituminous roof sealing layer 1 cm
Thermal insulation polyurethane 20 cm
Bituminous vapor barrier
concrete sealing 28cm
Round holes in boarding d=9.4cm
Vertical textile blind
Wooden window with metal coating
Z-shaped steel profil
Flat polycarbonate glass sheet 8mm
Outer Wall (outside to inside)
Wood boarding tongue
and groove 22/140mm
with perforations d=9.4cm
near window areas
Horizontal battens 30/50mm
(trapezoid section) a=0.5m
Vertical battens 30/50mm a=0.65m
Moisture-diffusing sealing layer
Vertical post 60/140mm a=0.65m
Mineral-fibre thermal
insulation 160mm
Variable distance ca. 20mm
Limestone wall 150mm
Floor (top to bottom)
Solid concrete floor 36.5cm,
varnished
Soffit:boarding category 2
(all boards arranged in same direction)
Insulated window element,
Boarding mounted onto exterior
Flat metal frame
Floor (top to bottom)
Sealed granolithic screed 40mm
concrete ceiling 35 cm,
Soffit:boarding category 2
(all boards arranged in same direction)
angled steel console
Aluminum Folding Window System
outward opening
Floor (top to bottom)
Sealed granolithic screed 70mm
Thermal insulation Polyurthane 40mm
Concrete floor 25 cm
Moisture Barrier
Perimeter Thermal Insulation XPS
Vertical Section of Facade
0
0.1
0.2
0.5
1

Detail

Former Haystack Refurbishment, Mora, Spain

OOIIO Architecture

BEFORE_

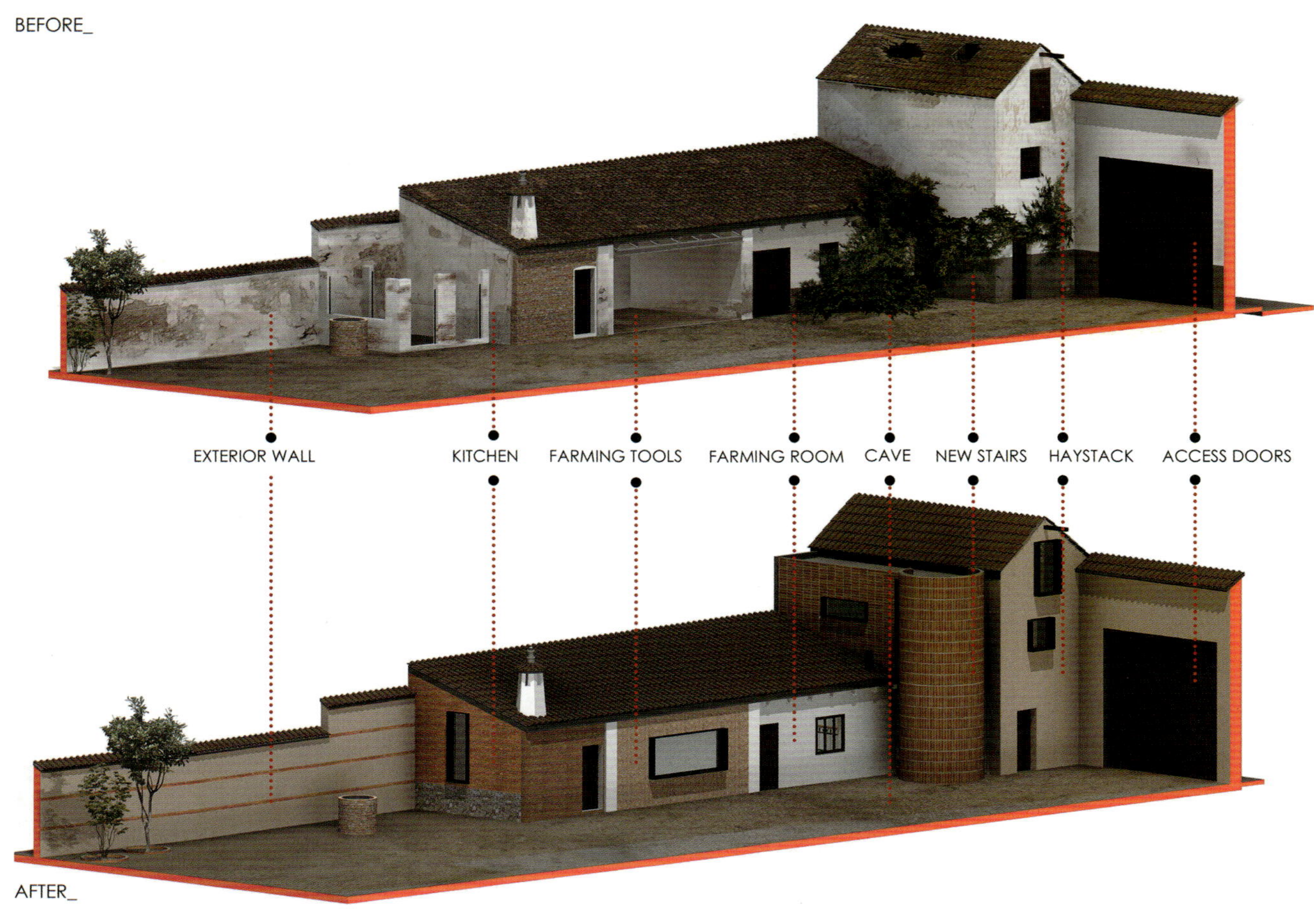

AFTER_

KITCHEN_BEFORE

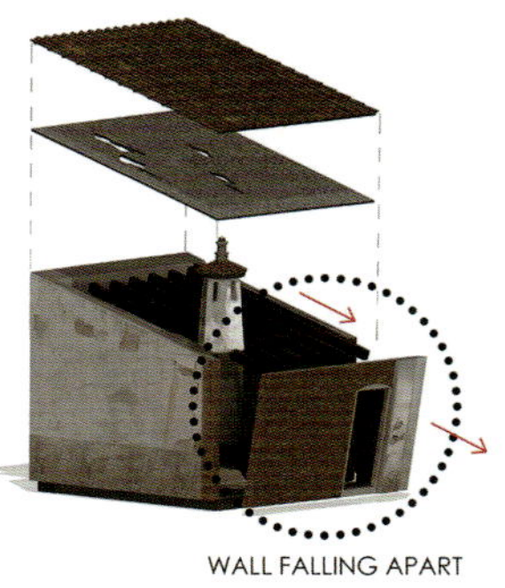

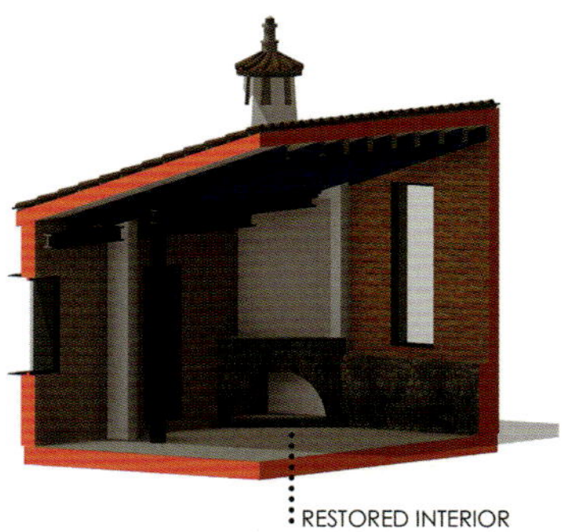

KITCHEN_AFTER

CAVE_BEFORE
STAIRS TOO STEEP
SHABBY INTERIOR FINISHES
CAVE_AFTER
CAVE_BEFORE
ENLARGED OPENING
ADDITIONAL STEPS
STAIRS TO CAVE_AFTER
CAVE_AFTER
FARMING TOOLS ROOMS_BEFORE
WORN-OUT ROOF
SHABBY OPENINGS
FINISH OF THE FACADE RUINED AND CLOSING ROOM
4. WORN DOWN INTERIOR
DEMOLISHED WALL
RESTORED ROOF TILE
NEW WINDOW
NEW WALL
4. RENOVATED CONSTRUCTIONS AND NEW FINISHES
FARMING TOOLS ROOMS_AFTER

© OOIIO Architecture, Josefotoinmo.

© OOIIO Architecture, Josefotoinmo.

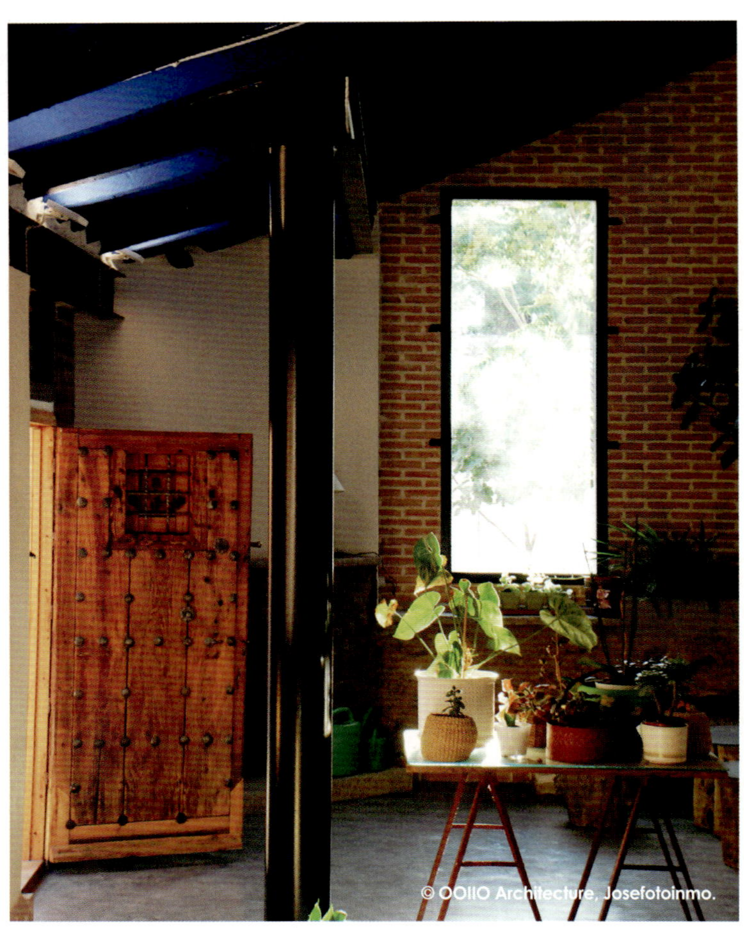

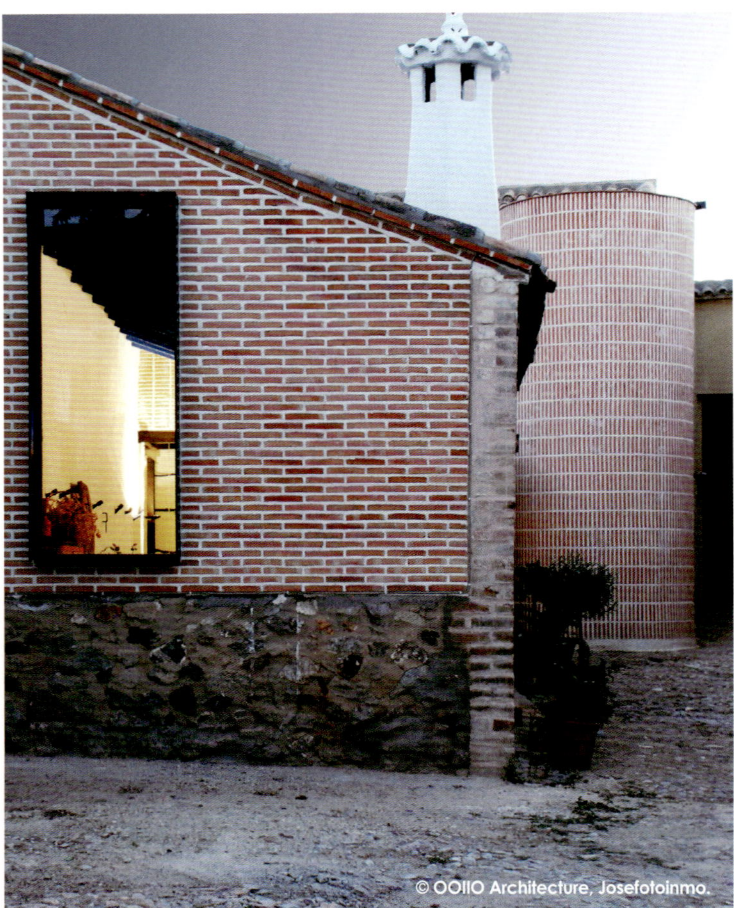

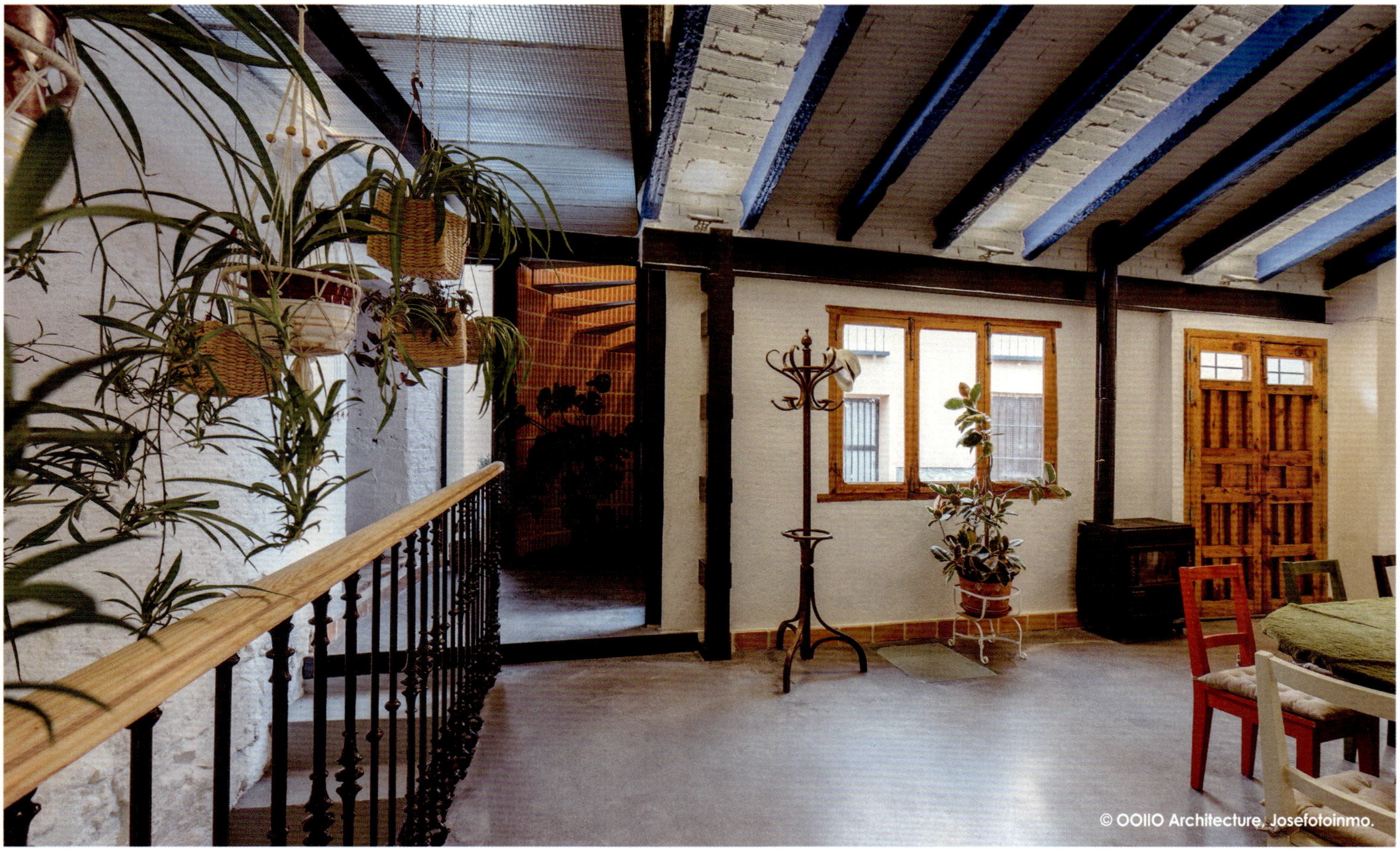
© OOIIO Architecture, Josefotoinmo.

HAYSTACK_BEFORE

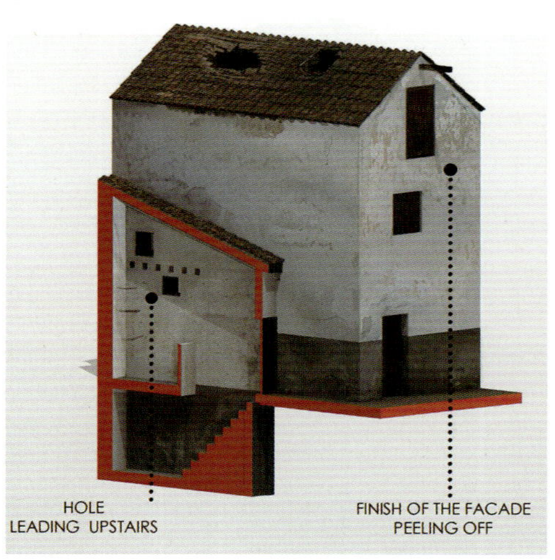

HAYSTACK_AFTER

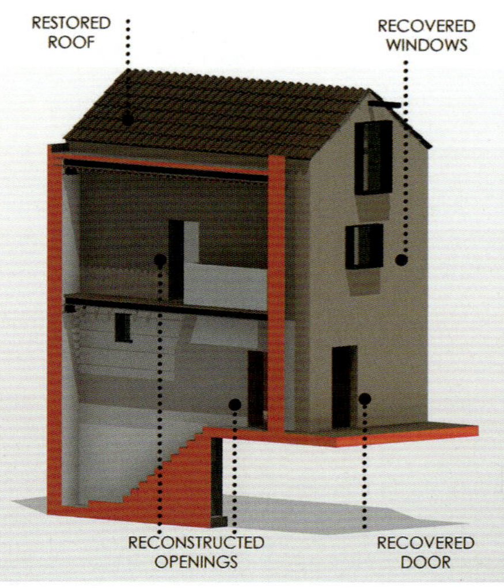

HAYSTACK_BEFORE

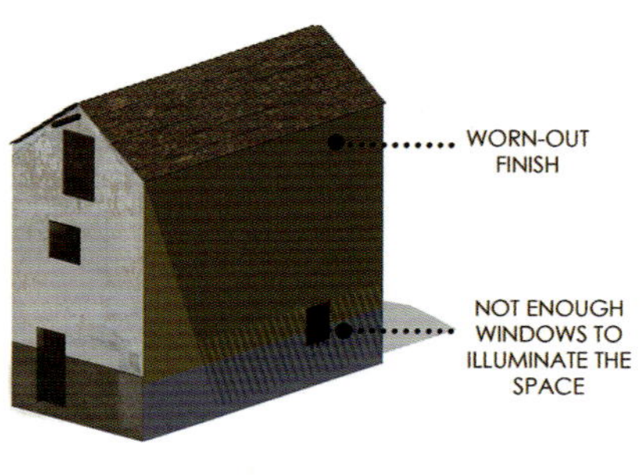

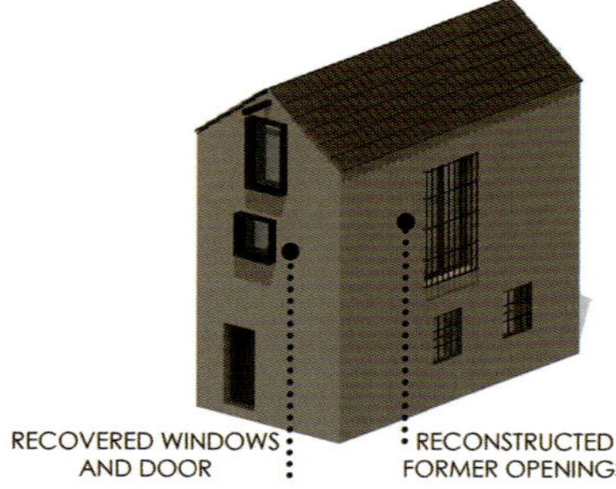

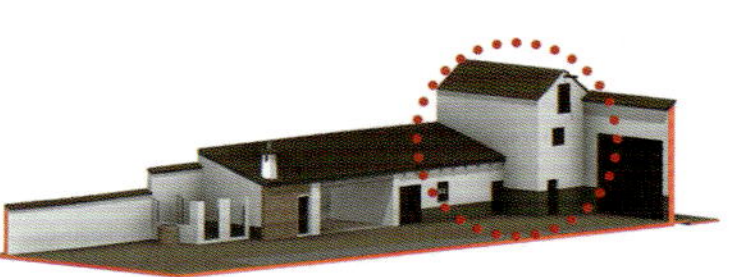

HAYSTACK_BEFORE

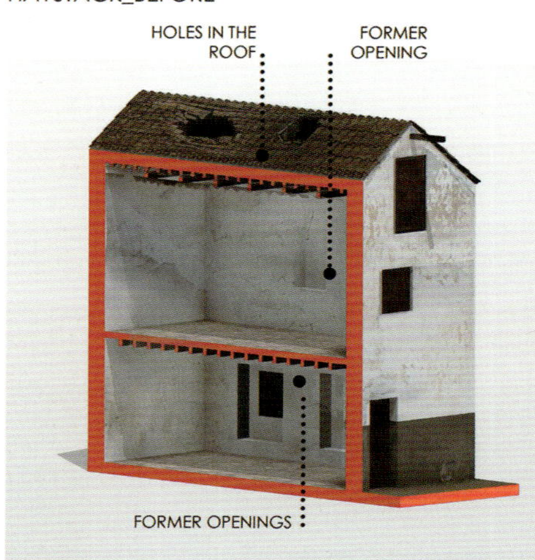

HAYSTACK_AFTER

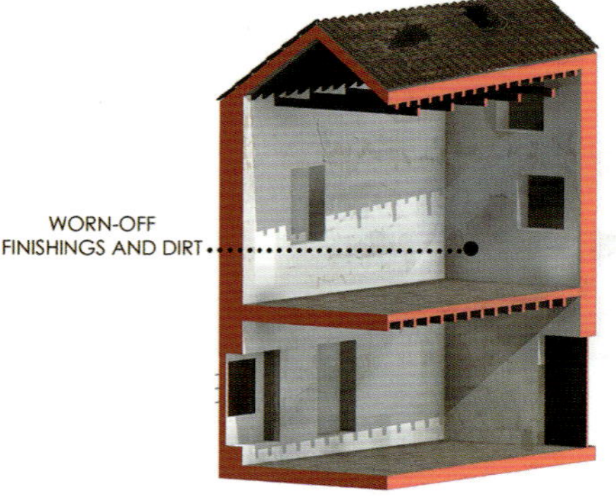

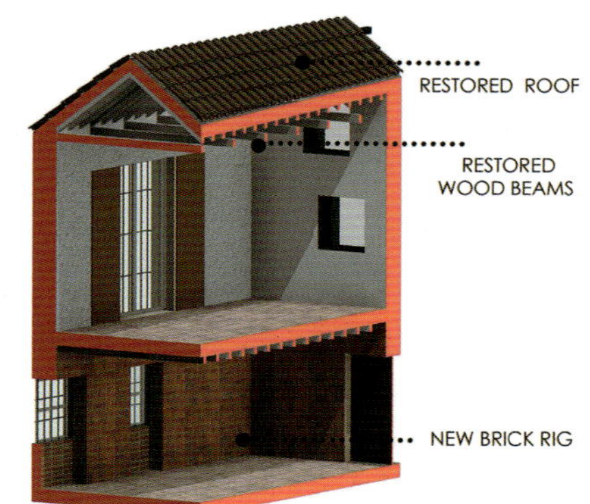

HAYSTACK_AFTER

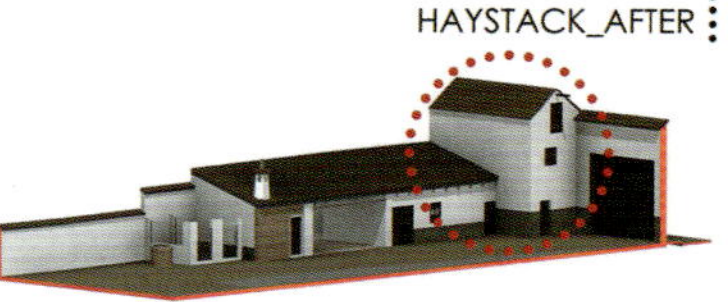

1.

2.

3.

4.

5.

6.

7.

8.

9.

Construction Process

Construction Process

© OOIIO Architecture, Josefotoinmo.

© OOIIO Architecture, Josefotoinmo.

Home Outside, Zwolle, Netherlands

BOARD

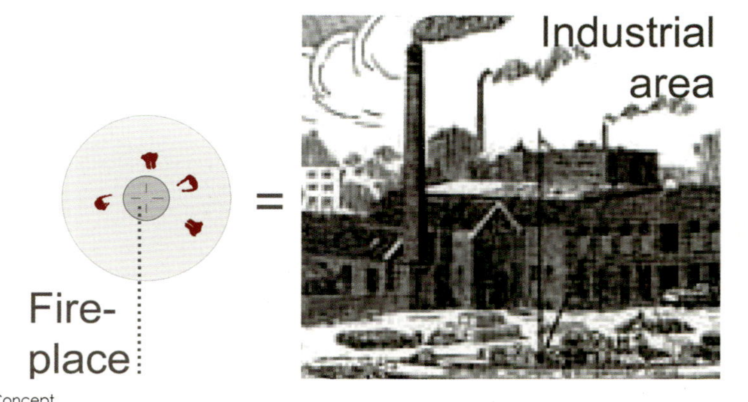

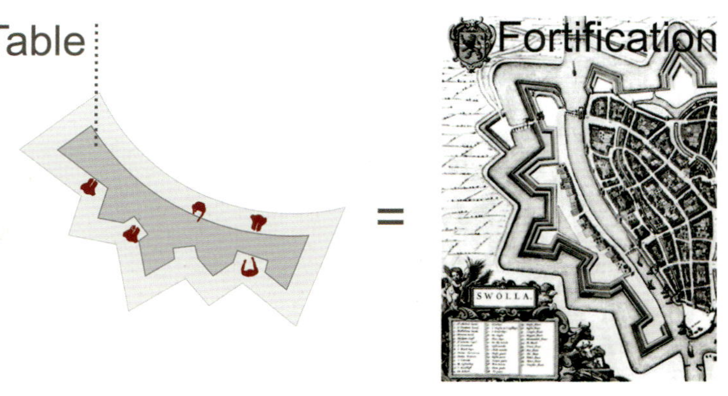

Concept

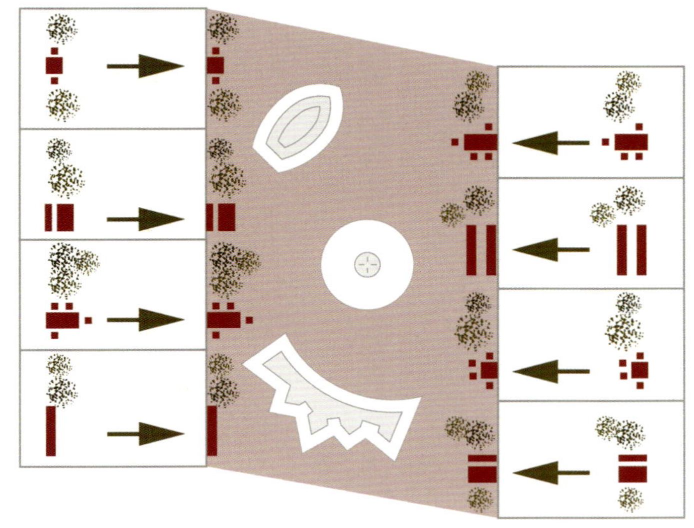

Landscape Architecture

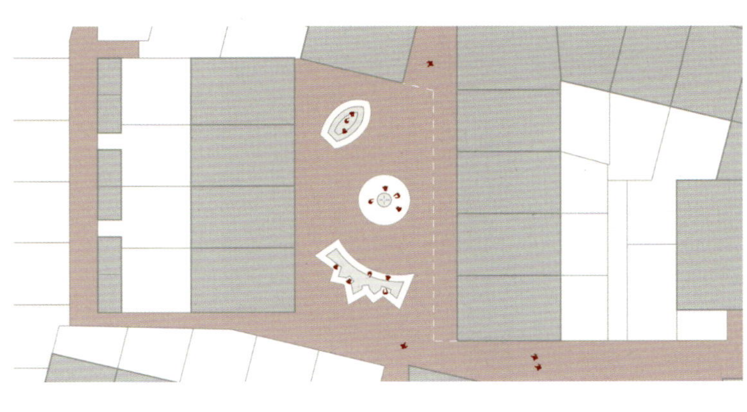

Floor Plan

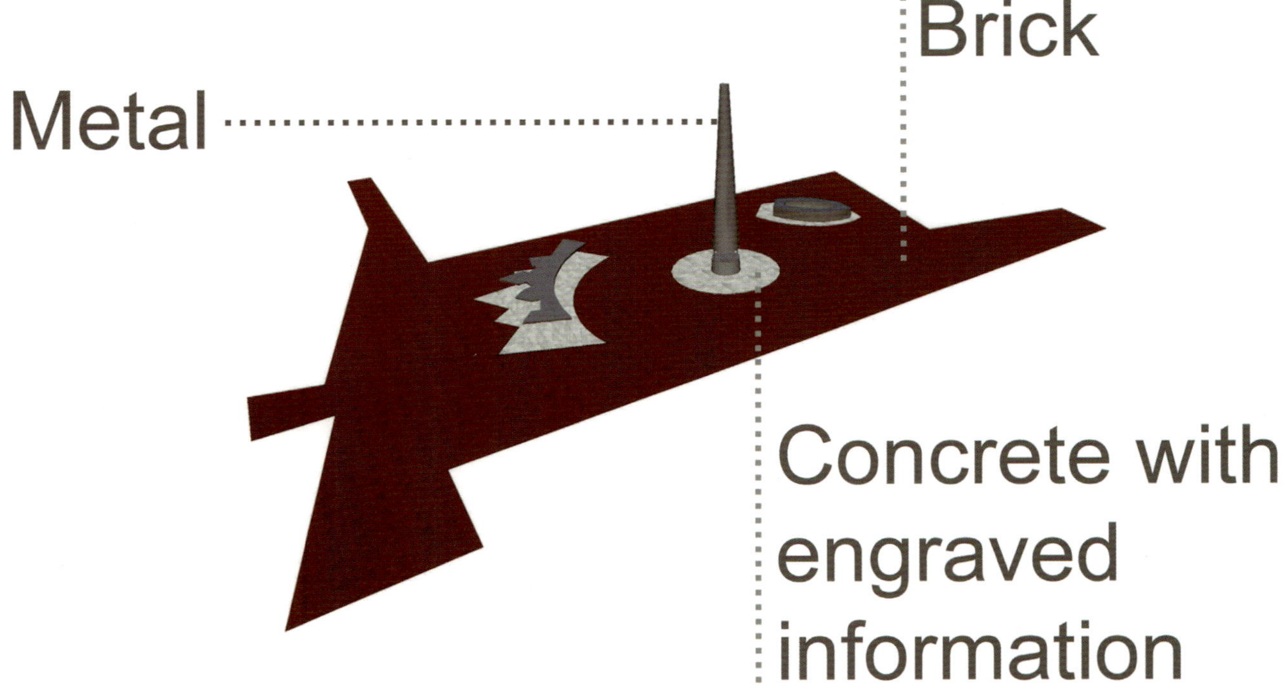

Axonometric

New National Museum of Afghanistan, Kabul , Afghanistan

Donner Sorcinelli Architecture

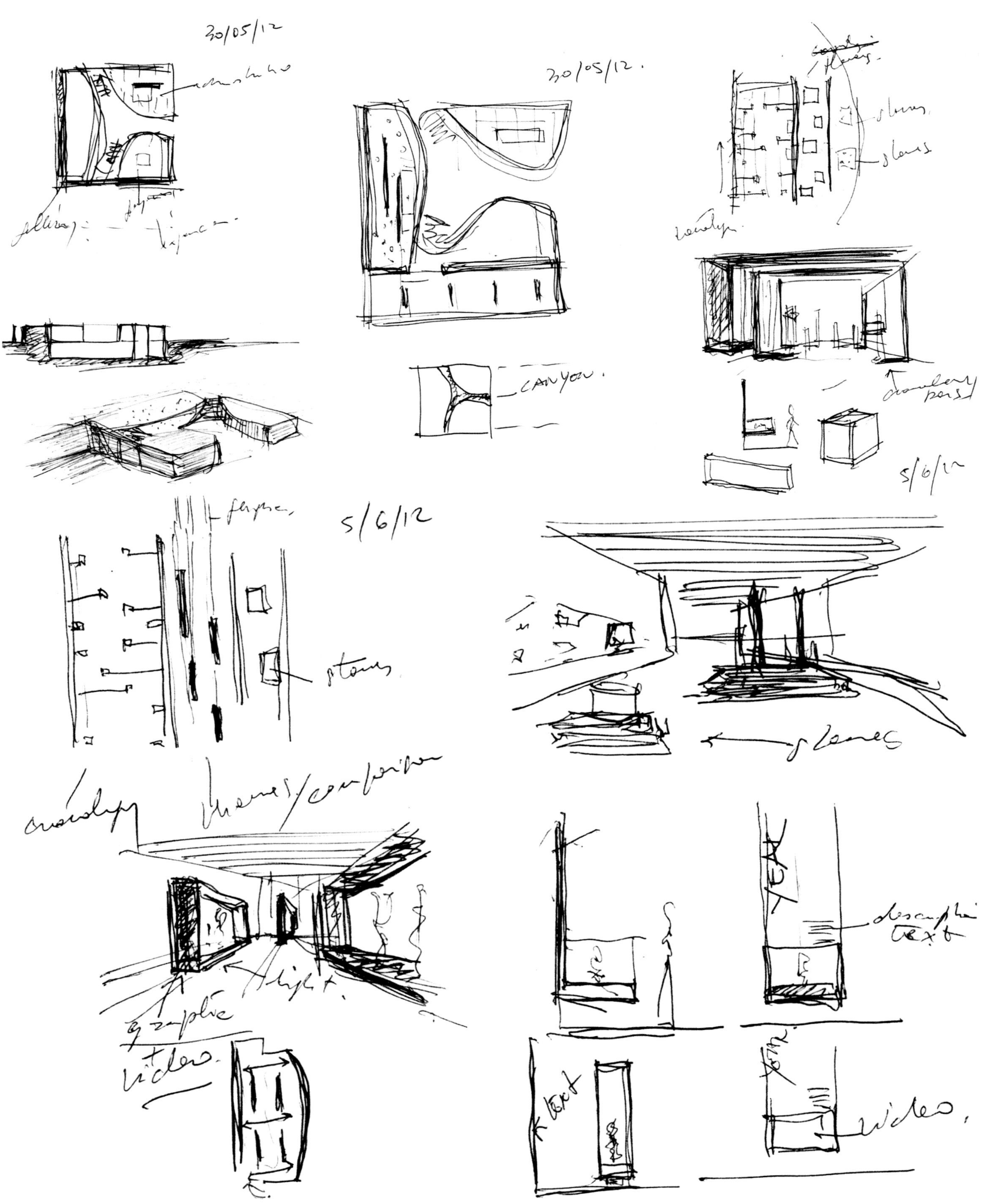

Sketch

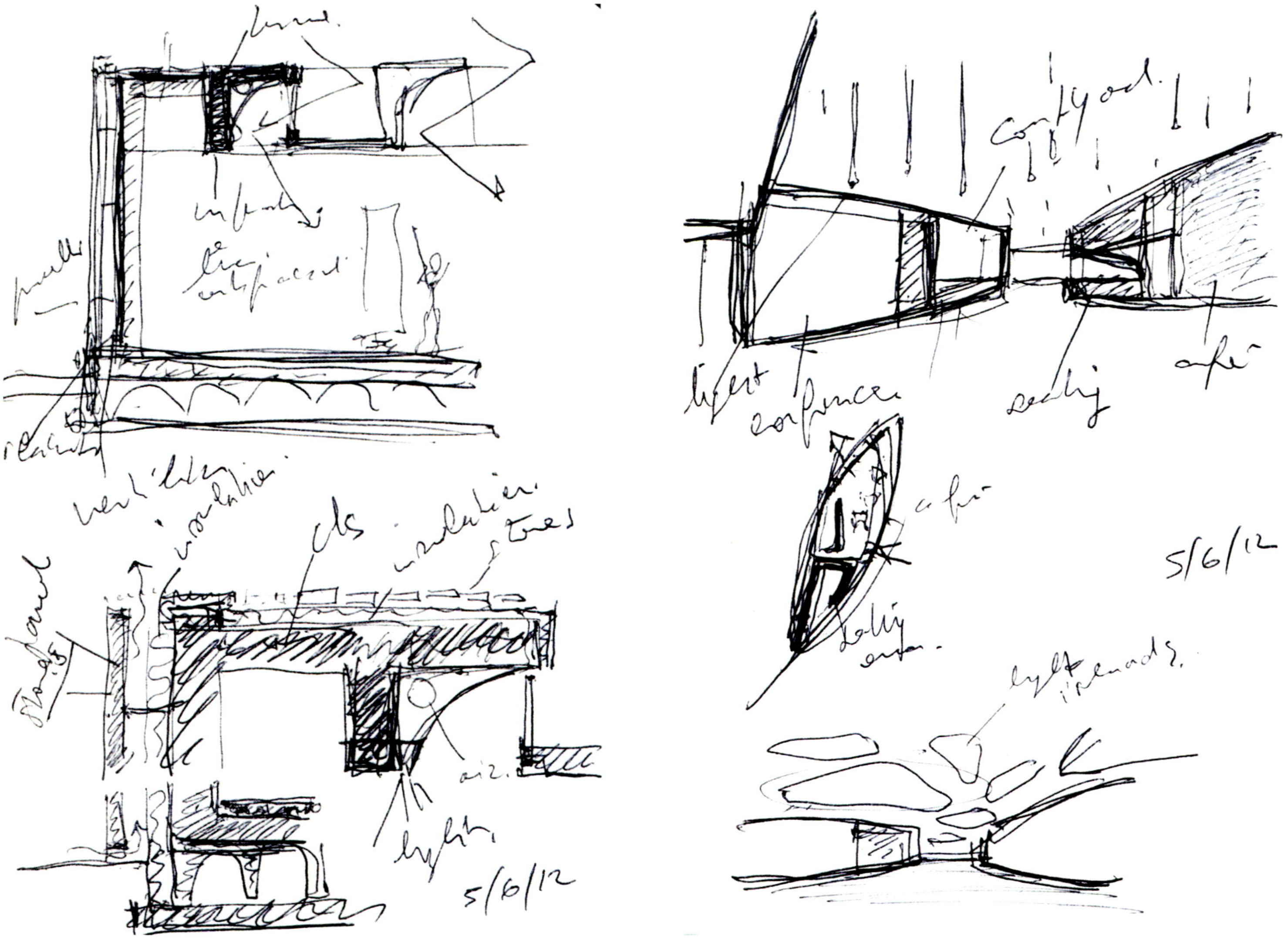

Sketch

existing building - cultural centre for historical studies
garden
new pavilion
plant room
embankment

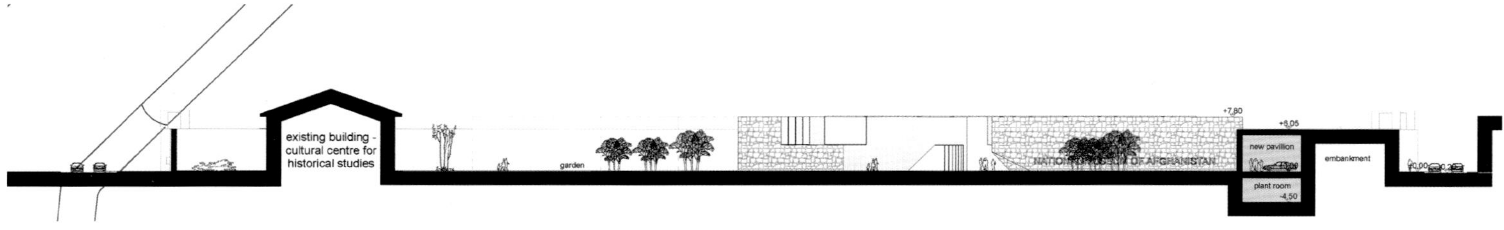

Facade

facade detail
waterproof membrane
vapor barrier
concrete
natural light
photovoltaic panels
gravel
air ducts
light
chronology
100 BC
local stone
air
insulation
local bricks
plasterboard
gravel
video
theca
detail facade
scale 1/25

Facade Detail

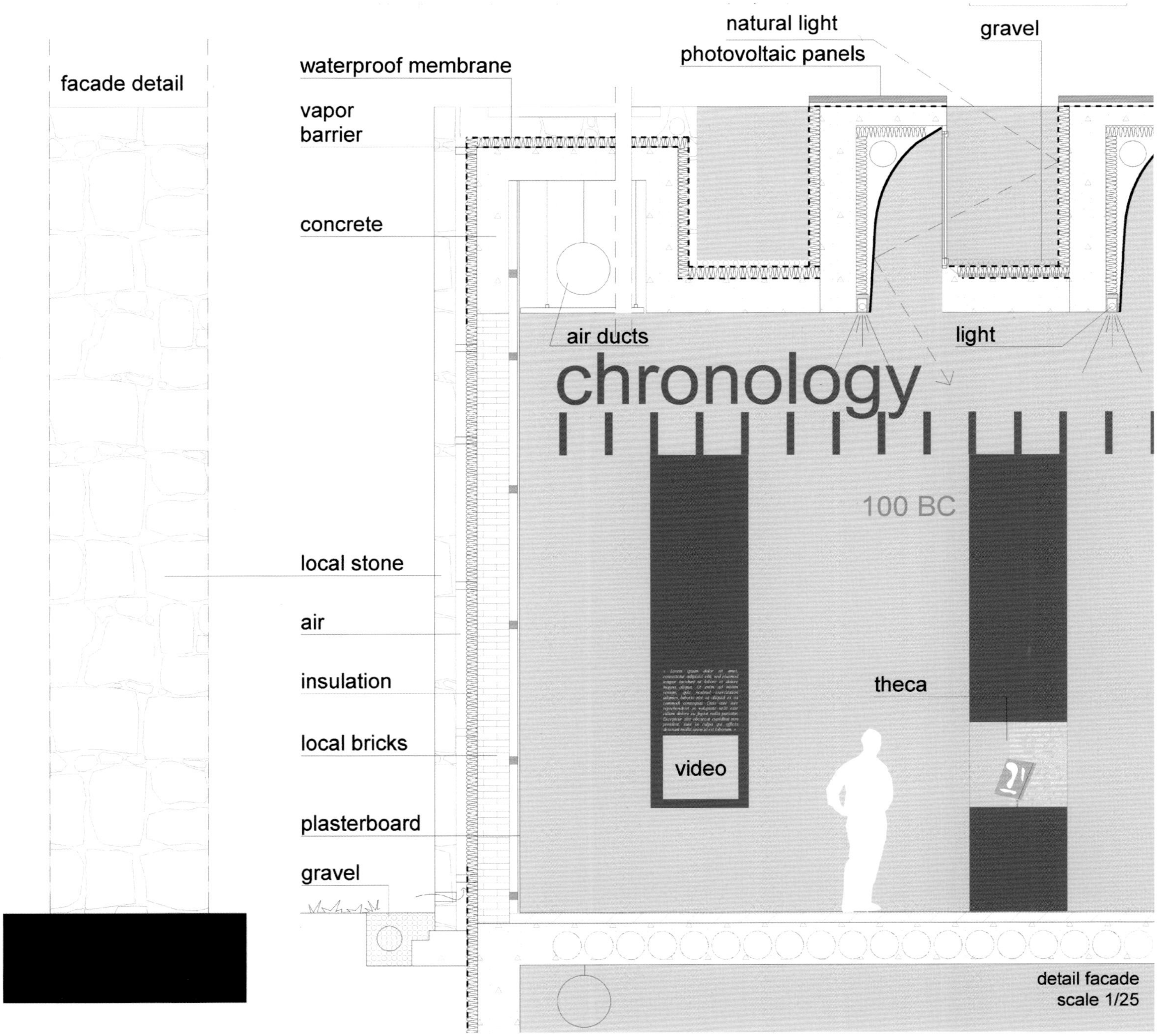

Gymnasium Hausburgviertel, Berlin, Germany

Rebecca Chestnutt, Robert Niess

© Werner Huthmacher

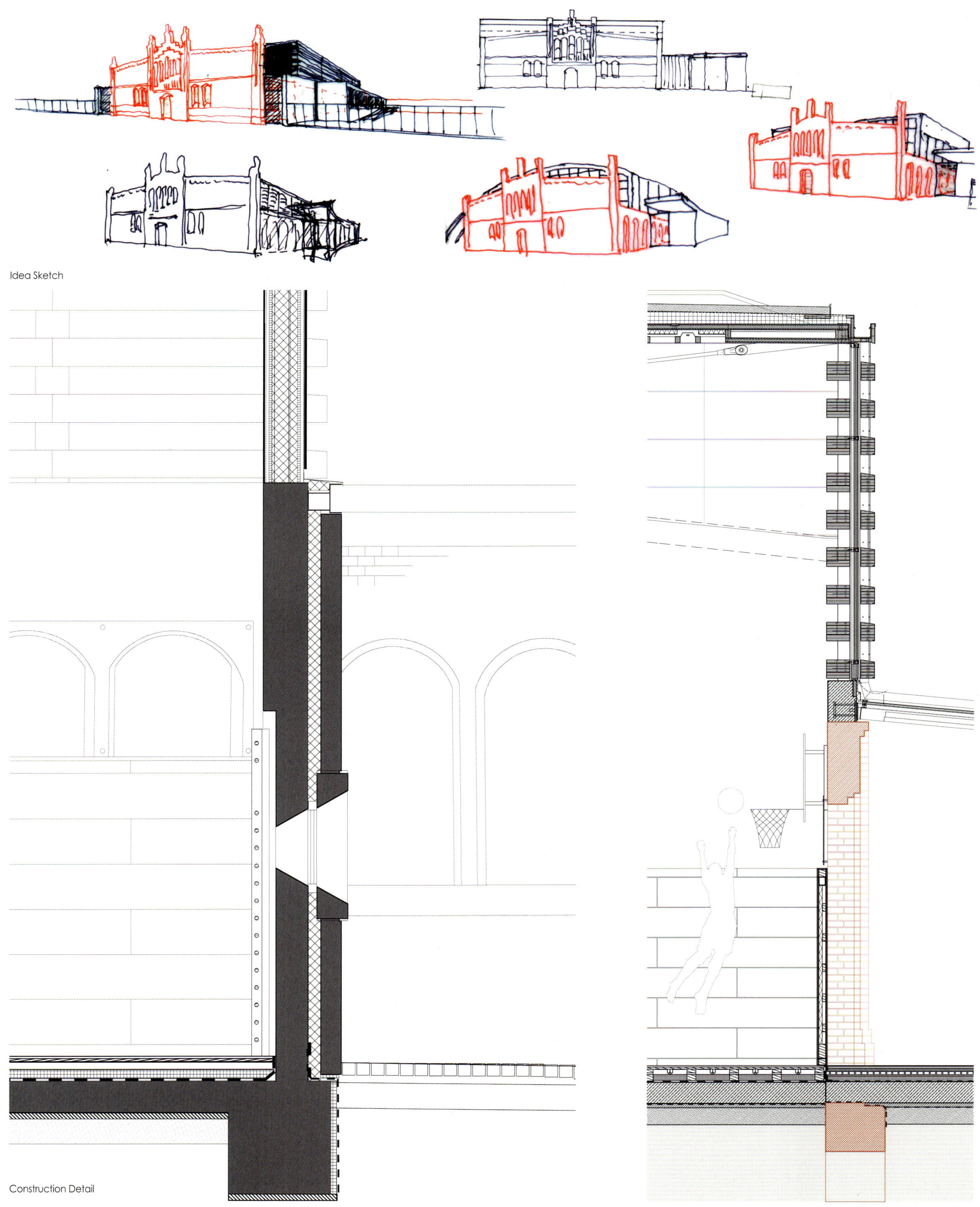

Idea Sketch

Construction Detail

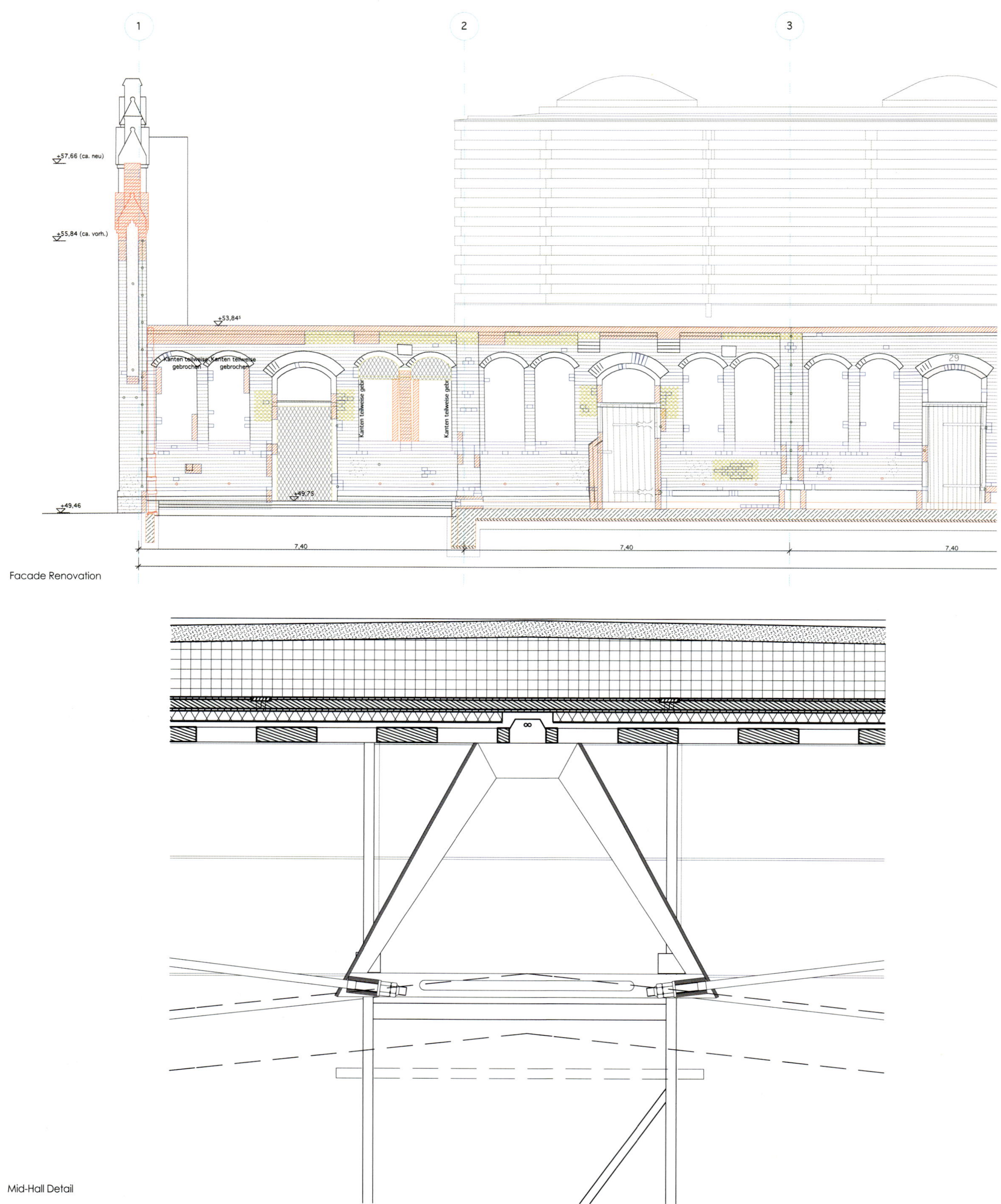

Facade Renovation

Mid-Hall Detail

Construction Process

© Werner Huthmacher

Architectural Material Series

To be Continued